COMPUTATIONAL QUANTUM PHYSICS

AIP CONFERENCE PROCEEDINGS 260

COMPUTATIONAL QUANTUM PHYSICS

NASHVILLE, TN 1991

EDITORS:

A. S. UMAR
V. E. OBERACKER
VANDERBILT UNIVERSITY

M. R. STRAYER
C. BOTTCHER
OAK RIDGE NATIONAL LABORATORY

American Institute of Physics   New York

Authorization to photocopy items for internal or personal use, beyond the free copying permitted under the 1978 U.S. Copyright Law (see statement below), is granted by the American Institute of Physics for users registered with the Copyright Clearance Center (CCC) Transactional Reporting Service, provided that the base fee of $2.00 per copy is paid directly to CCC, 27 Congress St., Salem, MA 01970. For those organizations that have been granted a photocopy license by CCC, a separate system of payment has been arranged. The fee code for users of the Transactional Reporting Service is: 0094-243X/87 $2.00.

© 1992 American Institute of Physics.

Individual readers of this volume and nonprofit libraries, acting for them, are permitted to make fair use of the material in it, such as copying an article for use in teaching or research. Permission is granted to quote from this volume in scientific work with the customary acknowledgment of the source. To reprint a figure, table, or other excerpt requires the consent of one of the original authors and notification to AIP. Republication or systematic or multiple reproduction of any material in this volume is permitted only under license from AIP. Address inquiries to Series Editor, AIP Conference Proceedings, AIP, 335 East 45th Street, New York, NY 10017-3483.

L.C. Catalog Card No. 92-71777
ISBN 0-88318-933-X
DOE CONF-9105238

Printed in the United States of America.

Contents

Preface vii
Lattice QCD in the 1990s 1
  Carleton DeTar
Designing a 2 Black Hole Code 12
  Richard A. Matzner
Chaos → Ergodicity → Isothermal Dynamics 23
  John H. Sloan, Dimitri Kusnezov, and Aurel Bulgac
Microscopic Calculations of the Ground State Structures, Collective Excitations, and the Dynamic Structure Function of ^{4}He Clusters 35
  S. A. Chin and E. Krotscheck
Particle Production through Two-Photon Processes in Relativistic Heavy-Ion Colliders 47
  J.-S. Wu, C. Bottcher, M. R. Strayer, and C. M. Shakin
Optimal Methods for Large-Scale Scientific Database and Sparse Matrix Applications 59
  P. Rochford, S. C. Park, J. P. Draayer, and S.-Q. Zheng
Siegert's Curse: Taming and Domesticating Divergent Wave Functions 72
  Peter Winkler
Exotic Gamow States of Atoms, Nuclei, and Particles 83
  Rubin H. Landau
Atomic Negative Ions 94
  Tomas Brage
Lorentz Covariant Multiple Scattering with Application to Kaon–Nucleus Scattering 109
  C. M. Chen and D. J. Ernst
Exact Monte Carlo for Few-Electron Systems 122
  Shiwei Zhang and M. H. Kalos
Many-Body Calculations of Photoionization Cross Sections 131
  Hugh P. Kelly
Computation Applied to Particle Acceleration Simulations 142
  W. B. Herrmannsfeldt and Yiton T. Yan
A Discussion of Computational Methods and Applications in Molecular Physics 149
  Kate P. Kirby
A Dynamical Picture of Hadron–Hadron Collisions with the String-Parton Model 159
  D. J. Dean, A. S. Umar, J.-S. Wu, and M. R. Strayer
High T_c Superconductors: Numerical Studies of the Hubbard Model 172
  Elbio Dagotto
The Classical Scattering of Waves: Some Analogies with Quantum Scattering ... 180
  Michael F. Werby
The R-Matrix Approach to Atomic Collisions 207
  N. R. Badnell, M. S. Pindzola, and D. C. Griffin

Nonperturbative Electromagnetic Muon-Pair Production with Capture in Peripheral Relativistic Heavy-Ion Collisions 215
    J. C. Wells, V. E. Oberacker, A. S. Umar, C. Bottcher, M. R. Strayer, and J.-S. Wu

Quark–Gluon Transport Theory: A Monte Carlo Simulation 228
    Klaus Geiger and Berndt Müller

Renormalization Group Approach to Thermal Quantum Field Theory 244
    Eric Braaten

Electromagnetic Lepton-Pair Production in Relativistic Collisions 255
    C. J. Albert, D. J. Ernst, M. R. Strayer, and C. Bottcher

Laser-Assisted Molecular Dissociation and Recombination of Diatomic Molecules 267
    Zi-Min Lu, Michel Vallières, and Jian-Min Yuan

Real-Time Evolution in Lattice Gauge Theory 280
    A. Trayanov and B. Müller

Monte Carlo Simulation of a Dynamical Fermion Problem: The Light $q^2\bar{q}^2$ System 292
    G. Grondin

Conceptual Framework for High-Spin Hadronic Physics 309
    D. V. Ahluwalia and D. J. Ernst

List of Participants 317

Author Index 321

PREFACE

This conference is the third in a series of workshops and summer schools to foster computational physics education across the disciplines of atomic, nuclear, and particle physics. With the rapid growth in massively parallel architectures and associated developments in distributed computing and visualization, there is an urgent need to raise the computational literacy of young scientists and students. A workshop of approximately 40 to 60 people provides a comfortable environment for the exchange of views, ideas, and methodologies. The disciplines of theoretical atomic, nuclear, and particle physics share many common concepts and methods of calculation, thus providing an ideal setting for students to broaden their perspective on physics.

The organization of the workshop as a collaboration between Oak Ridge National Laboratory, Vanderbilt University, the Science Alliance of the State of Tennessee, Duke University, Texas A&M University, and Auburn University illustrates the fruitful synergism between laboratories and universities. We would like to thank Vanderbilt University as hosts of this workshop.

Organizing Committee

Lattice QCD in the 1990 s

Carleton DeTar
Department of Physics, University of Utah, Salt Lake City, UT 84112

I give a brief sketch of the current status of QCD and of prospects over the next few years, anticipating the development of teraflop scale computers.

I. INTRODUCTION

More complete discussions of the status and prospects of lattice QCD can be found in the proceedings of the recent annual Lattice conferences.[1–3]

Our current understanding of the fundamental processes in Nature, i.e. the currently accepted field theory of the strong, weak, electromagnetic, and gravitational interactions, is embodied in the so called "Standard Model". Although this model has so far passed experimental muster exceedingly well, it is fundamentally incomplete and contains tantalizing hints of discoveries still to come. Because of the success of the Model, it could be said that our understanding has reached a plateau over the past decade.

To push beyond, physicists have embarked upon two very ambitious efforts. One is an experimental program, leading to the construction of gigantic new accelerators at the end of this decade, namely the US Superconducting Supercollider and the European Large Hadron Collider. The other is an ambitious theoretical effort to advance the power of high performance computing technology and to apply it to a study of the Standard Model, simultaneously sharpening the predictive power of the model, so that it can be confronted with experiment, and enlarging our ability to extract useful information from experimental results. The US initiative is called the "QCD Teraflop Project". Teraflop efforts are also underway in Japan and Europe.

A numerical approach to field theory has become the method of choice in recent years when conventional analytic techniques have failed. Particularly, it is needed in order to tackle problems exhibiting highly nonlinear behavior. This need has arisen in every facet of the Standard Model:

- Quantum chromodynamics

- Confinement. Demonstrating that confinement follows from first principles.
- Mass spectrum of hadrons. Explaining the mass ratios from first principles.
- High temperature phase transition. Characterizing the phase transition and exploring the properties of the high temperature phase from first principles. These properties will be investigated at the Relativistic Heavy Ion Collider, currently under construction at Brookhaven National Laboratory.

• Electroweak theory

- Bounds on undiscovered Higgs particle mass. Discovering limitations imposed by the strong-coupling features of the electroweak theory.
- Baryon number violation in early universe.

• Quantum Electrodynamics

- New behavior at very short distances.

• Quantum Gravity. Studying the complicated theory of random surfaces.

For the rest of this talk I will focus on quantum chromodynamics (QCD), for which the numerical effort is most advanced.

II. QUANTUM CHROMODYNAMICS

Quantum Chromodynamics is the widely accepted quantum field theory of the strong interactions. Here, we mention briefly the features that are important for understanding the numerical analysis. The theory describes the interactions of quarks and gluons. The quarks are permanently confined in specific combinations, forming the hadrons (strongly interacting particles), such as the proton, neutron, and the various mesons. Since it is a field theory, the quarks and gluons are described in terms of quark fields $q^i(x)$ and gluon fields $A_\mu^a(x)$. Just as with an ordinary electric field, the fields have a value at every point in space and time (denoted by x). The quark field is a fermion, like an electron, and so obeys the Pauli exclusion principle. As we shall see, this requirement adds complexity to the numerical algorithms. Both fields carry color indices ($i = 1, 2, 3$ and $a = 1, \ldots, 8$). The color degree of freedom is associated with an SU(3) gauge symmetry, so the gluon field can be represented alternatively by an SU(3) matrix $U_\mu(x)$. The field theory is local, so changes in the field at one space-time point have an immediate effect only in the close vicinity of that point. The adjustable parameters of the theory are the quark masses (m_u, m_d, m_s), the number of quark species ("flavors") included, and the coupling contant g between the quarks and gluons.

Among the features of QCD to be explained are these:

- Spectroscopy. For example, why is the ratio of the proton to the rho meson mass equal to 938/770, and why is the ratio of the rho meson to the pi meson mass equal to 770/140?

- Confinement. Does QCD really have the feature that quarks cannot be isolated? Related to this question is the problem of determining the potential energy of interaction between a quark and an antiquark, as they are separated.

- Weak decays. For example, one of the great puzzles of the Standard Model is the pattern of weak interactions between the various quark flavors. In Fig. 1 is shown a process leading to the decay of a B meson, which contains a "b" quark and and anti "d" quark. The crux of the interaction is the circled part in which the "b" quark turns into a "u" quark. We are especially interested in knowing the probability for such a transformation. Although the decay can be observed and measured in an experiment, the analysis of the experimental result is hampered by the fact that strong interactions (shown by wiggly lines) distort the initial and final states. Without a numerical simulation of these strong interactions, we will not be able to complete the analysis of the experimental measurement and isolate this essential parameter of the Standard Model.

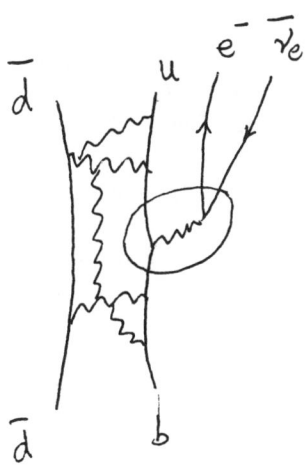

FIG.1. Weak decay of "b" quark (in oval) obscured by strong interactions (wiggly lines).

- High temperature and high pressure behavior. At extreme temperature and/or pressure, conditions are favorable for the formation of a new phase of matter, called the quark-gluon plasma. Such conditions may have occurred in the early universe, may be found at the cores of compact stars, and may be created in relativistic heavy ion collisions. Numerical simulations are needed to determine the properties of this phase.

III. NUMERICAL METHOD

A. Lattice and Continuum

The numerical analysis of QCD proceeds first by putting the field theory on a lattice in space and time, following pioneering work of Wilson.[5] In order for a lattice approach to be effective, it is essential that the continuum theory be recoverable as the lattice spacing a is taken to zero. Because QCD is an asymptotically free renormalizable theory, it has the important feature that in order to reduce the lattice constant a, it suffices to decrease the coupling constant g at the same time. The correlation between these quantities is ultimately prescribed by perturbation theory.

Figure 2 shows the temperature of the phase transition in the theory without quarks, determined from hadron masses, as a function of inverse lattice spacing.[?] These values should approach a constant. The extent to which they do is a measure of how closely this limit is being approached. The lattice spacing for the points labeled $1/a = 8$ GeV ($a \sim 0.2$ F) and above correspond to $\beta = 6/g^2 = 6.0$ to 6.5. (A proton is roughly one fermi in diameter).

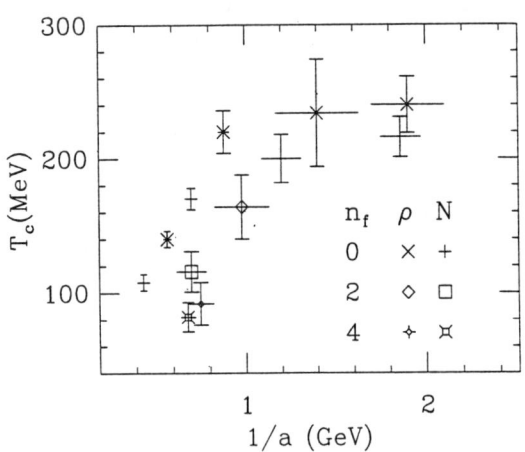

FIG.2. Continuum scaling of T_c.

Of course, it is easy enough to make a as small as we like by simply increasing $\beta = 6/g^2$. But the problem is that we must also keep the box size big enough that we don't cramp and distort the hadrons. So to make a small, we must increase the number of lattice points, and increase the computational labor accordingly. That is, the box size should be at least a few times the longest correlation length in the problem. Many computations that include quarks reduce this correlation length artificially by keeping the quark masses too large. Thus what is needed is the ability to calculate both with small a *and* small quark mass m, while keeping the box size larger than a few correlation lengths.

B. Feynman Path Integration

The numerical analysis then proceeds by carrying out the Feynman path integration of the field theory. Feynman's path integral is a formalism for calculating in the quantum mechanical Gibbs ensemble of the theory at a finite temperature T. Such a formalism permits the simulation of the field theory at finite temperature in thermal equilibrium, and also permits the determination of static ground state properties, obtained in the limit of zero temperature. Important finite temperature properties include the characterization of the phase transition and the high temperature phase. Important static properties include the determination of hadron masses, wave functions, and weak matrix elements. Every one of these properties is obtained by calculating the expectation value of an appropriate operator $\langle O \rangle$ on the ensemble. All operators are expressible in terms of the basic quark and gluon fields.

In brief the Feynman path integral formalism is based on the classical action $S[q]$, which depends on the values of the fields at each point in space and time (here represented by the single variable q). The time variable is complex and ranges from 0 to $1/T$, the inverse temperature. To be explicit, the expectation value of the operator O is given by

$$\langle O \rangle = \text{Tr}\left(Oe^{-H/kT}\right) = \int dq\, O(q) \exp(-S[q]) / \int dq \exp(-S[q]).$$

In the lattice version of this formalism, the integration dq is taken over all possible values of the fields at each lattice point in space and time. For QCD there are about 40 field variables for each lattice site, so for a 100^4 lattice the number of variables in the integration is about 4×10^9.

C. Numerical Integration

The rather formidable task of carrying out this integration is reduced to manageable proportions by Monte Carlo techniques. This stroke of great fortune was first discovered by Creutz, Jacobs, and Rebbi.[6] The Monte Carlo approach generates a series of points in this multidimensional space. Let q be one of these points, and call the series of points q_i. Each point corresponds to a specification of the value of the field at each space-time lattice site. Thus it is called a lattice "configuration". The series is generated by a random process, devised so that the probability of encountering a configuration q is given by $\exp(-S[q])$. Then the required expectation value is just the average of $O(q)$ on the generated sites.

$$\langle O \rangle = \lim_{N \to \infty} \sum_{i=1}^{N} O(q_i)/N.$$

The task is then to devise such a random process. The elementary step is the move from the configuration q_i to the configuration q_{i+1}. This move generates a Markov chain. Most popular algorithms for this move involve changing the integration variables at a single site and testing the effect of this change on the probability weight $\exp(-S[q])$. What makes the problem tractable is that QCD, like many other field theories, is a local field theory. For the theory without quarks (fermions) this means that the effect on the probability weight of a change at a single site can be computed by knowing the values of the integration variables on just the neighboring lattice sites. Thus the cost per site of assessing the effect of the change does not increase with the lattice size. When quarks are included in the calculation, the Pauli exclusion principle introduces complications that force upon us a degree of nonlocality. The result is that each move requires the solution of a large sparse matrix problem. The fortunate sparseness of this matrix is a legacy of the locality of the quark-gluon interaction.

The challenge in developing a good algorithm is to provide efficient and thorough coverage of the important regions of the integration space.

The most popular current method for generating the move in q is called the molecular dynamics algorithm (MDA), developed by the Illinois group.[7] This algorithm introduces a momentum variable p for each original variable q and generates a trajectory in (p,q) phase space according to a fictitious classical Hamiltonian dynamics. The expectation value of an operator O is, as before, the average of its value over the trajectory in q. The method is improved by introducing random changes in p at intervals along the deterministic molecular dynamics trajectory.

As with all numerical integration strategies, the most important question is, how many steps must we take before we can be confident that we have a good expectation value $\langle O \rangle$? In order to get a small variance for $\langle O \rangle$, we must have a large sample of statistically independent configurations q. The basic problem is that algorithms based on local changes in the coordinates tend to build in long autocorrelations in the Markov chain. The effect of local changes propagates only slowly across the lattice. Thus as the lattice constant a and the quark mass m is decreased, the molecular dynamics trajectories must go farther before a "statistically independent" configuration is reached. This widespread problem is called "critical slowing down". Much of the current effort to improve upon the popular algorithm are based on strategies to make nonlocal moves.

IV. STATUS OF QCD

Numerical simulations for QCD have been in progress for over one decade. What has been accomplished over this time? We have space to list only a few major results:

Confinement Most practitioners would agree that lattice simulations have given the most convincing demonstration to date that the color forces of QCD are confining. Shown in Fig. 3 is a the confining potential between a heavy quark and antiquark pair as a function of their separation.[8] The results of a numerical simulation are compared with a form determined phenomenologically from the experimentally measured bound-state spectrum in the J/ψ and Υ families. The agreement is very encouraging.

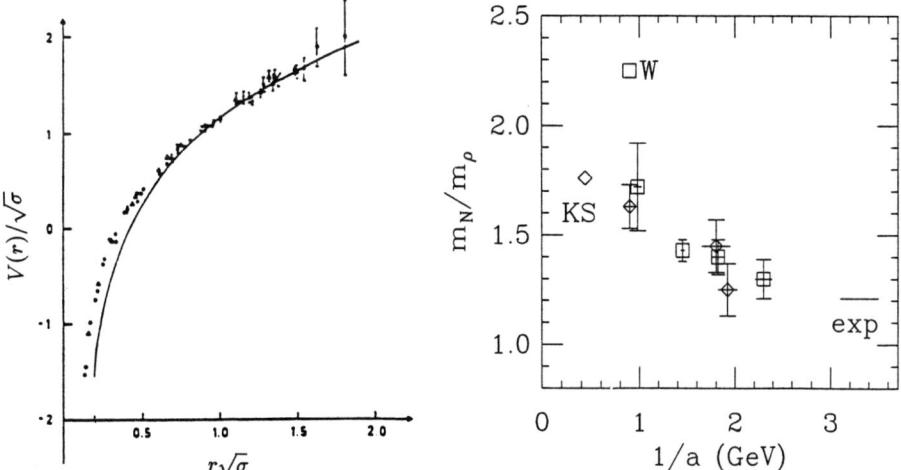

FIG.3. Confining potential between heavy quarks.

FIG.4 Continuum limit of the ratio of the nucleon mass to the rho meson mass.

Mass spectrum The mass spectrum of the hadrons appears to be on track, but not yet on target for want of lattices large enough to permit a small lattice spacing and quark mass. Figure 4 shows that the nucleon to rho meson mass ratio appears to be approaching the experimental value as the lattice spacing decreases.[4] Figure 5 gives another popular way to show where we stand with both the nucleon to rho meson mass ratio and the rho to pi meson mass ratio.[9] The point at the left is the experimental value, which should be reached by decreasing the quark mass parameter. Of course, one could have simply dialed in the correct experimental value for the pi to rho mass ratio by decreasing the quark mass. However, small quark masses require larger box sizes and much bigger lattices to avoid serios finite size effects. Thus all the simulations shown have kept the quark mass too high, leading to an honest, if unrealistic simulation.

Weak amplitudes A good illustration of the current state of the art is the measurement of the parameter B_K, which is needed in the analysis of the very important CP violating decays of the K meson, potentially the Achilles heel of the Standard Model. Figure 6 shows how this parameter varies with lattice size.[10] With simulations of this quality it should be possible to achieve good accuracy in the extrapolation to $a = 0$.

8 Lattice QCD in the 1990s

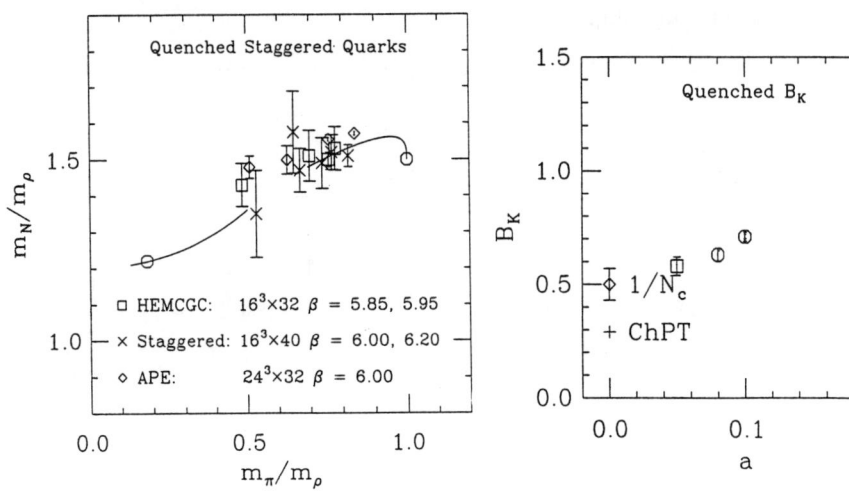

FIG.5. Hadron mass ratios. Solid line is based on perturbation theory in small quark masses.

FIG.6. Continuum limit of the CP-violation B_K parameter scaled to 1 GeV. Preliminary data, "Staggered" collaboration.

High temperature phase transition The issue here is whether the phase transition is strongly first order, only weakly so, or even absent altogether. It has been suggested that the nature of the phase transition depends on the quark mass and the number of quark species (flavors). Shown in Fig. 7 is a summary of results of simulations with varying numbers of flavors n_f and a varying ratio of quark mass to temperature.[11] Since in nature, $n_f \sim 2$ and $m/T \sim 1/10$, it is becoming evident from these simulations that the phase transition, if it occurs at all, is very weak. These results have great significance for the interpretation of heavy ion experiments intended to produce this phase transition.

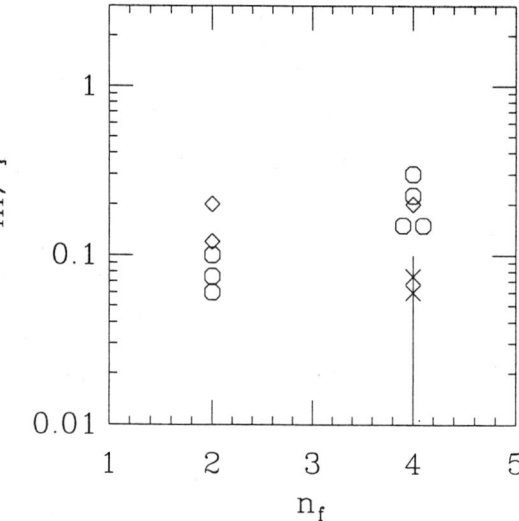

FIG.7. Gap in the phase boundary. Results from various simulations with $N_t > 4$ are presented. Crosses: a simulation found evidence for a first order transition. Octagons: a simulation found no phase transition. Diamonds: inconclusive.

In summary one could say that the major successes of lattice QCD have been important, but largely qualitative. Our agenda for this decade is to turn lattice QCD into a *quantitative* tool for the study of the Standard Model.

V. WHAT PRICE THE CONTINUUM?

To elevate the simulations to quantitative respectability, we must be able to reduce the lattice spacing further, and decrease the quark mass at the same time. What is the price? I give a conservative answer based on current well-understood algorithms: the computational effort scales approximately as a^{-6} for calculations without quarks and as a^{-8} for calculations with quarks included. In these exponents a power of four comes from simply increasing the number of lattice points in a four-dimensional space-time. The additional powers come from critical slowing down and, with quarks included, also the added costs of solving the sparse matrix problem. Thus with a thousandfold increase in computational power, we could reduce the lattice constant by a factor of about three without quarks and somewhat less that three with quarks included. Obviously, improvements in algorithms that return a power of a or more are urgently needed. For this reason much work has been devoted to developing nonlocal strategies.

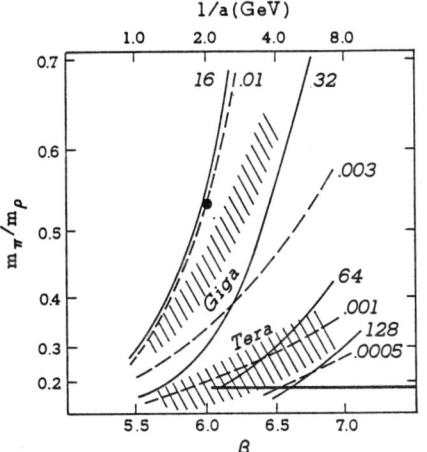 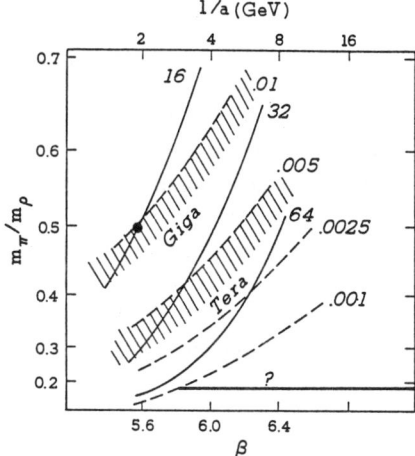

FIG.8. Computational effort required to reach physical mass ratios and small lattice spacings for QCD without quarks.

FIG.9 Computational effort for QCD with quarks.

How close will we come to our goal with a Teraflop-scale machine? Shown in Figs. 8 and 9 are rough estimates of the computational effort required in order to achieve this goal. Figure 8 is for the theory without quarks and Fig. 9 is for the theory with quarks included. The vertical scale shows the pi to rho mass ratio.

The horizontal scale gives the inverse lattice spacing. The black dot shows roughly where present simulations reach. The bold black horizontal bar at the lower right of the plot shows the range of parameters we would like to be able to reach. That range puts the pion mass at its experimental value and puts the lattice constant comfortably close to the continuum. The two cross hatched regions indicate where one should be able to get with a Gigaflop and a Teraflop computer. It is evident that the promised land is quite accessible at a Teraflop for calculations without quarks. With quarks present we may still be able to get close enough that we can finish the job with a perturbative expansion in small quark masses.

VI. QCD TERAFLOP PROJECT

Encouraged by the feasibility of Teraflop-scale architectures in the next couple of years and by the significant potential of QCD simulations, the Columbia University group, headed by Norman Christ, has spearheaded the formation of the QCD Teraflop Collaboration, with a current membership of 37 American high energy physicists. We are submitting a proposal to the US Department of Energy to enter into a partnership with Thinking Machines Corporation to construct a special purpose parallel machine capable of a sustained Teraflop in QCD simulations. The machine would go into operation in 1994. It would evolve into a national facility.

VII. CONCLUSION

I have argued that QCD simulations will come of age in the 1990's. We will use results of numerical simulations to increase the predictive power of the Standard Model and make it possible to extract vital information from experiments. What is needed to reach this goal is a combination of significant advances in raw computer power, computer architecture, and in algorithms. We can achieve a QCD Teraflop with computer power available in the near future, if the architecture is tuned to the requirements of QCD. And with continued efforts to improve upon algorithms we expect to make the most of this capability.

REFERENCES

[1] A.S. Kronfeld and P.B. Mackenzie eds. *Lattice 1988*, Fermi National Laboratory, 22-25 September 1988, Nucl. Phys. B (Proc. Suppl.) **9** (1989);
[2] R. Petronzio, et al., eds., *Lattice '89*, Capri, Italy, September 18-21, 1989, Nucl. Phys. B (Proc. Suppl.) **17** (1990).
[3] Int. Symp. *Lattice '90*, Tallahassee, FL, Oct., 1990, to be published by Nucl. Phys. (Proc. Suppl.)
[4] A. Ukawa in Proceedings, 25th International Conference on High Energy Physics, Singapore, August, 1990 (World Scientific, Singapore, 1991)
[5] K.G. Wilson, Phys. Rev. D **10**, 2445 (1975).
[6] M.Creutz, L. Jacobs, and C. Rebbi, Phys. Rev. Lett. **42**, 1390 (1979); M. Creutz, Phys. Rev. Lett. **43**, 553 (1979); M. Creutz, Phys. Rev. D **21**, 2308 (1980).
[7] S. Duane and J.B. Kogut, Nucl. Phys. B **275**, 398 (1986); Phys. Rev. Lett. **55**, 2774 (1985).
[8] R. Gupta in *"Lattice Gauge Theory Using Parallel Supercomputers"*, Beijing, 1987.
[9] The APE collaboration, Nucl. Phys **B317**, 509 (1989); *ibid.*,**343**, 228 (1990); E. Marinari talk given at *Lattice '89*, Capri, Italy, Nucl. Phys. **B (Proc. Suppl.) 17**, 431 (1990); R. Gupta, talk given at **Lattice '89**, Capri, Italy, Nucl. Phys. **B (Proc. Suppl.) 17**, 70 (1990); R. Gupta, G. Guralnik, G. W. Kilcup, S. R. Sharp, Phys. Rev. **D43**, 2003 (1991); The High Energy Monte Carlo Grand Challenge Phys. Rev. Lett. **65**, 2106 (1990), Phys. Rev. **D42**, 3794 (1990).
[10] I am grateful to Stephen Sharpe for providing this preliminary data.
[11] F.R. Brown et al, Phys. Lett. B **251**, 181 (1990); S. Gottlieb et al, Phys. Rev. D **41**, 662 (1990); R. Gavai et al, Phys. Lett. B **241**, 567 (1990); E. Kovacs et al, Phys. Rev. Lett. **58**, 751 (1987); MILC Collaboration, unpublished.

DESIGNING A 2 BLACK HOLE CODE

Richard A. Matzner
Center for Relativity, University of Texas, Austin, TX 78712

ABSTRACT

Among the large problems for which teraflop (in principle massively parallel) machines will provide the means of solution are those involving the prediction of gravitational wave amplitudes from interacting strong field gravitational situations. Examples are neutron star/neutron star, neutron star/black hole, and black hole/black hole interactions, collisions, and mergers.

While the weak field, long distance behavior of the gravitational field can be well treated by approximate methods, the strong field interaction and radiation regimes is full 3-dimensional numerical analysis, apparently on grids of order $(1000)^3$. Variable count amounts typically to about 80 variables per zone. Based on smaller scale evolutions running at about 0.2 to 0.3 Gflop/sec (Cray Y-MP) which require $\sim 2 \times 10^{-5}$ seconds per zone cycle, one deduces $\sim 2 \times 10^4$ seconds per cycle ~ 6 hours/cycle for $(1000)^3$ (7×10^{11} Bytes of memory required) runs. Because a typical run would require $\sim 10^4$ cycles, CPU times in decades would be required at Y-MP CPU rates. However, teraflop sustained machines would reduce this to the 6–10 hour range, and algorithm improvements would be expected to lower this by another factor of 10 (at least). Since there are a number of feasible designs for teraflop/sec machines, it appears the 2-black hole problem will be double within the next five years or so. This paper discusses an outline for designing the code, discusses various alternatives in approach, and drafts a possible path through the forest of alternatives.

INTRODUCTION

The study of Numerical Relativity began in the 1960's with May and White[1] who considered spherical collapse (supernovae). Hahn and Lindquist[2] studied the evolution of a 2-black hole system, but lack of computational power, plus lack of a fully developed algorithmic resource meant that they were only partly successful. The first really successful nonspherical numerical relativity was by Smarr[3] and Eppley[4] who ran axisymmetric (head on, nonspinning) black hole collisions. Centrella,[5] did numerical cosmology, and Piran[6] studied a number of differencing schemes in cylindrical symmetry. The two black hole problem languished until recently.

Active workers in the field now are: Oohara and Nakamura,[7,8] Nakamura, Oohara, and Kojima,[9] Seidell,[10] Choptuik,[11] Abrahams,[12] Evans,[13] Laguna,[14] Kurki-Suonio,[15] Thornburg,[16] Cook,[17] Wilson,[18] among others. Almost all of these are working on the two-black-hole problem in the general, nonsymmetric case of interaction of spinning black holes. The two-black-hole problem is in a real sense the ideal Relativity situation, since it is essentially unaffected by exterior astrophysics and should have a uniquely determined gravitational wave production which is a function of a small number of parameters.

A number of developments are driving this problem at this time. There has been a substantial improvement in our understanding of the analytical properties for evolving black-hole spacetimes, particularly for horizon properties,[19] for the initial value problem[17,20,21]; and a simultaneous improvement of numerical techniques and algorithms.[18] There are a number of proposals for large, teraflop/sec machines—massively parallel designs, but numerical relativity is a field theory which can be adapted (is being adapted) to parallel coding—to be developed in the next few years. For instance, Oak Ridge National Laboratories has an ongoing development strategy in this regard.[22] And finally there is the possibility that a very sensitive set of detectors using laser interferometry to measure relative separations of $\sim 10^{-22}$, or less, will become operational within the decade, providing a unique opportunity for prediction, and for comparison to observation.[23]

Weak gravitational interactions are computable via various weak-field approximations. It is only strong field interactions, like the two-black-hole case, that require the full nonlinear computation. These are however the ones which produce the strongest and most unique wave signatures.

There are a number of choices to be made in designing a two-black hole code, and I will describe some of the choices, and eventually plot the computationally simplest, though probably not the best, path through them. In the next section I first give the analytical background for the two black hole problem.

THE EINSTEIN EQUATIONS

For a full exposition, see any standard Relativity text, e.g., Misner, Thorne, Wheeler.[24]

The general form of the Einstein equations is usually given in a very compact 4-dimensional representation

$$G_{\mu\nu} \equiv R_{\mu\nu} - \frac{1}{2}g_{\mu\nu}R = 8\pi G T_{\mu\nu}. \qquad (1)$$

Here μ, ν label components of 4-dimensional objects ($\mu, \nu = 0, 1, 2, 3$) where

'0' is time and 1, 2, 3 are spatial coordinate labels. The quantity $G_{\mu\nu}$, called the Einstein tensor, is a 4 × 4 symmetric object, and thus has 10 free components. It is defined in terms of the Ricci curvature $R_{\mu\nu}$, and the trace R of the Ricci curvature; the Ricci curvature itself is a nonlinear combination involving the fundamental *metric* tensor $g_{\mu\nu}$, second derivatives of $g_{\mu\nu}$ and quadratic terms involving first derivatives of $g_{\mu\nu}$. $g_{\mu\nu}$ is a function in principle—and in practice for realistic systems—of all three spatial coordinates, and of time: $g_{\mu\nu} = g_{\mu\nu}(x^\alpha) = g_{\mu\nu}(t, x, y, z)$. The final quantity in this equation is $T_{\mu\nu}$, which is formed from matter variables, e.g., densities and pressures. Here a slight simplification appears, for we treat black holes as *pure gravitational* entities. No matter appears, $T_{\mu\nu} = 0$, and we are dealing with what are called the vacuum equations.

Equation (1) is a 4-dimensional representation, but our knowledge of physics is instead based very much on 3-dimensional intuition about things evolving in time; that is the way we typically compute. So we introduce a splitting which respects our intuition. We assume there is a global definition of time, t, such that t = constant defines 3-dimensional spaces. The choice of t, and of these 3-spaces, is arbitrary, except that the 3-spaces must be *space-like*: No physical signal can propagate from one point to another point on any such 3-space. Each such 3-space thus corresponds to our Newtonian idea of *space at some instant of time*.

We are of course dealing with relativity, and so we want to incorporate a generalization of the special Relativity metric:

$$ds^2 = -dt^2 + dx^2 + dy^2 + dz^2.$$

(This incorporates the Euclidean 3-space, x, y, z; and it incorporates a time t, which is *proper time* for objects at rest in the x, y, z coordinates: if $dx = dy = dz = 0$ then $ds^2 = -(d(\text{proper time}))^2 = -dt^2$.)

Our generalization is to write

$$ds^2 = -\alpha^2 dt^2 + \gamma_{ij}\left(dx^i + \beta^i dt\right)\left(dx^j + \beta^j dt\right) \tag{2}$$

where α, β^i, and γ_{ij} are all functions of the x^i and of t (i, j run over 1, 2, 3, i.e., spatial coordinates xyz). Also there is the inevitable relativistic implied sum over repeated indices. If $\alpha = 1$, $\beta^j = 0$ and $\gamma_{ij} = \delta_{ij}$ the δ symbol, then (2) reduces to the special relativity form (1). In general, we see that the constant-t spaces are given by some distortion of Euclidean geometry [If $dt = 0$ then $ds^2 = \gamma_{ij} dx^i dx^j$], and in Relativity it is this distortion that carries the information about the gravitational field. The quantity β^i, which we mostly ignore henceforth, allows the "same" point at two different times to have different spatial coordinates; i.e., it allows the spatial coordinates to slide around as a function of time. This arbitrariness is an important feature of the covariance implied by Relativity, but for now set $\beta^i = 0$. Finally α gives the relation (point for point) between our arbitrary coordinate time t, and the proper time of a particle at that point:

$$ds^2 = -\alpha^2 \, dt^2 \,,$$

if $\beta^i = 0$ and $dx = dy = dz = 0$.

Much work has established that the Einstein equations are in fact a constrained hyperbolic system. Constrained: certain consistency conditions among the components γ_{ij} and their momenta must be satisfied on one "initial" constant-time space; these consistencies are then maintained in the subsequent hyperbolic evolution. As implied by the way we constructed Eq. (2), the α and β^i depend on an arbitrary choice of time function t, and are not dynamical quantities.

CONSTRAINTS

The constraint equations are relations between the metric γ_{ij} and its momentum π_{ij} (essentially its time derivative):

$$H \equiv \gamma^{-1/2} \left(\pi^{ik} \pi_{ik} - \frac{1}{2} \left(\pi^k{}_k \right)^2 \right) - \gamma^{1/2} \, {}^3R = 0 \tag{3}$$

$$H^i = -2\pi^{ij}{}_{|j} = 0 \,. \tag{4}$$

The second of these says that the divergence of the momentum π^{ij} must vanish ($|$ denotes a covariant space derivative using γ_{ij} as the metric). The first equation involves $\gamma \equiv \det \gamma_{ij}$, and ${}^3R =$ the trace of the Ricci tensor formed from γ_{ij}. Index raising is done using γ_{ij} and its inverse; for instance $\pi_k^k = \gamma^{kl} \pi_{kl}$, with implied sums, where γ^{kl} is pointwise the matrix inverse to γ_{lm}: $\gamma^{kl} \gamma_{lm} = \delta^k{}_m$.

Because ${}^3R$ involves second spatial derivatives of γ_{ij}, the first constraint is an elliptic equation for γ_{ij}, if the momentum π_{ij} is given. The second condition is a 'transverse' condition on π_{ij}. The standard way[21] to solve these equations is to suppose t is chosen so that π_k^k vanishes initially. Then the metric is rewritten involving a conformal factor ψ:

$$\gamma_{ij} = \psi^4 \tilde{\gamma}_{ij} \,.$$

For two black holes we choose the background metric $\tilde{\gamma}_{ij} = \delta_{ij}$, i.e. Euclidean, but the topology is not Euclidean: there are "holes" in the space. By choosing also a conformal mapping for π_{ij}, one can write the linear equation $\pi^{ij}{}_{|j} = 0$ in terms of a potential. This can be solved essentially analytically—a technique similar to the method of images.[20] The obtained A_{ij}, the conformally related π_{ij}, then appears in nonlinear equation (3) as a source for $\nabla^2 \psi$:

$$-8\nabla^2 \psi = \left(A_{ij} A^{ij} \right) \psi^{-7} \,. \tag{5}$$

This has to be solved in a 3-space with holes (two holes, for the two black hole case). We proceed by linearizing and iteration. We use some "old" initial guess ψ_0 on the right side of (5) and write $\psi^{-7} = (\psi_0 + \delta\psi)^{-7} \approx \psi_0^{-7}\left(1 - \frac{7(\psi_{\text{new}} - \psi_0)}{\psi_0}\right)$, solve for a new ψ by solving the linear equation resulting, use the new ψ as the "old" and iterate. Since this has been reduced to a straightforward linear equation, there are a number of techniques to carry out this solution. We currently have an ADI (Alternating Direction Implicit) solver for this problem. In order to solve the elliptic equation we need a boundary condition at the black hole 'throat.' This, in the Euclidean space of the calculations, is

$$\left. \frac{n^i \partial \psi}{\partial x^i} \right|_{\text{Hole Surface}} = -\frac{\psi}{2a_{\text{Hole}}} \quad (6)$$

where a_{Hole} is the (Euclidean) radius of the cutout hole. This leads to finite value and finite slope at the hole surface. At the outer edge of the computational grid, we use the fact that ψ varies as $1 +$ Newtonian potential:

$$\psi \sim 1 + \frac{C}{r} + \cdots \quad (7)$$

where C is an (a priori undetermined) constant, and r is the distance from the center of mass. The mixed *Robin* boundary condition

$$r \frac{\partial \psi}{\partial r} + (\psi - 1) = 0 \quad (8)$$

guarantees this behavior without demanding an a priori knowledge of C.

EVOLUTION

The evolution equations for this (second order) hyperbolic system are usually written as a set of coupled first order equations. The evolution equation for γ_{ij} is simplest:

$$\frac{\partial \gamma_{ij}}{\partial t} = 2\alpha \gamma^{-1/2}\left(\pi_{ij} - \frac{1}{2}\gamma_{ij}\pi_k^k\right) + \beta_{i|j} + \beta_{j|i}. \quad (9)$$

The appearance of the covariant derivatives of β on the right is required because β moves the point of observation on the next constant-time space surface, where the 'new'—and the 'old' γ_{ij} are different. Concentrate on the first term. It has a factor α, because α large means larger *proper time* step for a given coordinate time step Δt, meaning we expect larger changes in γ_{ij}. Otherwise we have only a tensor generalization of $\dot{x} = p$; the rate of change of the coordinate (here γ_{ij}) is given by the momenta (here π_{ij}).

The remaining vacuum field equation, which I shall not write explicitly, gives the time evolution of π_i,

$$\frac{\partial \pi_{ij}}{\partial t} = \left(\text{complicated nonlinear expression involving } \pi_{ab}, \gamma_{cd}, {}^3R_{ef}, \alpha, \beta^i\right). \quad (10)$$

Recall that $^3R_{ef}$ involves second derivatives of γ_{ab}, so this equation is like a wave equation:
$$\frac{\partial \pi_{..}}{\partial t} \sim \frac{\partial^2 \gamma_{..}}{\partial t^2} \cong \frac{\partial^2 \gamma_{..}}{\partial x^2}.$$
However it is nonlinear. It also depends on our arbitrarily specifiable constant time spaces, hence on α, β^i.

The usual way to solve the equation is to specify the topology (holes, here), specify the momentum and spin of the holes (the analytical solution of the momentum constraint, Eq. (4) accomplishes this) and solve the initial value problem for ψ; then construct all the initial fields.

Then make some choice which implicitly or explicitly specifies β^i and α, and evolve the dynamical equations (9)–(10) into the future. A complicating feature is that while the constraint (initial value) solution is maintained at every time in an analytical evolution, there is no guarantee this will be the case for arbitrarily finite differenced evolution. To date no general constraint-maintaining finite method has been developed. One can then either just monitor the maintenance of the constraints as a validity check, or can instead force their solution at every line. Because the constraints are elliptic in character, such repeated solution has a stabilizing effect on typical evolutions. It may also be excessively dissipative, and in any case is very time costly.

OPTIONS

The discussion so far has touched on some of the decision points—and the resulting decisions—made in a first attack on the two-black-hole problem. In fact there are a number of such binary choices, and I list here those that come to mind.

- Spacelike Time Slices or Null Time Slices?
- Rectangular Coordinates or Adapted (generalized bispherical)?
- Explicit Differencing for Evolution or Implicit Differencing or Adaptive Mixed Differencing?
- Unconstrained evolution or constrained evolution?
- Wave extraction via spacelike slices or via null slices?
- Questions of elegance and delicacy.
- Do we have a structure for development?

The group at the University of Texas now has a code under development which chooses among these alternatives. I will indicate and try to explain the alternatives in sequence.

- I have already given an explanation of the decomposition of a 4-dimensional space-time into a sequence of spaces, evolving in time. We required that these spaces be *spacelike* with no two events causally connected in them. This is the choice made in the Texas code, though it should be noted that spaces that become lightlike, at least near spatial infinity, may be more suitable for wave problems. Since gravitational waves travel at the speed of light, it can be expected that they will asymptotically lie almost exclusively in one of the light-like spaces. In the spacelike formulation we use instead, the radiation moves outward through very many slices. The lightlike option is at least interesting, and should be investigated.

- There is an obvious attraction to using generalized bispherical or other adapted coordinate systems to compute the two black hole problem. The ability finely to fit the boundary conditions at the hole surface and at the outer boundary is the strongest reason to use them. Nonetheless, there are also advantages to rectangular coordinates. In particular, Relativity involves a number of tensor quantities which can be computed, in rectangular coordinates, in a repetitive way. We have written several programs which produce FORTRAN code by this process, and rectangular coordinates are the choice we take for the Texas code.

- Explicitly differenced hyperbolic codes must satisfy some version of the Courant condition: the timestep can be at most a fraction of the time it takes for light (or other disturbance) to propagate across a computational cell. Otherwise the code develops numerical instabilities. Explicit methods are often quite easy to program and debug. Implicit methods, on the contrary, tend to suppress short time fluctuations, and to drive the system to a long-term, relaxed state. They are however often much more difficult, and numerically more costly, to solve than are explicit methods. The Texas code (a variant of the Laguna/Kurki-Suonio/Matzner[25] cosmology code, see also Matzner[26]), in the interest of quickly obtaining working results, uses an explicit method. It will be interesting to compare this to implicit (Evans,[13] Allen[27]) or mixed implicit/explicit (Demkowicz, Oden, and Rachowicz[28]) codes.

- In discussing the initial data (constraint) problem I pointed out that the *numerical* maintenance of the constraints has not been accomplished in any code. One *may* prefer to evolve only some of the dynamical variables and resolve the "initial" value problem at every timestep. This is certain to be more time-consuming than simply monitoring the evolution of the constraints—they should stay zero in an explicit evolution. The constraints are four conditions on a total of 12 variables—the γ_{ij} and the π_{ij}—and are simply one measure of accuracy of evolution. Nonetheless Hobill (1990) has convincingly shown that if, for instance the Hamiltonian (nonlinear) constraint produces a value that corresponds to an effective negative energy density, then the system does behave as if it has a negative energy source, with all the instability that implies. Hence it is important the constraints be maintained. It may be possible to devise difference algorithms which explicitly maintain the constraints (numerically), but no one has yet succeeded in that. The Texas code uses unconstrained evolution but monitors the constraint.

- 'Wave extraction' refers to the determination of the gravity wave profile and amplitude at infinity, given all the dynamics of the strong field black hole region. The gravitational radiation will be strongly concentrated over a limited time of reception by a distant observer, and because the radiation propagates at the speed of light, this suggests that 'light cones' or at least an asymptotic lightlike surfaces would be the best way to monitor the radiation 'at infinity.' The radiation field would then be strong only for a limited range of light surfaces—those that cross infinity with the wave. In contrast, if one has spacelike evolution slices, then the wave slowly propagates outward in successive time slices. Since the point is to measure the wave far from the source, this means that a large number of successive evolution slices have to be computed, just to get the wave to 'infinity.' Errors will accumulate, and this may be a limiting feature of the technique.

Although some theoretical work has been done on the lightlike space wave extraction technique, no numerical application has been made. The Texas code in particular is not yet so well developed as to require addressing this question.

- Questions of elegance and delicacy arise in black hole problems because of the need to describe the large scale Newtonian-like background of the colliding black holes together with the oscillatory, propagating radiation. Differencing methods that describe one field are not well adapted to describing the other. The straightforward solution to this problem appears to be to use a large number of zones in the mesh. Consistent with other kinds of 3-dimensional systems evolving in time, (e.g., hydrodynamics) we find from 2-dimensional experiments (Hobill[29]) that $\sim (500)^3$ is the minimum acceptable size for 3-dimensional evolutions; below we take $(1000)^3$ as typical. In addition there is a perhaps related delicacy question, in that the usual way of numerically evolving black holes does not in fact produce a static representation of even a single black hole, not even asymptotically. This has the result that no one has evolved even a single black hole space time 'forever,' and this question must be attacked before we can confidently predict 2-black-hole spacetimes.

- Do we have a structure for development? In the technical sense, the Texas code is a possible structure for development. It's features:

  - rectangular coordinates
  - spacelike slicing
  - Alternating Direction Implicit for solution of Initial Value Problem
  - Unconstrained evolution with constraint monitoring
  - Modular code, allowing easy modification of algorithms and of choices for α, β^i, and for boundary algorithms.

The initial value problem for 2-black holes is running. The evolution code has not yet run on the black hole case, but has been tested for simpler topologies in cosmology.

Except for its modularity, almost every feature of this code may be considered tentative, and other choices possible. Some timings have been done with the cosmology code. It requires about 2×10^{-5} seconds per zone-cycle (Cray Y-MP, about 2×10^8 floating point operations per second $\equiv$ 0.2 Gflop/sec) and requires about 80 words (640 Bytes) of variable storage per zone. Thus a 10^9 zone run would require $\sim$ 3 hours per cycle, and 80×10^9 Cray words = 7×10^{11} Bytes of storage. The largest existing Cray memory size is $\sim 0.5 \times 10^9$. Hence memory currently precludes running even one cycle of a $(1000)^3$ code. For full evolution, one expects [from 2-d codes, Hobill[29]] to run $\sim 10^4$ cycles; the elapsed time of evolution becomes years, regardless of the memory requirements.

However, the hardware developments now in sight offer the way out. There are a number of feasible proposals for massively parallel machines, which will sustain 10^{12} floating point operations per second (1 Teraflop/sec): "teraflop machines." These machines must also contain massive memories to make them feasible. Their speed factor $\sim 10^4$ over Y-MP processors reduces the evolution time for one of the proposed black hole codes to $\sim$ hours. It has been suggested[30] that algorithmic developments will lead to an additional factor $\sim$ 100 in speed improvements. While I have no doubt that there will be substantial speed developments, I prefer to conservatively discount them against the complexity of the 2-black-hole problem, and estimate $\sim$ hours (teraflop machine) for accurate 2-black-hole computations in 1996.

The complexity of the 2-black hole problem and the number of possible options and difficulties remaining means that a very substantial effort remains to reach such a 1996 goal. At present there are perhaps five or six groups in the United States, and one or two outside, who are actively approaching this problem. It is our hope that close collaboration and cooperation can be maintained, and that the problem will lead to developments in algorithms, hardware and software, as well as new results in black-hole and gravitational wave physics.

ACKNOWLEDGMENTS

This work is supported by NSF grant PHY 88-06567, by a Cray Research Grant, and by The University of Texas Center for Relativity. Much of this work was done in collaboration with P. Laguna (Los Alamos National Laboratories) and H. Kurki-Suonio (Lawrence Livermore National Laboratories).

REFERENCES

1. M. May and R. H. White, Phys. Rev. **141**, 1232 (1966).

2. S. G. Hahn and R. Lindquist, Ann. Phys. (NY) **29**, 304 (1964).

3. L. L. Smarr, Ph.D. dissertation, The University of Texas at Austin (1975).

4. K. Eppley, Ph.D. dissertation, Princeton University (1975).

5. J. Centrella, Astrophys. J. **241**, 875 (1980).

6. T. Piran, Journal of Computational Physics **35**, 254 (1980).

7. K.-I. Oohara and T. Nakamura, Progress in Theoretical Physics **81**, 360 (1989).

8. K.-I. Oohara and T. Nakamura, in *Frontiers in Numerical Relativity*, Evans, Finn, and Hobill, eds. (Cambridge University Press, Cambridge, 1989).

9. T. Nakamura, K.-I. Oohara, and Y. Kojima, Progress in Theoretical Physics Suppl. **90** (1987).

10. E. Seidell, in Texas Symposium on 3-Dimensional Numerical Relativity, R. Matzner, ed., Relativity Center "Armadillo," The University of Texas at Austin (1990).

11. M. Choptuik, in Texas Symposium on 3-Dimensional Numerical Relativity, R. Matzner, ed., Relativity Center "Armadillo," The University of Texas at Austin (1990).

12. A. Abrahams, in Texas Symposium on 3-Dimensional Numerical Relativity, R. Matzner, ed., Relativity Center "Armadillo," The University of Texas at Austin (1990).

13. C. Evans, in Texas Symposium on 3-Dimensional Numerical Relativity, R. Matzner, ed., Relativity Center "Armadillo," The University of Texas at Austin (1990).

14. P. Laguna, in Texas Symposium on 3-Dimensional Numerical Relativity, R. Matzner, ed., Relativity Center "Armadillo," The University of Texas at Austin (1990).

15. H. Kurki-Suonio, in Texas Symposium on 3-Dimensional Numerical Relativity, R. Matzner, ed., Relativity Center "Armadillo," The University of Texas at Austin (1990).

16. J. Thornburg, Classical and Quantum Gravity **4**, 1119 (1987).

17. G. Cook, preprint, Cornell University (1991).

18. J. Wilson, in Texas Symposium on 3-Dimensional Numerical Relativity, R. Matzner, ed., Relativity Center "Armadillo," The University of Texas at Austin (1990).

19. S. W. Hawking, Phys. Rev. Lett. **26**, 1344 (1971).

20. A. Kulkarni, L. C. Shepley, and J. York, Phys. Lett. **96A**, 228 (1983).

21. J. York, in *Sources of Gravitational Radiation*, L. L. Smarr, ed. (Cambridge University Press, Cambridge, 1979).

22. T. H. Dunigan, Performance of the INTEL iPSC/860 Hypercube, ORNL/TM-11491, Oak Ridge National Laboratory (1990).

23. R. Isaacson, in Texas Symposium on 3-Dimensional Numerical Relativity, R. Matzner, ed., Relativity Center "Armadillo," The University of Texas at Austin (1990).

24. C. Misner, K. Thorne, and J. A. Wheeler, *Gravitation* (Freeman, San Francisco, 1973).

25. P. Laguna, H. Kurki-Suonio, and R. Matzner, submitted to Phys. Rev. (1991).

26. R. Matzner (1991), Annals New York Academy of Science (in press).

27. G. Allen, in Texas Symposium on 3-Dimensional Numerical Relativity, R. Matzner, ed., Relativity Center "Armadillo," The University of Texas at Austin (1990).

28. L. Demkowicz, J. T. Oden, and W. Rachowicz, Computer Methods in Applied Mech. and Engineering **84**, 275 (1990).

29. D. Hobill, in Texas Symposium on 3-Dimensional Numerical Relativity, R. Matzner, ed., Relativity Center "Armadillo," The University of Texas at Austin (1990).

30. L. Smarr, Talk at Toronto Numerical Relativity Workshop (1991).

CHAOS → ERGODICITY → ISOTHERMAL DYNAMICS

John H. SLOAN,
Department of Physics, Ohio State University, Columbus, OH 43210,
Dimitri KUSNEZOV[†] and Aurel BULGAC
Department of Physics and Astronomy and
National Superconducting Cyclotron Laboratory,
Michigan State University, East Lansing, MI 48824.

ABSTRACT

We present a cursory review of the new developments of the finite temperature molecular dynamics approach along with some selected applications.

ISOTHERMAL MOLECULAR DYNAMICS

In recent years molecular dynamics (MD) techniques have been improved and applied to a wide range of physical systems: fluids, gases, molecular systems, atomic clusters and in some cases lattice gauge calculations. The extension of these methods to non-equilibrium phenomena is an unbeatable advantage of molecular dynamics methods over other existing approaches. One can safely say that MD techniques are on the rise and their potential has not yet been fully exploited and explored. We shall not try to present a review of what was already done in this field, but instead we shall concentrate on presenting our recent developments in isothermal MD.

In 1984 Nosé[1] presented a very elegant method of simulating the coupling of a classical system with N degrees of freedom to a thermal bath. He showed that, by adding only one additional degree of freedom to the system under study, one can emulate the action of an infinite number of degrees of freedom of the thermostat. Subsequently, Hoover[2] simplified the approach by replacing the additional coordinate and momentum introduced by Nosé with a single pseudofriction coefficient, thus and eliminating a rather unintuitive method of time sampling of the phase space. Jellinek[3] pointed to some other potentially interesting ways to generalize Nosé's initial scheme. However, in all these formulations of the isothermal MD scheme it was assumed that the resulting equations of motion generate an ergodic trajectory in the enlarged phase space, not merely a chaotic one. Ergodicity implies that the integral over the phase space, with the appropriate Boltzmann factor in the measure, is identical to a time average over the trajectory followed by the system. This is just the opposite of what the forefathers of statistical mechanics told us to do. The absence of ergodicity in some relatively simple cases was observed almost right away[2], but ways to cure this undesireable feature were only found recently[4,5].

In Refs. 4,5 we proposed the following equations of motion in order to study

24 Chaos→Ergodicity→Isothermal Dynamics

a system coupled with a thermostat at temperature T

$$\dot{q}_i = \frac{\partial H(p,q)}{\partial p_i} - h_2(\xi)F_i(q,p), \quad \dot{p}_i = -\frac{\partial H(p,q)}{\partial q_i} - h_1(\zeta)G_i(q,p), \quad (1a,b)$$

$$\dot{\zeta} = \alpha \sum_{i=1}^{N}\left[\frac{\partial H(p,q)}{\partial p_i}G_i(p,q) - T\frac{\partial G_i(p,q)}{\partial p_i}\right], \quad (2a)$$

$$\dot{\xi} = \beta \sum_{i=1}^{N}\left[\frac{\partial H(p,q)}{\partial q_i}F_i(p,q) - T\frac{\partial F_i(p,q)}{\partial q_i}\right], \quad (2b)$$

where $H(p,q)$ is the Hamiltonian of the envisaged system, $G_i(p,q)$, $F_i(p,q)$, $h_1(\zeta)$, $h_2(\xi)$ are some arbitrary functions, α, β are constants and T is the absolute temperature at which one is interested in studying the system. We shall call the extra variables ζ and ξ pseudofriction coeficients and their introduction into the equations of motion is aimed at simulating the coupling to the thermostat.

It can be shown that in the extended phase space $(q_1,...,q_N,p_1,...,p_N, \zeta, \xi)$ the equations of motion (1,2) are compatible with the existence of the following invariant measure

$$d\mu = \mathcal{N} \exp\left\{-\frac{1}{T}\left[H(p,q) + \frac{1}{\alpha}g_1(\zeta) + \frac{1}{\beta}g_2(\xi)\right]\right\} dq_1...dq_Ndp_1...dp_Nd\zeta d\xi, \quad (3)$$

where $dg_1(\zeta)/d\zeta = h_1(\zeta)$ and $dg_2(\xi)/d\xi = h_2(\xi)$ and $\mathcal{N}$ is a normalization constant. Consequently, if Eqs. (1,2) generate an ergodic trajectory, one can replace the phase space integral over the measure (3) with a time average integral.

The equations of motion (1-2) have a rather unexpected structure. First of all, the symplectic structure has been lost in the augmented phase space $(q_1,..., q_N, p_1,..., p_N, \zeta, \xi)$ or, in other words, the equations of motion no longer have a Hamiltonian character. As a consequence, the phase space volume is not conserved, since

$$\frac{\partial \dot{q}_i}{\partial q_i} + \frac{\partial \dot{p}_i}{\partial p_i} \neq 0, \quad (i = 1,...,N). \quad (4)$$

However, the time or phase space average of the r.h.s. of Eq. (4) vanishes. Consequently, under the influence of the thermostat the phase space "breathes". The additional terms in Eqs. (1) are referred to as pseudofriction for a simple reason. A friction force has a dissipative character, it leads to a loss of energy of the system. The pseudofriction terms have the same formal appearance as a dissipative force. However, since the pseudofriction variables ζ, ξ take both positive and negative values as functions of time, energy is either extracted or given to the system in such a way as to achieve thermalization, and the motion never ceases. On this augmented phase space one can introduce a phase space probability[2] $f(q_1,...,q_N, p_1,...p_N, \zeta, \xi, t)$ which can be easily shown to satisfy the following generalized Liouville equation

$$\frac{\partial f}{\partial t} + \sum_{i=1}^{N}\left[\frac{\partial(f\dot{q}_i)}{\partial q_i} + \frac{\partial(f\dot{p}_i)}{\partial p_i}\right] + \frac{\partial(f\dot{\zeta})}{\partial \zeta} + \frac{\partial(f\dot{\xi})}{\partial \xi} = 0. \quad (5)$$

It can be easily checked that the weight of the measure $d\mu$ (3) is a stationary solution of Eq. (5). This is the only time-independent solution of the generalized Liouville equation (5) if and only if the system described by Eqs. (1-2) is ergodic. Neither the form nor the number of the pseudofriction terms is unique or prescribed by "theory". This should be looked upon as an advantage, allowing the possibility of modeling different types of couplings to a thermostat. The rate of energy exchange between the system and the thermostat can also be controlled in this way. The rather particular and to some extent peculiar type of coupling we have chosen in Eqs. (1) is to some extent accidental. Since at equilibrium the distribution of coordinates and momenta are uncorrelated, we have chosen this form of coupling to the thermal bath merely to achieve thermalization as quickly as possible. If someone is interested in dynamical properties of a thermalized system (e.g. autocorrelation functions for different quantities) the specific form of the coupling might be crucial. We shall not dwell anymore on these questions here and refer the reader to Ref. 5 for a detailed discussion of this.

SHORT DESCRIPTION OF SEVERAL APPLICATIONS

As a simple illustration of the effectiveness of this approach we present in Fig. 1 the time evolution of the relative deviations of the computed distributions for variables (q, p, ζ, ξ) for the case of a simple 1D harmonic oscillator $H(q,p) = (p^2 + q^2)/2$ coupled to a thermal bath. This example is important for several reasons: it is one of the simplest examples in which nonergodicity of the Nosé-Hoover scheme was observed; at low temperatures essentially every system can be approximately described by coupled harmonic oscillators; the harmonic oscillator is in a way "very regular" and relatively hard to make chaotic; numerical simulations can be easily compared to exact results. We describe the coupling to the thermostat most of the time by the following functions (or slight modifications of these):

$$h_1(\zeta) = \zeta^3, \quad g_1(\zeta) = \zeta^4/4, \quad h_2(\xi) = \xi, \quad g_2(\xi) = \xi^2/2, \quad (6a)$$

$$F_i(q,p) = q_i^3, \quad G_i(q,p) = p_i, \quad (i = 1, ..., N). \quad (6b)$$

We link the ability of this type of approach to generate canonical distributions with the high degree of chaoticity of the trajectories. In Fig. 2 we present the time evolution of a circle in the phase space (q,p) for a 1D harmonic oscillator using the present approach as well as the earlier Nosé-Hoover scheme. The calculated Lyapunov exponents[5] are rather big, which fact ensures a rapid exploration of the phase space, starting essentially from any point in the phase space. The fact that the motion is chaotic (positive Lyapunov exponents) is in itself not a sufficient property of this scheme to describe a system in thermal equilibrium. Ergodicity implies chaoticity but the reverse is generally not true. It can be shown as well that the equations of motion (1-2) do not have stable fix points.

26 Chaos→Ergodicity→Isothermal Dynamics

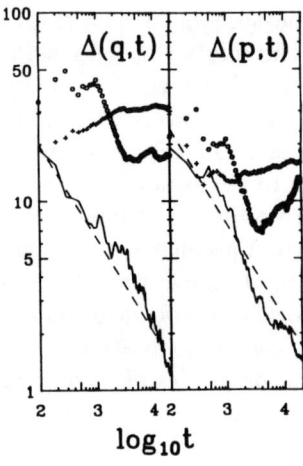

Fig. 1. The time evolution of the integrated absolute difference between the computed and the exact thermal distribution (solid) for the specified variables, $\Delta(\phi,t)$, ($\phi = q,p$). Circles and crosses are results from the Nosé-Hoover scheme[2]. The dashed line corresponds to $\Delta(\phi,t) = const/t^{1/2}$.

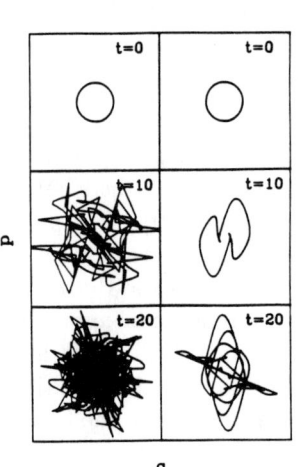

Fig. 2. Time evolution of a circular set of 1000 initial points using the couplings of Eqs. (6) (left) compared to Nosé-Hoover[2] for a 1D harmonic oscillator.

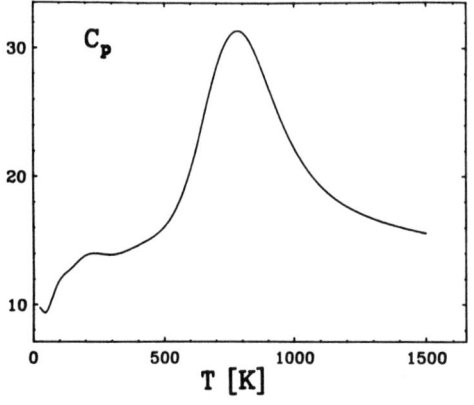

Fig. 3. Specific heat for the potential energy for a Na_8 microcluster.

Recently[6], we have applied this technique to the study of thermal properties of Na microclusters. One can treat the interaction among the ions and the effect of valence electrons in a Na microcluster by some suitably defined effective many-particle (not two-body) interactions among atoms only. In the range of temperatures we were interested in, the motion of the atoms can be treated fairly well at the classical level and consequently the procedure we sketched above can be applied in a straightforward manner. We have computed a wide range of properties, such as shape, bond lengths, momenta of inertia, rotational, vibrational and potential energies, density of states, specific heats, etc. in order to fully characterize the behaviour of the microcluster upon changing the temperature. We shall limit ourselves here to displaying only a couple of them. In this case we have introduce six pseudofriction coefficients, three for coordinates and three for the momenta (one for each cartesian direction). The motivation for this is to ensure that both the total angular momentum and its direction are not conserved quantities, and a sufficiently high randomness was induced. Also the couplings (6) were slightly modified in such a way as not to affect the center of mass coordinate and total momentum of the system, since we were only interested in the intrinsic properties of the cluster. The equations of motion in this case read (only for the x-components here)

$$\dot{x}_i = \frac{p_{xi}}{m} - \xi_x \left[x_i^3 - \frac{1}{N} \sum_{k=1}^{N} x_k^3 \right], \quad \dot{p}_{xi} = -\frac{\partial H}{\partial x_i} - \zeta_x^3 \left[p_{xi} - \frac{1}{N} \sum_{k=1}^{N} p_{xk} \right], \quad (7a, b)$$

$$\dot{\xi}_x = \beta \sum_{i=1}^{N} \left[\frac{\partial H}{\partial x_i} x_i^3 - 3\frac{N-1}{N} T x_i^2 \right], \quad \dot{\zeta}_x = \alpha \left[\sum_{i=1}^{N} p_{xi}^2 - (N-1)T \right]. \quad (7c, d)$$

In spite of the relatively small number of particle in a microcluster its properties are rather complex. One of the most spectacular features is the coexistence of liquid and solid phases and the occurrence of phase transitions. In Fig. 3 the specific heat (for the potential energy U only) for a Na_8 cluster is shown. One can see the presence of two peaks, a smaller, less defined one around 200 K, corresponding to a transition to a "molten" or "glassy" state, and a bigger one around 700-1000 K, corresponding to the "melting" of the cluster. From the runs at about ten different temperatures, ranging from 25 to 1500 K we extracted the density of states (here for the potential energy only) in a rather wide range of excitation energies of the cluster, see Fig. 4. In each run we had 10^6 "measurements" (the total time of the trajectory between 200 psec for high T and 2000 psec for low T). Using the fact that the distribution of the potential energy in a canonical ensemble is $P(U,T) = \rho(U)\exp(-U/T)/Z(T)$, where $\rho(U)$ is the density of states and $Z(T)$ is the partition function (for potential energy only) one can easily "measure" $\rho(U)$ in a run, up to an arbitrary multiplication factor. One can fix the normalization of the density of states in several ways if that is required as well. One million configurations allowed us to define the density of states over essentially twelve orders of magnitude, in an energy interval around

28 Chaos→Ergodicity→Isothermal Dynamics

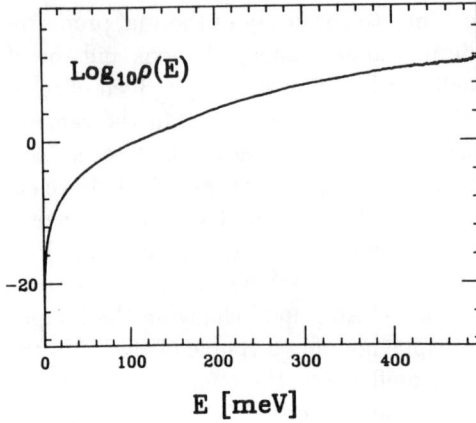

Fig. 4. The unnormalized density of states (potential energy only) for a Na_8 microcluster as extracted from isothermal MD. At energies near 150 meV the logarithm of $\rho(E)$ has a slight negative curvature. This is responsible for the occurence of the phase transition.

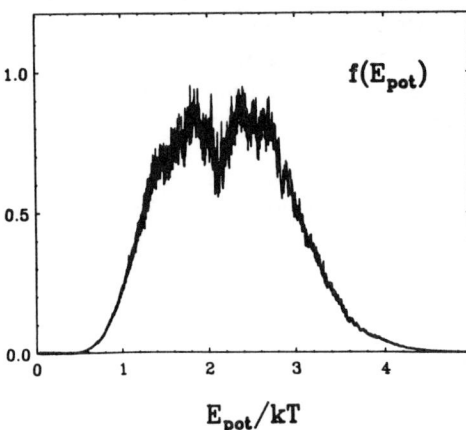

Fig. 5. The distribution of the potential energy at $T = 820\ K$ for Na_8. This type of double-humped structure is characteristic of the phase transition in microclusters and corresponds to the coexistence of a liquid and a solid phase.

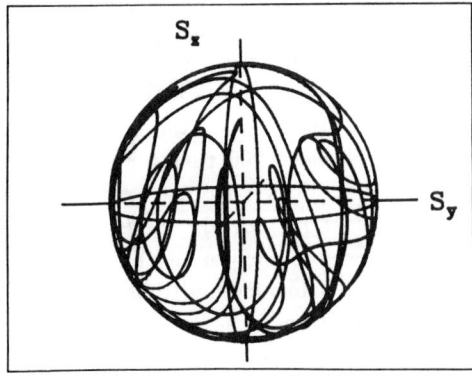

Fig. 6. Chaotic trajectory of a classical spin (SU(2)) coupled to a thermal bath on the sphere S_2.

the average $U(T)$. By piecing together different portions of $\rho(U)$ we defined the density of states with quite high accuracy for over more than 20 orders of magnitude. The same thing can be done rather accurately actually using only about half of these runs, over the same range of temperatures and of the same length. The knowledge of the density of states allows one to predict the thermal behaviour of the system. In contradistinction to an infinite system, the phase transitions are not sharp, however, they are quite well defined and amenable to observation. The distribution of (potential) energy can have a double-humped structure at temperatures around the phase transition, see Fig. 5.

In Refs. 4,10 we presented a generalization of the isothermal MD method to systems characterized by a compact phase space (e.g. classical limit of Lie algebras). The classical mechanics formalism in this case looks a bit unusual, there are "topological conserved" quantities, the Poisson brackets have the structure of Dirac brackets and the theory has a beautiful gauge structure. In spite of their exotic character, compact phase spaces appear quite often in different physical situations. In particular quantum mechanics can be cast into such a framework and this has proved extremely useful in some circumstances (e.g. large amplitude nuclear collective motion). In the case of a compact phase space one cannot define canonical variables globally, but instead one has to use different atlases. A naive attempt to directly apply the Nosé-Hoover scheme or our generalization of it to this case, by using local canonical variables, failed. Instead we have chosen a slightly different route[2,10] and managed to describe thermal properties of classical SU(2), SU(3) algebras and classical spin chains in this way. One possible form of the equations of motion which we investigated is

$$\dot{X}_i = J_{ij}(X)\left[\frac{\partial H(X)}{\partial X_j} - h(\zeta)D_j(X)\right], \quad (i=1,...,N), \quad (8a)$$

$$\dot{\zeta} = \alpha J_{ij}(X)\left[\frac{\partial H(X)}{\partial X_i}D_j(X) - T\frac{\partial D_j(X)}{\partial X_i}\right], \quad (8b)$$

where X_i are the (noncanonical) phase space variables, $J_{ij}(X)$ is the Poisson tensor, which defines the Poisson brackets, $H(X)$ is the Hamiltonian of the system. The functions $D_i(X)$ satisfy the constraints

$$\frac{\partial \dot{X}_i}{\partial X_i} = -h(\zeta)J_{ij}(X)\frac{\partial D_j(X)}{\partial X_i} \neq 0. \quad (9)$$

These equation of motion are compatible with the conservation of the following measure on the augmented phase space $(X_1,...,X_N,\zeta)$

$$f(X,\zeta) = \mathcal{N}\exp\left\{-\frac{1}{T}\left[H(X) + \frac{g(\zeta)}{\alpha}\right]\right\}, \quad (10)$$

where $h(\zeta) = dg(\zeta)/d\zeta$. We display in Fig. 6 the chaotic trajectory of a classical spin, coupled to a thermal bath, on a S_2 sphere. The time average of this trajectory reproduces the expected canonical distribution. As before, the number of pseudofriction coefficients is arbitrary and should be chosen in such a way as to ensure ergodicity. However, the number of pseudofriction coefficients can influence rather strongly the convergence properties of the method. In Fig. 7 we have plotted the average internal energy as a function of the time of the simulation for an SU(3) Hamiltonian for the case of two and three pseudofriction coefficients. The scheme with three pseudofriction coefficients leads to a significantly quicker thermalization of the system.

The so-called XY-model is a spin lattice model and has been studied quite often in the past[7]. The potential energy for the XY-model is

$$V(\theta) = - \sum_{<i,j>} \cos(\theta_i - \theta_j), \qquad (11)$$

where θ_i is the angle made by the spin i with some arbitrary direction in the plane and the sum is over all nearest-neighbor pairs on a square lattice with periodic boundary conditions. The equations of motion describing the isothermal dynamics in this case are)here we have introduce a simple redefinition of (ζ, ξ)

$$\dot{\theta}_i = p_i - \frac{\beta}{NT} \xi \sin^3 \theta_i, \quad \dot{p}_i = -\frac{\partial V(\theta)}{\partial \theta_i} - \frac{\alpha}{NT} \zeta^3 p_i, \qquad (12a,b)$$

$$\dot{\zeta} = \alpha \left[\frac{1}{NT} \sum_i p_i^2 - 1 \right], \quad \dot{\xi} = \frac{\beta}{N} \left[\frac{1}{T} \sum_i \frac{\partial V(\theta)}{\partial \theta_i} \sin^3 \theta_i - \sum_i 3 \sin^2 \theta_i \cos \theta_i \right]. \qquad (12c,d)$$

In the case when $\alpha, \beta \simeq O(1)$ the frequency of motion of the pseudofriction coefficients is comparable with the with the highest frequency mode of the spin system. We have investigated the cases $\alpha, \beta \simeq O(1), O(\sqrt{N})$ and $O(N)$, where N is the total number of spins for 16×16 and 64×64 lattices. The algorithm described above was slightly modified by introducing a refreshing of the momenta every $t = 64$ for the 64×64 lattice[9]. The main reason for us to do this was the fact that in this particular type of coupling to the thermal bath the zeroth mode is not thermalized properly and one should modify the Eqs. (12) in the same spirit as we have done for the Na microclusters[6], see Eqs. (7). We compared the results of the isothermal MD with standard hybrid Monte Carlo calculations[8a]. In Fig. 9 we present the autocorrelation function for the total magnetization at a temperature $T = 1/4$, below the phase transition, for the standard hybrid Monte Carlo method and the isothermal MD. The autocorrelation time versus the correlation length for the 64×64 lattice for the total magnetization (Fig. 9) and one isolated spin (Fig. 10) were computed for the same two approaches. As one can judge from the presented results, the isothermal MD approach seems to perform extremely well. We also intend to compare these results with the improved hybrid Monte Carlo algorithm suggested in Ref. 8b.

One extension of these methods to real quantum problems is straightforward. In the summer of 1990, two undergraduate students worked under our supervision on the applicability of isothermal MD to the finite temperature path integral

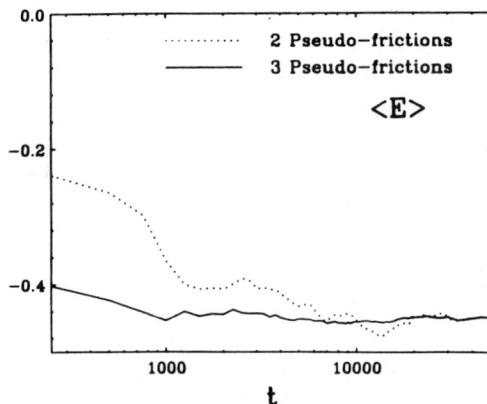

Fig. 7. Time evolution of the average energy $<E>$ for an $SU(3)$ Hamiltonian using two (dots) and three (solid) pseudo-frictions. In general, convergence to the canonical ensemble can be achieved much sooner by using additional pseudo-frictions.

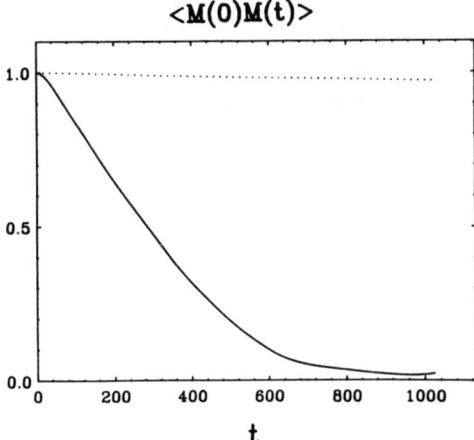

Fig. 8. XY-model auto-correlation function for the total magnetization using isothermal MD (solid) and standard hybrid Monte Carlo (dots). These results are for $\beta = 4$ on a 64^2 lattice with 160K statistics.

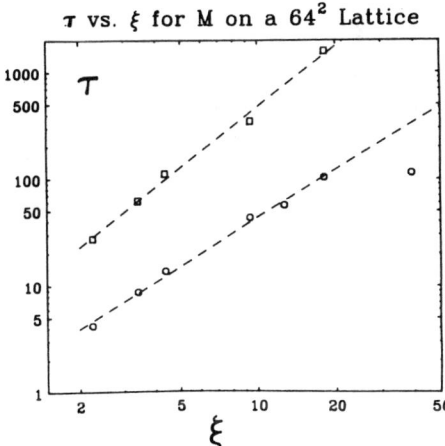

Fig. 9. Auto-correlation time versus correlation length for a 64^2 lattice. Isothermal MD (circles) methods have a noticeably smaller critical exponent and prefactor than standard hybrid Monte Carlo methods (boxes). (Results are preliminary, see Ref. 9 for details.)

description of the properties of a simple Ne_2 system[11]. Quantum propagators expressed through a Feynman path integral become equivalent to a many-body classical partition function, and consequently the above methods can be applied without any modifications. It is likely that one can devise a completely different approach as well, by modifying the time dependent Schrödinger equation through the addition of a properly chosen "random" field and/or making it nonlinear and/or relinquishing the hermiticity of the Hamiltonian. These modifications of the Schrödinger equation are in the same spirit as the modifications of the classical Hamilton equations of motion we described earlier.

We did not discuss another extremely interesting extension of the MD techniques, namely the study of non-equilibrium phenomena. We shall only present an extension of the present ideology to the study of Brownian motion[12]. Almost anyone is familiar with the phenomenology of the Brownian motion in the Langevin description. The Newton equation of motion is modified by adding a dissipative term (friction) and a random force (usually white noise). Both these additional terms lead to an irreversible time evolution. There is a long standing argument on whether the equation of motion for a Brownian particle should have a dissipative and irreversible character; it does not seem that there is a consensus yet. In a way the presence of the two additional terms looks like overdoing things. The random force should account for whatever the surrounding molecules do to the Brownian particle and the friction term looks a bit superfluous. MD ideology should in principle be able to describe Brownian motion as well, especially if one intends to study transport and nonequilibrium phenomena. In Ref. 12 we have shown that one can develop a deterministic and time-reversal invariant description for a Brownian particle. Instead of the celebrated Langevin equation, we introduced the following equations of motion for a one-dimensional Brownian particle (unit mass $m = 1$)

$$\dot{q} = p, \quad \dot{p} = -\alpha\zeta^3 p - \beta\xi(p^2 - a) - \gamma\varepsilon p^3, \qquad (13)$$

$$\dot{\zeta} = p^2 - T, \quad \dot{\xi} = p(p^2 - a) - 2Tp, \quad \dot{\varepsilon} = p^4 - 3Tp^2, \qquad (14)$$

where, as before, T is the absolute temperature in energetic units and a, α, β, γ are some arbitrary constants. These equations display a time-reversal invariance if the phase space coordinates change as $q, p, \zeta, \xi, \varepsilon \rightarrow q, -p, -\zeta, \xi, -\varepsilon$ when $t \rightarrow -t$. The right hand side of the equation for the momentum can be interpreted as the "thermal" force exerted by the medium on the particle. In Fig. 11 we show a characteristic time dependence of this force. As one can see, it looks pretty random. Namely this feature ensures that the rms radius of a particle (initially at the origin) behaves like $1/\sqrt{t}$ over long periods of time, see Fig. 12. At the same time, the momentum has an Boltzmann distribution as expected for a Brownian particle. We find it remarkable that a diffusion process can be described without any recourse to dissipation (time irreversibility) and/or random noise (non-deterministic element).

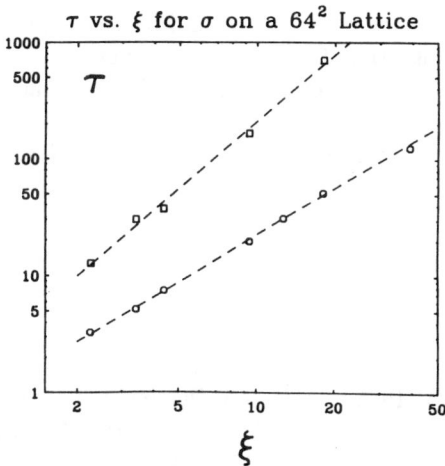

Fig. 10. Same as figure 9, except that auto-correlation times τ are extracted from the single-spin auto-correlation function. As before, Isothermal MD (circles) methods have a noticeably smaller critical exponent and prefactor than standard hybrid Monte Carlo methods (boxes). (Results are preliminary, see Ref. 9 for details.)

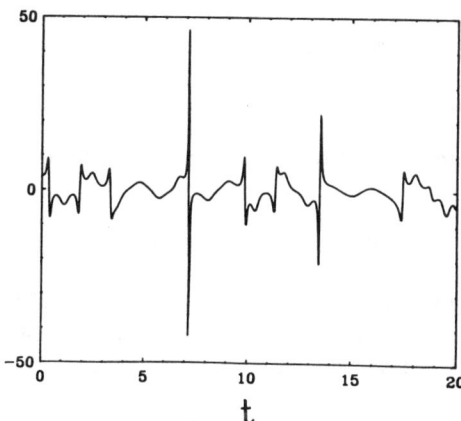

Fig. 11. Characteristic behavior of the (deterministic) friction force $\dot{p} = F(t)$ for a Brownian particle, using equations (13).

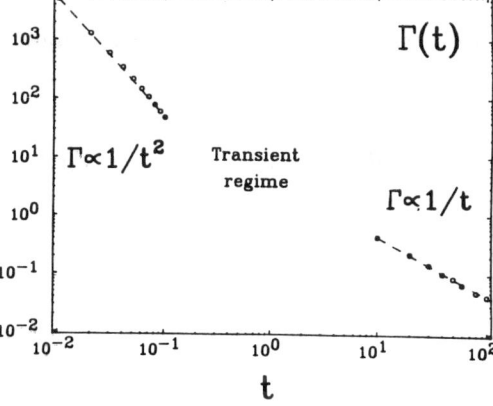

Fig. 12. Time dependence of the inverse width of the distribution $f(q,t) = \sqrt{\Gamma(t)/\pi} \exp[-\Gamma(t)q^2]$ obtained by fitting to spatial distributions at selected times. The mean spreading velocity reaches its asymptotic behavior $\sim \sqrt{t}$ for long times, characteristic for the Brownian particle.

We hope that our extremely short exposure of the isothermal MD techniques, gave the reader an idea about the huge potential of the MD techniques.

The present work was partially supported by the Department of Energy and the National Science Foundation.

† Address after September 1-st, 1991, Center for Theoretical Physics, Sloane Laboratory, Yale University, New Haven, CT 06511.

References

1. S.Nosé, J.Chem.Phys. **81**, 511 (1984); Mol.Phys. **52**, 255 (1984).

2. W.G.Hoover, Phys.Rev. **A 31**, 1695 (1985); *Molecular Dynamics*, Lecture Notes in Physics (Springer-Verlag, New York, 1986), Vol. 258; H.A.Posch, W.G.Hoover and F.J.Veseley, Phys. Rev. **A 33**, 4253 (1986).

3. J.Jellinek, J.Chem.Phys. **92**, 3163 (1988).

4. A.Bulgac and D.Kusnezov, Phys. Rev. **A42**, 5045 (1990).

5. D.Kusnezov, A.Bulgac and W.Bauer, Ann. Phys. **204**, 155 (1990).

6. A.Bulgac and D.Kusnezov, *Thermal properties of Na microclusters*, to be submitted.

7. R.Gupta et al. Phys. Rev. Lett. **61**, 1996 (1988) and references therein.

8. a)S.Duane, A.D.Kennedy, B.J.Pendleton and D.Roweth, Phys. Lett. **195B**, 216 (1987) and references therein; b)S.Duane, Nucl.Phys. **B257** [FS14], 652 (1985), A.D.Kennedy and B.Pendleton, SCRI preprint (1990).

9. J.Sloan, D.Kusnezov and A.Bulgac, *XY-model and isothermal dynamics*, in preparation.

10. D.Kusnezov and A.Bulgac, *Canonical ensembles from chaos II: constrained dynamics*, to be submitted.

11. R.Kumon and G.Rogers, REU MSUNSCL report (1990); R.Kumon, G.Rogers, D.Kusnezov and A.Bulgac, *Isothermal dynamics and path integrals*, in preparation.

12. A.Bulgac and D.Kusnezov, Phys. Lett. **A151**, 122 (1990).

MICROSCOPIC CALCULATIONS OF THE GROUND STATE STRUCTURES, COLLECTIVE EXCITATIONS, AND THE DYNAMIC STRUCTURE FUNCTION OF ^{4}He CLUSTERS

S. A. Chin and E. Krotscheck

Center for Theoretical Physics and Department of Physics,
Texas A&M University, College Station, TX 77843

ABSTRACT

By sampling the exact many-body ground state via a second order Diffusion Monte Carlo algorithm, we show how the collective excitation spectrrum, transition densities, and the dynamic structure functions of ^{4}He clusters can be determined variationally by solving a generalized Feynman eigenvalue equation.

I. INTRODUCTION

Helium clusters are unique in that they are the only known *bosonic*, but fully quantum, finite systems. Because of the simplicity of the ^{4}He-^{4}He potential, Helium clusters are excellent laboratories for testing our understanding of finite quantum many-body systems. In this case, unlike that of nuclei, it is possible to calculate intrinsic many-body effects microscopically unhampered by the complexity of the interaction. With the advent of supercomputers, we have the necessary raw computational power for elucidating the physics of clusters from first principles. What is needed, however are theoretical methods and efficient algorithms that can harness this power for the extraction of physical results.

In this work, we describe our efforts to understand the static and dynamic structure of ^{4}He clusters by exact Monte Carlo methods[1,2]. We show that by directly solving a generalized Feynman eigenvalue equation with inputs of exact one- and two-body ground state densities, we not only determine the *optimal* excitation functions and collective excitation energies[3] simultaneously, but also obtain a parameter-free dynamic structure function[4]. Further study of sum rules[5] then allows us to precisely determine the extend of collectivity of each state. Due to space limitation, we will focus mainly on presenting results of our investigation, interested readers can consult Refs.3-5 for more details and references.

II. GROUND STATE RESULTS

Employing the Aziz[6] potential, we evolve the product state $\Psi_0 \Phi_0$ of a N-atom Helium cluster by a second order Diffusion Monte Carlo (DMC) algorithm, DMC2b, as developed by one of us in Ref. 2. The variational trial function Φ_0

used is of the McMillian form

$$\Phi_0 = \prod_{i<j} \exp\{-\frac{1}{2}(a/r_{ij})^5\} \prod_i \exp\{-\frac{1}{2}b^{-2}(\mathbf{r}_i - \mathbf{r}_{cm})^2\}. \quad (1)$$

The ground state properties thus obtained for Helium clusters of size $N = 40$, 70, and 112 are tablulated in Table 1 and compared with the GFMC result of Ref. 7. The parameter a is fixed at $a = 3.0$. Our cluster radii and energies are both systematically below previous GFMC results.

TABLE 1. Ground state properties of ^{4}He clusters. ε_0 is the ground state energy per particle in degree K, $r_0 = (5/3)^{1/2} \langle r^2 \rangle^{1/2} N^{-1/3}$ is the unit radius in unit of Å.

N	b	ε_0^{DMC}	ε_0^{GFMC}	r_0^{DMC}	r_0^{GFMC}
40	3.6	-2.525(3)	-2.487(3)	2.551(2)	2.57
	3.5	-2.529(3)		2.532(2)	
70	3.9	-3.188(2)	-3.12(4)	2.441(2)	2.47
	3.8	-3.170(2)		2.461(3)	
112	4.3	-3.702(3)	-3.60(1)	2.399(2)	2.44
	4.2	-3.705(3)		2.391(2)	

To obtain the ground state expectation value of operators other than the Hamiltonian, such as the one- and two-body densities,

$$\rho_1(\mathbf{r}) = \langle \Psi_0 | \sum_i \delta(\mathbf{r}_i - \mathbf{r}_{cm} - \mathbf{r}) | \Psi_0 \rangle,$$
$$\rho_2(\mathbf{r}, \mathbf{r}') = \langle \Psi_0 | \sum_{i \neq j} \delta(\mathbf{r}_i - \mathbf{r}_{cm} - \mathbf{r})\delta(\mathbf{r}_j - \mathbf{r}_{cm} - \mathbf{r}') | \Psi_0 \rangle, \quad (2)$$

we use the perturbative estimate

$$\langle \Psi_0 | O | \Psi_0 \rangle = 2\langle \Psi_0 | O | \Phi_0 \rangle - \langle \Phi_0 | O | \Phi_0 \rangle. \quad (3)$$

Figs. 1-3 show the corresponding one-body ground state densities. Despite differences in the Variational Monte Carlo (VMC) densities (from just sampling Φ_0^2), there is excellent agreement between the two DMC calculations using the perturbative estimate (3). While all VMC densities are generally smooth, DMC densities generally show persistent oscillations. In particular, the $N = 112$ case clearly shows that two completely independent DMC calculations, using different trial functions, yielded similar patterns of oscillation.

S. A. Chin and E. Krotschek 37

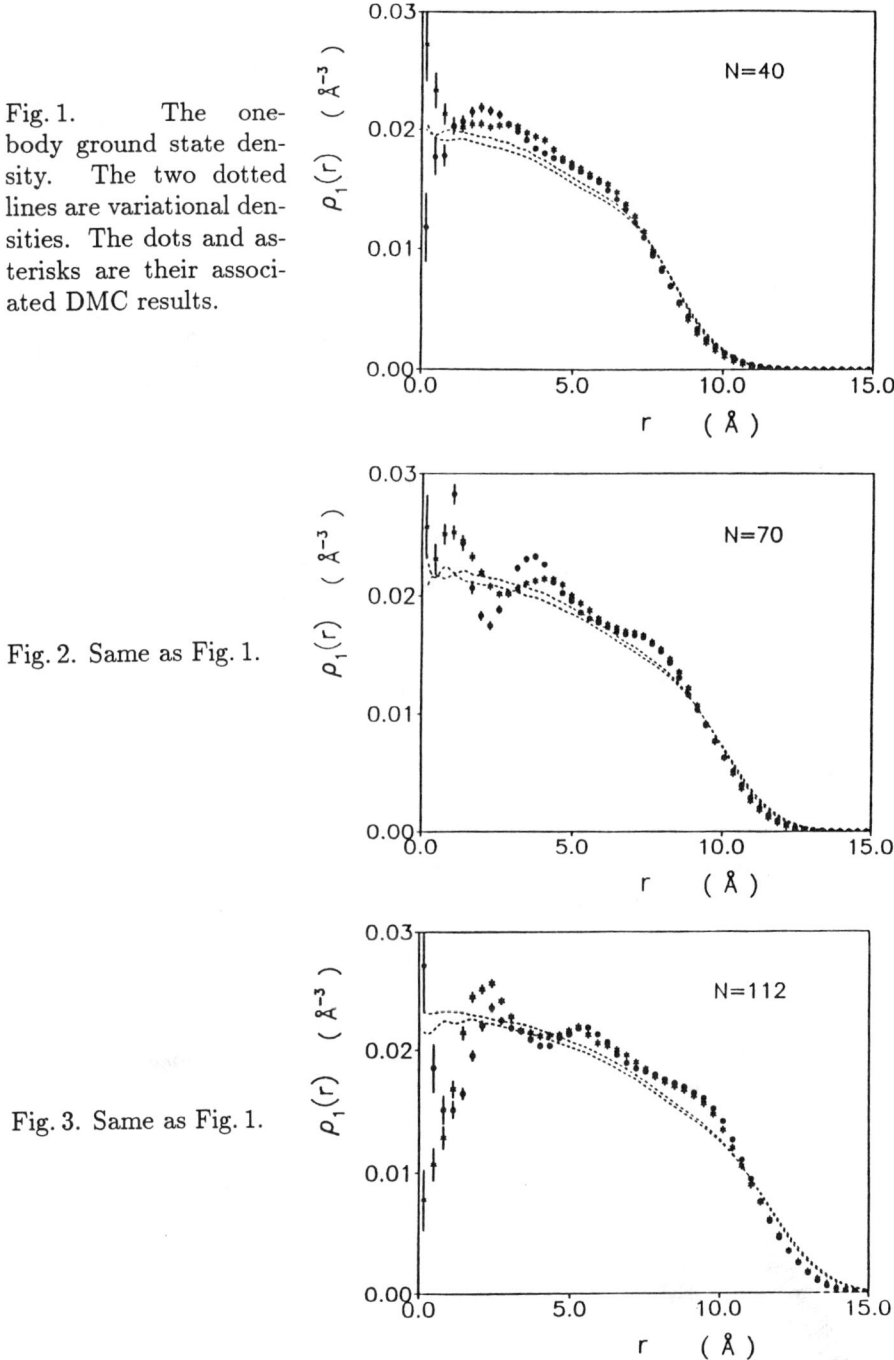

Fig. 1. The one-body ground state density. The two dotted lines are variational densities. The dots and asterisks are their associated DMC results.

Fig. 2. Same as Fig. 1.

Fig. 3. Same as Fig. 1.

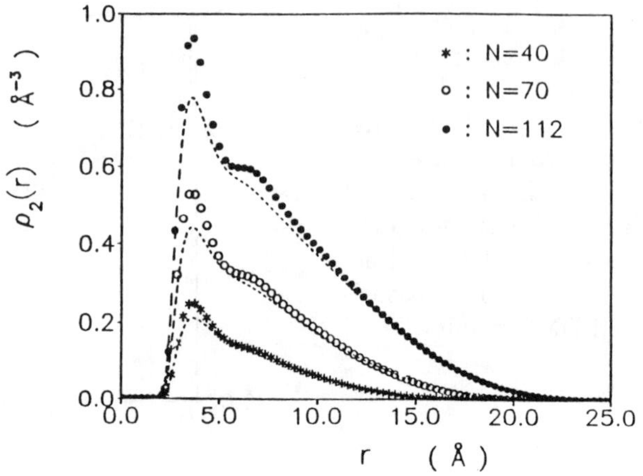

Fig. 4. The pair-distance density for three cluster size as defined by Eq.(4).

Fig. 4 compares our VMC and DMC results for the pair-distance distribution $\rho_2(r)$ defined by

$$\rho_2(r) = \int d^3r_1 \, d^3r_2 \delta(r - |\mathbf{r}_1 - \mathbf{r}_2|)\rho_2(\mathbf{r}_1, \mathbf{r}_2). \tag{4}$$

In binning the two-body density $\rho_2(\mathbf{r}_1, \mathbf{r}_2)$, it is useful to bin its partial wave amplitudes $\rho_2^{(\ell)}(r, r')$ defined by

$$\rho_2^{(\ell)}(r, r') \equiv \int d\Omega \rho_2(\mathbf{r}, \mathbf{r}') P_\ell(\hat{\mathbf{r}} \cdot \hat{\mathbf{r}}'). \tag{5}$$

rather than the two-body density itself. Figs. 5-7 give the $\ell = 0, 1, 2$ partial wave amplitude for $N = 112$ in the normalized form of

$$g_\ell(r, r') = \frac{(2\ell + 1)}{4\pi} \frac{\rho_2^{(\ell)}(r, r')}{\rho(r)\rho(r')}. \tag{6}$$

One clearly sees in Fig. 5 that $g_0(r, 0)$ has the expected behavior of a pair correlation function familiar from bulk liquid helium.

III. COLLECTIVE EXCITATIONS

We generalizing Feynman's original theory[8] of collective excitation for bulk liquid ^{4}He to the case of finite clusters by writing the trial excited state as

$$\Psi_F(\mathbf{r}_1, \ldots \mathbf{r}_N) = F(\mathbf{r}_1, \ldots \mathbf{r}_N)\Psi_0(\mathbf{r}_1, \ldots \mathbf{r}_N), \tag{7}$$

where $F(\mathbf{r}_1, \ldots \mathbf{r}_N) = \sum_{i=1}^{N} \left[f(\mathbf{r}_i - \mathbf{r}_{cm}) - \langle f_i \rangle \right]$, $\langle f_i \rangle = \langle \Psi_0 | f(\mathbf{r}_i - \mathbf{r}_{cm}) | \Psi_0 \rangle$, and $\mathbf{r}_{cm}$ is the center-of-mass coordinate. A rigorous upper bound to the excitation

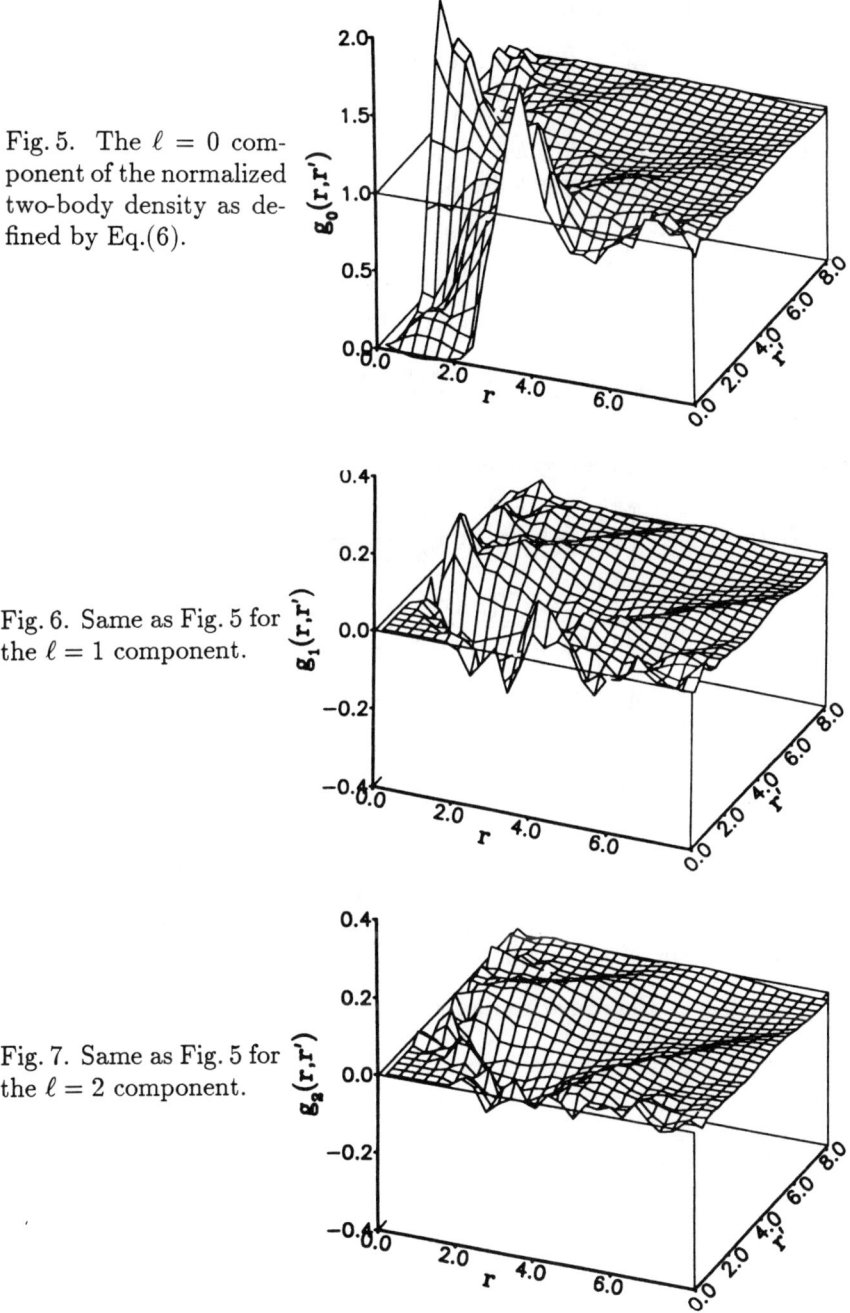

Fig. 5. The $\ell = 0$ component of the normalized two-body density as defined by Eq.(6).

Fig. 6. Same as Fig. 5 for the $\ell = 1$ component.

Fig. 7. Same as Fig. 5 for the $\ell = 2$ component.

energy can then be estimated by

$$E_F - E_0 = \frac{1}{2}\frac{\langle \Psi_0 | [F,[H,F]] | \Psi_0 \rangle}{\langle \Psi_0 | F^2 | \Psi_0 \rangle}. \tag{8}$$

Denoting the excitation energy $E_F - E_0 = \hbar\omega$ and define $u(\mathbf{r}) = \sqrt{\rho_1(\mathbf{r})}f(\mathbf{r})$, Eq. (8) can be written symmetrically as

$$\hbar\omega = \frac{\int d^3r d^3r' u(\mathbf{r})H_1(\mathbf{r},\mathbf{r}')u(\mathbf{r}')}{\int d^3r d^3r' u(\mathbf{r})S(\mathbf{r},\mathbf{r}')u(\mathbf{r}')}, \tag{9}$$

with the coordinatate space representation of the static structure function

$$S(\mathbf{r},\mathbf{r}') = \delta(\mathbf{r}-\mathbf{r}') + \frac{\rho_2(\mathbf{r},\mathbf{r}') - \rho_1(\mathbf{r})\rho_1(\mathbf{r}')}{\sqrt{\rho_1(\mathbf{r})\rho_1(\mathbf{r}')}}, \tag{10}$$

and the non-local kinetic energy operator

$$H_1(\mathbf{r},\mathbf{r}') = -\left[1-\frac{1}{N}\right]\delta(\mathbf{r}-\mathbf{r}')\frac{\hbar^2}{2m}\frac{1}{\sqrt{\rho_1(\mathbf{r})}}\nabla\rho_1(\mathbf{r})\cdot\nabla\frac{1}{\sqrt{\rho_1(\mathbf{r})}}$$
$$-\frac{\hbar^2}{2mN}\frac{\nabla_\mathbf{r}\cdot\nabla'_\mathbf{r}\rho_2(\mathbf{r},\mathbf{r}')}{\sqrt{\rho_1(\mathbf{r})}\sqrt{\rho_1(\mathbf{r}')}}. \tag{11}$$

Minimizing $\hbar\omega$ with respect to the excitation function $u(\mathbf{r})$ yields the generalized Feynman eigenvalue equation

$$\int d^3r' H_1(\mathbf{r},\mathbf{r}')u(\mathbf{r}') = \hbar\omega \int d^3r' S(\mathbf{r},\mathbf{r}')u(\mathbf{r}'). \tag{12}$$

With inputs of one- and two-body densities, this equation can be solved by partial-wave expansion to determined the optimal excitation function and the lowest excitation energies. The resulting monopole ($\ell = 0$) and the quadrupole ($\ell = 2$) excitation energies[3] are tabulated in Table 2.

TABLE 2. Collective excitation energies $\hbar\omega_\ell$ of ^{4}He clusters in units of K.

N	b	$\hbar\omega_0^{VMC}$	$\hbar\omega_0^{DMC}$	$\hbar\omega_2^{VMC}$	$\hbar\omega_2^{DMC}$
40	3.6	3.44	3.60	1.77	1.37
	3.5	3.58	3.53	1.84	1.50
70	3.9	3.38	3.94	1.52	1.50
	3.8	3.53	3.96	1.60	1.54
112	4.3	2.97	3.92	1.27	1.76
	4.2	3.16	4.27	1.34	1.62

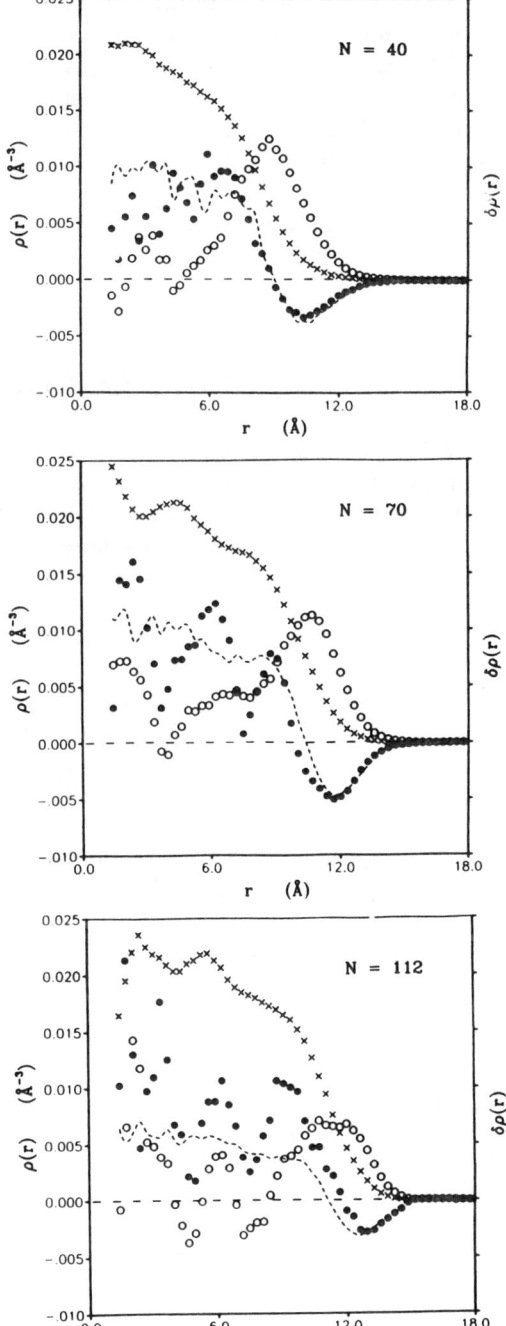

Fig. 8. The monopole (dots) and the quadrupole (circles) transition density for a $N = 40$ Helium cluster. The dotted line is the variational monopole transition density. For comparsion, The one-body density is denoted by crosses.

Fig. 9. Same as Fig. 8 for a $N = 70$ Helium cluster.

Fig. 10. Same as Fig. 8 for a $N = 112$ Helium cluster.

The VMC results are obtained by solving the eigenvalue equation (12) with inputs of one- and two-body densities sampled from the trial function (1). Both our VMC and DMC excitation energies differs significantly from the liquid drop model[9]. Our DMC results do not even begin to share the same trend as LDM predictions. We conclude therefore that, for cluster sizes considered, the detail dynamics of Helium clusters is far from the liquid drop limit.

Once the excitation function is known, the corresponding transition density can be obtained via

$$\delta\rho(\mathbf{r}) = \langle \Psi_F | \sum_i \delta(\mathbf{r}_i - \mathbf{r}_{cm} - \mathbf{r}) | \Psi_0 \rangle,$$
$$= \rho_1(\mathbf{r})f(\mathbf{r}) + \int d^3r' [\rho_2(\mathbf{r},\mathbf{r}') - \rho_1(\mathbf{r})\rho_1(\mathbf{r}')] f(\mathbf{r}'). \quad (13)$$

Figs. 8-10 show the resulting transition densities. Both the monopole and the quadrupole transition density show marked oscillations with wavelength approximately equal to the avarage particle separation, suggesting that they are connected with the geometric, hard-sphere like, shell structure of the droplets. By contrast, the variational monopole transition density shows no such oscillation.

IV. DYNAMIC STRUCTURE FUNCTION

The scattering cross-section for a weak probe scattering from a composite system is given in Born approximation by[10]

$$\frac{d^2\sigma}{d\Omega d\omega} = \left(\frac{d^2\sigma}{d\Omega d\omega}\right)_0 S(\mathbf{k},\omega), \quad (14)$$

where $(d^2\sigma/d\Omega d\omega)_0$ is the double differential scattering cross-section for scattering from a single constitutent, and

$$S(\mathbf{k},\omega) = \sum_{n\neq 0} |\langle \Psi_n | \rho(\mathbf{k}) | \Psi_0 \rangle|^2 \delta(E_n - E_0 - \omega) \quad (15)$$

is the dynamic structure function. The calculation of $S(\mathbf{k},\omega)$ thus requires knowledge of both the excitation spectrum $E_n - E_0$ and the complete set of transition densities $\langle \Psi_n | \rho(\mathbf{k}) | \Psi_0 \rangle$. As a first step toward an exact calculation of $S(\mathbf{k},\omega)$, one can replace the exact spectrum and transition densities by those of the Feynman theory, as describe previously. Thus for a given ω, (12) can be solve to give the transition density (13), and from which one obtains a parameter-free model dynamical structure function,

$$S(\mathbf{r},\mathbf{r}';\omega) = \delta\rho_\omega(\mathbf{r}) \delta\rho_\omega(\mathbf{r}'), \quad (16)$$

or its momentum-space equivalent,

$$S(\mathbf{k},\mathbf{k}';\omega) = \int d^3r d^3r' e^{i(\mathbf{k}\mathbf{r}-\mathbf{k}'\mathbf{r}')} S(\mathbf{r},\mathbf{r}';\omega). \quad (17)$$

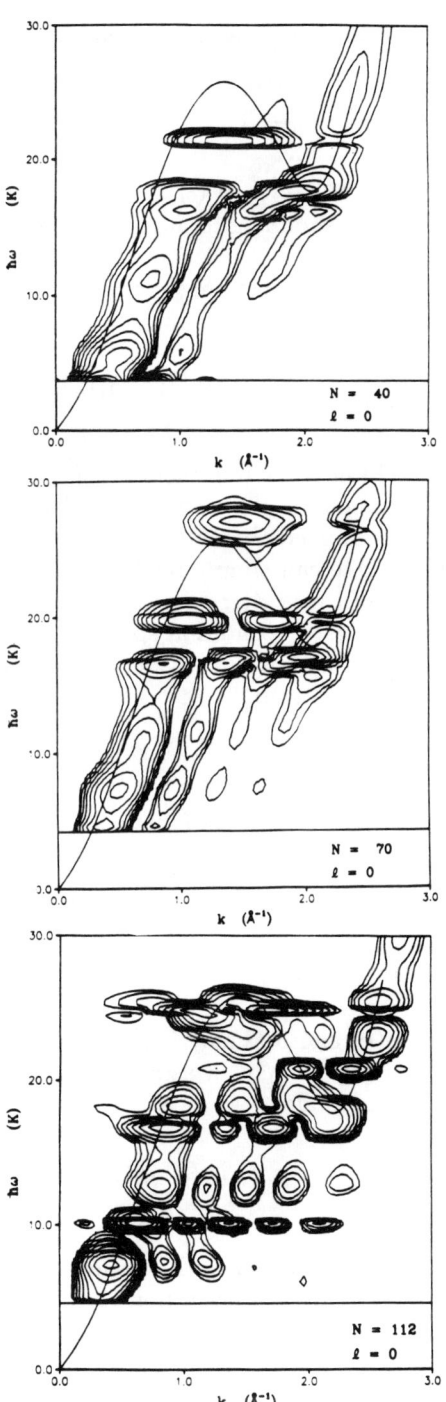

Fig. 11. A contour plot of the $\ell = 0$ component of the dynamic structure function for $N = 40$. The solid line corresponds to the Feynman dispersion relation for bulk liquid Helium.

Fig. 12. Same as Fig. 11 for $N = 70$.

Fig. 13. Same as Fig. 11 for $N = 112$.

A partial wave expansion then yields

$$S(\mathbf{k},\mathbf{k}';\omega) = \sum_\ell \delta\rho_{\ell\omega}(k)\delta\rho_{\ell\omega}(k')P_\ell(\hat{\mathbf{k}}\cdot\hat{\mathbf{k}}') = \sum_\ell S_\ell(k,k';\omega)P_\ell(\hat{\mathbf{k}}\cdot\hat{\mathbf{k}}'), \quad (18)$$

where $\delta\rho_{\ell,\omega}(k) = \int d^3r\, \delta\rho_{\ell\omega}(r)j_\ell(kr)$ is the Fourier-transform of the transition density corresponding to angular momentum ℓ and energy $\hbar\omega$.

In inclusive scattering experiments, only the diagonal element of $S(\mathbf{k},\mathbf{k}';\omega)$ corresponding to the dynamic structure function is measured:

$$S(\mathbf{k},\omega) = S(\mathbf{k},\mathbf{k};\omega) = \sum_\ell S_\ell(k,k;\omega). \quad (19)$$

Figs. 11-13 show contour plots of $S_0(k,k;\omega)$ for cluster sizes $N = 40$, 70, and 112. In the continuum, the scattering peaks begins at discrete intervals in $\hbar\omega$, reflecting the quantized character of the collective excitation appropriate for a finite droplet. One can simply read off the higher excitation energies by noting the locations of the scattering peaks. Thus the dynamic structure function directly maps the cluster excitation spectrum and provides a graphic demonstration of how the continuous bulk dispersion relation develops out of the discrete cluster excitation spectrum.

V. SUM RULES

The static structure function $S(k)$ can either be expressed directly as

$$\begin{aligned} S(k) &= \frac{1}{N}\int d^3r\,d^3r'\,\sqrt{\rho_1(\mathbf{r})\rho_1(\mathbf{r}')}e^{i\mathbf{k}\cdot(\mathbf{r}-\mathbf{r}')}S(\mathbf{r},\mathbf{r}'), \\ &= 1 + \frac{1}{N}\left\{\tilde{\rho}_2(k) - |\tilde{\rho}_1(k)|^2\right\}, \end{aligned} \quad (20)$$

where $\tilde{\rho}_2(k)$ and $\tilde{\rho}_1(k)$ are Fourier transforms of the pair-distance and one-body densities respectively, or as a sum over integrated contribution of each angular momentum modes,

$$S(k) = \sum_\ell S_\ell(k) = \sum_\ell \frac{1}{N}\int_0^\infty d(\hbar\omega)S_\ell(k,\omega). \quad (21)$$

Comparing the latter with the former then allows us to analyze the specific contribution of each excitation to the full static structure function $S(k)$. This is shown in Figs. 14-16. Aside from the reduced overshoot of the cluster $S(k)$ around $k \approx 2\text{Å}^{-1}$, there appears a shoulder around $k \approx 0.5\text{Å}^{-1}$ which is absent in the bulk $S(k)$. This is primarily due the quadrupole mode, which accounts for 25%-30% of the overall excitation strength. By contrast, the monopole's

Fig. 14. The solid line is the cluster static structure function given by (20). The dotted line is the experimental bulk liquid Helium structure function. The contribution to $S(k)$ from the quadrupole collective state is given by the solid hump. The sum of contributions from all states in the three lowest angular momentum mode is given by the long dash line.

Fig. 15. Same as Fig. 14 for $N = 70$.

Fig. 16. Same as Fig. 14 for $N = 112$.

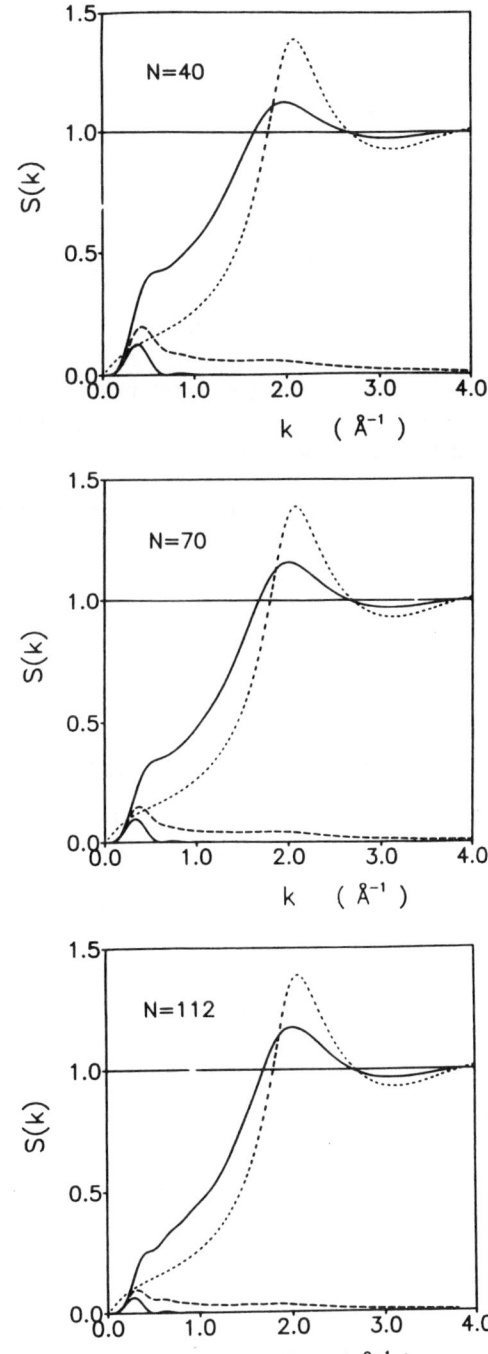

contribution is negligible (< 5%). In the limit of $N \to \infty$, the quadrupole (which is mainly a surface mode) ceases to be important, and the shoulder goes away.

VI. CONCLUSIONS

In this work, we have shown that by just sampling the ground state one- and two-body densities, one can extract a tremendous amount of physics about finite quantum many-body systems. By solving the generalized Feynman eigenvalue equation, one can simultaneously determine the collective excitation spectrum, transition densities, and a parameter- free dynamic structure struction. Once the one- and two-body densities for any finite system is known, the same theory can be applied. It is therefore of great interest to apply this generalized Feynman theory to Fermi systems as well as to improve the theory itself by including higher order ground state densities.

ACKNOWLEDGEMENT

This work was supported, in part, by the National Science Foundation grants PHY89-07986 (to SAC), PHY-8806265 (to EK), and the Texas Advanced Research Program under Grant 010366-012 (to EK). The simulations of the ground-state structure were performed on the CRAY-YMP of the Texas A&M University Supercomputer Center.

REFERENCES

1. D. Ceperly and M. Kalos, in *Monte Carlo Methods in Statistical Mechanics*, edited by K. Binder (Springer, New York, 1979).

2. S. A. Chin, Phys. Rev. **A42**, (1990) 6991.

3. S. A. Chin and E. Krotscheck, Phys. Rev. Lett. **65**, 2658 (1990).

4. S. A. Chin and E. Krotscheck, Chem. Phys. Lett. **178**, 435 (1991).

5. S. A. Chin and E. Krotscheck, Texas A&M University CTP preprint 51/91.

6. R. A. Aziz, V. P. S. Nain, J. C. Carley, W. L. Taylor, and G. T. McConville, J. Chem. Phys. **70**, (1979) 4330.

7. V. R. Pandharipande, J. G. Zabolitzky, S. C. Pieper, R. B. Wiringa, and U. Helmbrecht, Phys. Rev. Lett. **50** (1983) 1676.

8. R. P. Feynman, Phys. Rev. **94**, (1954) 262.

9. M. Casas and S. Stringari, J. Low. Temp. Phys. **79** (1990) 135.

10. L. Van Hove, Phys. Rev. **95**, (1954) 249.

Particle Production through Two-Photon Processes in Relativistic Heavy-Ion Colliders

J.-S. Wu, C. Bottcher, M. R. Strayer and C.M. Shakin*

*Center for Computationally Intensive Physics,
Physics Division, Oak Ridge National Laboratory, Oak Ridge, TN 37831*

Particle production via two-photon processes in relativistic heavy-ion colliders is discussed. We consider the coherent production, including its impact parameter dependence, and incoherent production. Particular attention is given to Higgs boson production in the energy regime of the LHC and the SSC.

I. INTRODUCTION

It was suggested several years ago that the highly peaked electromagnetic fields in ultra-relativistic heavy-ion collisions could be used as the sources to produce heavy particles coherently [1,2]. A variety of particle final states have been considered, including lepton pairs [3,4], W-pairs, Higgs bosons [5 - 8] and supersymmetric particles [9]. The advantage of production with heavy-ion beams is that the cross section of coherent production has Z^4 enhancement. At high energies the production is spread over a wide range of the impact parameter space of the colliding nuclei. The peripheral collisions can provide us with clean signals for detection.

Most of the calculations cited were carried out by using equivalent photon methods. However, in this approach the virtual photons are approximated by "on-shell" photons and the transverse momenta of the virtual photons are neglected. Therefore, the impact parameter dependence and transverse momentum spectra might be seriously wrong. Here is another approach which takes advantage of the rapidly developing high-performance computing resources. For the two-photon processes, production cross section can be written in terms of Feynman integrals in momentum space and integrated out numerically using Monte Carlo methods. It is straightforward to extend the approach to study the impact parameter dependence [8] and the incoherent production using structure functions [10].

We begin with the interaction Hamiltonian of quantum fields. The S-matrix can be defined as

$$S = \mathrm{T} \exp\{-i \int d^4 x \mathcal{H}_{\mathrm{int}}(x)\} , \qquad (1)$$

© 1992 American Institute of Physics

where the operator T gives the time-ordered product. The interaction Hamiltonian can be written as

$$\mathcal{H}_{int}(x) = \mathcal{J}_\mu(x) A^\mu(x) + H_{int}(x) , \qquad (2)$$

where $\mathcal{J}_\mu$ is the electromagnetic source current of the colliding heavy ions and $H_{int}(x)$ is the interaction Hamiltonian of the quantum fields. The interaction Hamiltonians in the standard model for a variety of two photon processes are summarized in Table I. Since the search for the Higgs boson or the mechanism responsible for electroweak symmetry breaking is one of the primary goals of the next generation of supercolliders, we mainly concentrate on Higgs boson production in relativistic heavy-ion colliders.

The Higgs boson is a neutral scalar particle and does not directly couple to electromagnetic fields. However, the production of Higgs bosons via two virtual photons can occur through the production of charged particles and their subsequent fusion. The effective interaction between the Higgs boson and the electromagnetic fields is given by

$$\mathcal{H}_{int} = \frac{e^2}{2} I : F^{\mu\nu} F_{\mu\nu} H : , \qquad (3)$$

where $F_{\mu\nu} = \partial_\mu A_\nu - \partial_\nu A_\mu$.

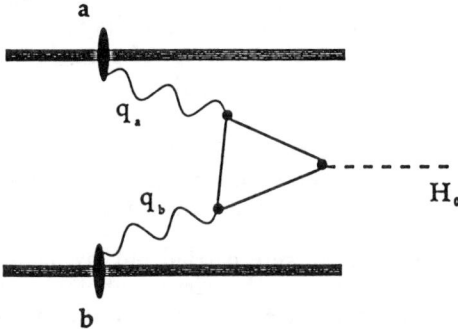

Fig. 1. The Feynman diagram for the effective coupling of the Higgs boson to two photons via a single loop. The loop sums over all combinations of charged fermions and bosons.

The lowest order Feynman diagram is shown in Fig. 1. In the figure the intermediate loops can be any charged particles, quarks, leptons or bosons. Their contributions can be written in terms of the above three-body vertex with an effective coupling I. In the standard model the effective coupling is given by

$$I = \frac{g}{8\pi^2 M_W} (A_q + A_\ell + A_W) , \qquad (4)$$

and where each amplitude A_i is only a function of the ratio of the masses, $\lambda_i = m_i^2 / M_H^2$. Here, the contributions of the amplitudes come from the quarks, charged leptons, and $W^\pm$ bosons. Their explicit form can be found in refs.[11] and [12]. Generally, these amplitudes are sensitive to the mass ratio, and the major contribution is from the vector bosons and the top quark. For the case of an intermediate-mass Higgs boson of mass less than 200 GeV, and the top quark mass between the limits set by recent experiments, i.e. 130±30 GeV [13], the effective coupling is not sensitive to the value of top quark mass. Thus, we choose $m_t = 100$ GeV and the standard values for other quark masses in the discussion.

II. COHERENT PRODUCTION

For the coherent production we treat the electromagnetic fields as classical fields arising from the motion of two colliding nuclei. The potentials can be derived from Maxwell equations and written as

$$A^\mu(x) = A_a^\mu(x) + A_b^\mu(x) , \qquad (5)$$

where a and b label the nuclei and

$$A_{a,b}^\mu(q) = 2\pi Z_{a,b}\, e\, \delta(q^0 \mp \beta q^z)\, \frac{F_{a,b}(-q^2)}{-q^2}\, \exp(\pm i\vec{q}_\perp \cdot \vec{b}/2) u_{a,b}^\mu , \qquad (6)$$

$$u_{a,b}^\mu = (1, 0, 0, \pm\beta) .$$

A few comments about the classical fields given in Eq. 6 are in order. $F_{a,b}(q^2)$ are the elastic form factors derived from the finite charge distribution of the nuclei. The velocity $u_{a,b}^\mu$ and the delta-function indicate that the fields can be obtained by boosting the nuclear Coulomb fields from the rest frame of the corresponding nucleus to the collider frame. The vector b displaces the straight-line trajectories of the two colliding nuclei and will be identified as the impact parameter.

The Higgs production cross section is obtained from the S-matrix to the first order in the Hamiltonian (3),

$$\sigma_{ab \to H_0} = \int \frac{d^3k}{(2\pi)^3 2\omega_k} \int d^2b \left| <k|S^{(1)}|0> \right|^2 , \qquad (7)$$

where

$$<k|S^{(1)}|0> = ie^2 I \int \frac{d^4q_1 d^4q_2}{(2\pi)^4} \delta^{(4)}(q_1 + q_2 - k)$$
$$\times [(q_1 \cdot q_2) A_\mu(q_1) A^\mu(q_2) - q_{1\mu} A^\mu(q_2) q_{2\nu} A^\nu(q_1)]. \qquad (8)$$

When evaluating the S-matrix element, the classical fields are c-numbers rather than field operators, and can be taken out of the ordered product.

The integation over impact parameters can be carried out analytically, leading to the cross section expression

$$\sigma_{ab \to H_0} = \frac{4 Z_a^2 Z_b^2 (4\pi\alpha)^4 |I|^2}{\beta^2} \int \frac{d^3k\, d^2q_{a\perp} d^2q_{b\perp}}{(2\pi)^5 2\omega_k} \delta^{(2)}(\vec{q}_{a\perp} + \vec{q}_{b\perp} - \vec{k}_\perp)$$
$$\times \frac{F^2(-q_a^2)}{(q_a^2)^2} \frac{F^2(-q_b^2)}{(q_b^2)^2} [(q_a \cdot q_b)(u_a \cdot u_b) - (q_a \cdot u_b)(q_b \cdot u_a)]^2 . \qquad (9)$$

The integral can be carried out numerically by the Monte Carlo method. Gauge invariance has been proved formally and tested numerically. This treatment also shows that both of the virtual photons q_a, q_b are space-like and considerably "off-shell", since $-q_a^2 = (\omega_a/\beta\gamma)^2 + (\vec{q}_{a\perp})^2 > 0$ and $-q_b^2 = (\omega_b/\beta\gamma)^2 + (\vec{q}_{b\perp})^2 > 0$.

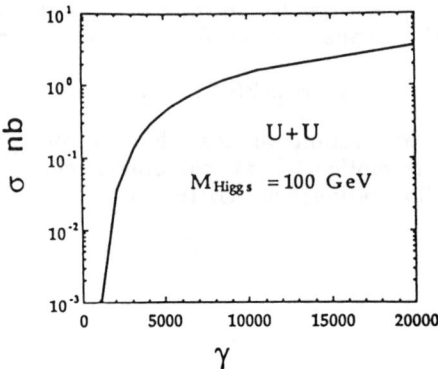

Fig. 2. The cross section for Higgs production in uranium collisions as a function of the Lorentz γ of the beams. The top quark mass is assumed to be 100 Gev.

Fig. 2 gives the cross section as a function of beam energy in uranium collisions for a Higgs mass $M_H = 100 GeV$. The qualitative behavior of the production cross section can be explained as follows. Consider the simple schematic model of a collision in which the nuclei each emit a virtual photon, q'_a, q'_b, in their respective rest frames, $q'_a = -q'_b = (0,0,0,q)$. The form factors $F(-q'^2)^2$ represent the probabilities that each nucleus will emit the corresponding photon and remain in their ground state. The form factors control the magnitude of the emission, which is strongly suppressed whenever the momentum transfer is much larger than the inverse nuclear size, i.e.,

$$F(-q'^2) \sim 0 \quad \forall \quad -q'^2 \gg 1/R. \tag{10}$$

In the collision, these momenta are boosted with velocities $\pm\beta$, respectively,

$$q_a^\mu = \gamma(\beta q, 0, 0, +q),$$
$$q_b^\mu = \gamma(\beta q, 0, 0, -q). \tag{11}$$

The square of the total available energy from the photons, $S = 4\gamma^2\beta^2 q^2$, is equal to the mass square of the produced Higgs boson, i.e.

$$M_H^2 = 4\gamma^2\beta^2 q^2, \tag{12}$$

while the "threshold" for the production,

$$q \geq \frac{M_H}{2\gamma\beta}. \tag{13}$$

For collisions at the LHC and the SSC the "thresholds" are about 14 and 6 MeV/c, respectively, and are comparable to the inverse nuclear size for heavy nuclei, $R^{-1} \sim$ 30 MeV/c. Thus, the coherent pair production can occur at these energies. For lower energies, the production is greatly suppressed by the form factor. As the energy is raised, the "threshold" climbs up the tail and the production cross section rises sharply. When the "threshold" reaches the plateau at small values of Q^2 the production dependence on energy saturates and goes over to its logarithmic asymptotic form. In the asymptotic regime the production is no longer suppressed by the form factor, and the magnitude of the cross section is not sensitive to the details of the form factor. It is only in this regime that the full Z^4 coherent enhancement is recovered.

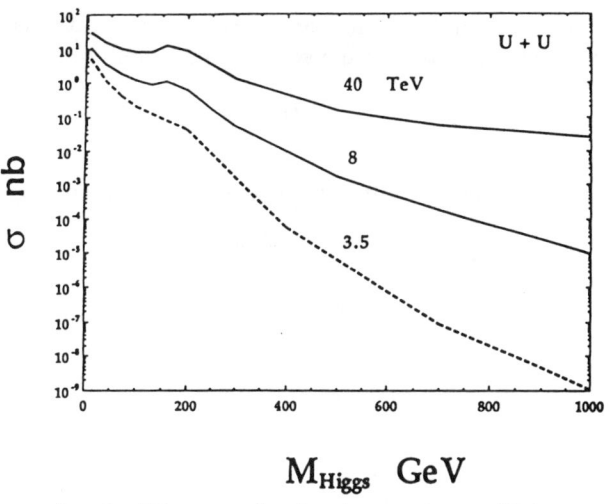

Fig. 3. The cross section for Higgs production in uranium collisions as a function of the Higgs mass at the LHC, the SSC, and the ELOISATRON. The top quark mass is assumed to be 100 GeV.

Fig. 3 gives the cross section in uranium collisions as a function of Higgs mass for the three beam energies $E = 3.5$, 8.0, and 40 GeV/u, corresponding to the LHC, the SSC, and the proposed ELOISATRON, respectively. We take the charge-mass ratio as 0.4 in the expected beam energy per nucleon. The luminosity in heavy-ion colliders is typically about 1 pb/year. We can estimate the Higgs boson production rates at these energies a few hundred to a few thousand per year for the Higgs mass $M_H = 100$ GeV. This figure also shows that at a large Higgs mass the production of the Higgs boson is very sensitive to the beam energies.

It can be shown from quantum field theory [10] that the above classical field approximation is especially good in the two-photon processes of heavy-ion collisions where the photon momentum transfer is small compared to the nuclear momenta. When the final nuclear momenta are approximated by their initial momenta in the expression for the production cross section, the classical field approximation can be recovered from the corresponding Feynman integral. As we shall discuss in the following section, the classical field approximation is particularly convenient for describing the dependence of particle production on the impact parameter.

III. IMPACT PARAMETER DEPENDENCE

We are mostly interested in the clean signals provided by the peripheral collisions. In the central collisions the nuclei will undergo very strong excitations which remove them from the elastic flux. Therefore, we have to consider the impact parameter dependence of the production and exclude the production from central collisions.

As we have seen, the classical fields depend explicitly on the impact parameter. The S-matrix elements for the two-photon processes can be written as

$$< f|S|0> = \int \frac{d^2 q_\perp}{(2\pi)^2} M_f(\mathbf{q}_\perp) \exp\{i(\mathbf{q}_\perp - \mathbf{k}_{f\perp}/2)\cdot \mathbf{b}\}, \qquad (14)$$

where M_f is independent of b, and the delta-functions in the classical fields have been integrated out. The total cross section can be reduced to the following form

$$\sigma_{ab\to H_0} = \int d^2 b \; s(b) \sum_f \left|< f|S^{(1)}|0>\right|^2, \qquad (15)$$

where $s(b)$ is a function excluding the flux from the central regime.

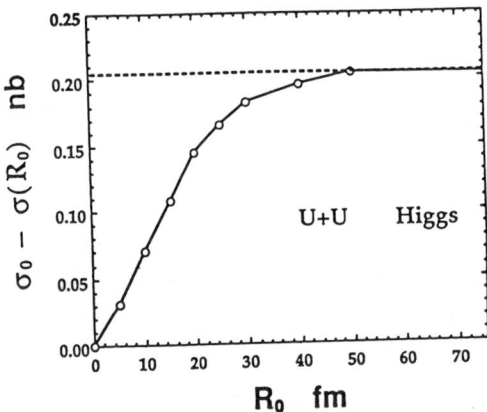

Fig. 4. The cross section $\sigma(0) - \sigma(R_0)$ for Higgs production in uranium collisions as a function of R_0, at beam energies per nucleon of 3.5 TeV. The top quark mass is assumed to be 100 Gev.

With an explicit form of $s(b)$, we may carry out the integration over impact parameters and obtain

$$\sigma_{ab\to f} = \sum_f \int \frac{d^2 q_{1\perp}}{(2\pi)^2} \frac{d^2 q_{2\perp}}{(2\pi)^2} M_f(\mathbf{q}_{1\perp}) M_f^*(\mathbf{q}_{2\perp}) S(\mathbf{q}_{1\perp} - \mathbf{q}_{2\perp}), \qquad (16)$$

where $S(\mathbf{q}_\perp) = \int d^2 b \, s(b) \exp\{i\mathbf{q}_\perp \cdot \mathbf{b}\}$. For convenience we may choose the simplest assumption, i.e.

$$s(b) = \theta(b - R_0), \qquad (17)$$

where R_0 is twice the nuclear radius, $2R$. In Fig. 4 we plot the cross section $\sigma(0) - \sigma(R_0)$ for Higgs production as a function of the distance R_0 for collisions of uranium at a beam energy of 3.5 TeV per nucleon.

From Fig. 4 we observe that a fairly large fraction of the cross section comes from impact parameters beyond peripheral collisions. Exclusion of central events at $R_0 = 2R$, where R is the nuclear radius, reduces the cross section for Higgs production by a factor of 1.9 at LHC (3.5 TeV), by a factor of 1.4 at SSC (8.0 TeV), and by a factor of 1.2 at ELOISATRON (40 TeV).

IV. HIGGS DECAY

For intermediate mass Higgs bosons, $M_Z < M_H < 2M_W$ and top quark masses $m_t > M_H/2$, 90 percent of the Higgs decay width is into b-quark pairs, and the lifetime of the Higgs boson is about 10^{-24} seconds. Therefore, in experiments, we expect to observe the Higgs through its decay into hadron jets. This decay has been suggested by other authors [5, 7], and the corresponding cross section for the b-jet distribution is obtained by including the decay process $H_0 \to b\bar{b}$, where

$$\mathcal{H}'_{int} = (\sqrt{2}G_F)^{1/2} m_f \bar{\psi}_f \psi_f H . \tag{18}$$

The resulting cross section is

$$\sigma_{ab \to H_0 \to f\bar{f}} = \frac{4Z_a^2 Z_b^2 (4\pi\alpha)^4 |I|^2}{\beta^2} C_f \int \frac{d^3 k_f d^3 k_{\bar{f}} d^2 q_{a\perp} d^2 q_{b\perp}}{(2\pi)^{10} 2E_f 2E_{\bar{f}}} \sum_{s_f, s_{\bar{f}}} |\mathcal{U}(k_f, s_f; k_{\bar{f}}, s_{\bar{f}})|^2$$

$$\times \delta^{(2)}(\vec{q}_{a\perp} + \vec{q}_{b\perp} - \vec{k}_{f\perp} - \vec{k}_{\bar{f}\perp}) \frac{F^2(q_a^2)}{(q_a^2)^2} \frac{F^2(q_b^2)}{(q_b^2)^2}$$

$$\times [(q_a \cdot q_b)(u_a \cdot u_b) - (q_a \cdot u_b)(q_b \cdot u_a)]^2 . \tag{19}$$

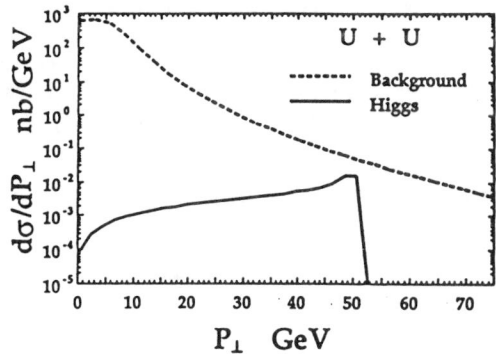

Fig.5. The transverse momentum distribution of the b-jets obtained from Higgs decay compared to the corresponding jet background obtained directly from two-photon production for uranium collisions at beam energies per nucleon of 3.5 TeV. The Higgs mass and the top quark mass are both assumed to be 100 GeV.

The $H_0 \to f\bar{f}$ vertex is given by

$$\mathcal{U}(k_f, s_f; k_{\bar{f}}, s_{\bar{f}}) = \frac{(\sqrt{2}G_F)^{1/2} m_f \bar{u}(k_f, s_f) v(k_{\bar{f}}, s_{\bar{f}})}{(k_f + k_{\bar{f}})^2 - M_H^2 + i\Gamma_H M_H}, \qquad (20)$$

while the colour degeneracies are $C_q = 3$ and $C_\ell = 1$, and Γ_H is the total Higgs decay width. Note that the b-quark momentum in (19) is integrated out. From these equations we have computed the b-jet transverse momentum distribution arising from the Higgs decay. This is shown in Fig. 5 for uranium collisions at beam energies per nucleon of 3.5 TeV. Also shown in this figure is the corresponding background jet distribution obtained from the peripheral production of $b\bar{b}$ pairs. We note that the background distributions are somewhat larger than those from the Higgs decay, indicating that intermediate mass Higgs bosons may be difficult to identify from the analysis of singles spectra.

One of the advantages of the Monte Carlo integration is that each Monte Carlo point can be treated as a weighted event. Any desired distribution may be accumulated by binning these events as the calculation proceeds. In general, more points are needed to calculate distributions than total cross sections.

V. INCOHERENT PRODUCTION

By considering coupling constants, it is expected that the cross section of two-photon production of Higgs bosons through inelastic events is smaller than gluon-gluon fusion by a factor of $(\alpha/\alpha_s)^4$, where α and α_s are the electromagnetic and strong coupling constants, respectively. However, it is worthwhile to make a quantitative comparison between the incoherent and coherent two-photon Higgs boson production using the direct Monte Carlo integration of the relevant Feynman diagrams.

For simplicity, we first consider the incoherent production in a p-p collision, in which the S-matrix can be factorized as the product of the proton current (inelastic) matrix elements and the two-photon production matrix elements. The production cross section is the incoherent sum of the inelastic processes, and summing over all possible final states yields an expression cast in terms of the proton structure function measured in the deep inelastic scattering.

The Higgs production cross section is obtained from the leading order term in the S-matrix, which is the third order in the Hamiltonian,

$$M = (-ie)^2(-ie^2 I)\frac{-i}{q_a^2}\frac{-i}{q_b^2} <X_a|\mathcal{J}_\mu|P_a s_a><X_b|\mathcal{J}_\nu|P_b s_b>$$
$$\times (q_{a\lambda}g_\sigma^\mu - q_{a\sigma}g_\lambda^\mu)(q_b^\lambda g^{\nu\sigma} - q_a^\sigma g^{\nu\lambda})(2\pi)^4\delta^4(P_{X_a} + P_{X_b} + k - P_a - P_b). \quad (21)$$

In the collider frame the cross section can be written as

$$\sigma = \frac{(4\pi\alpha)^4|I|^2}{2\beta}\int\frac{d^3k}{(2\pi)^3 2\omega_k}\frac{1}{2\omega_a}\int\prod_{X_a}\frac{d^3p_{X_a}}{(2\pi)^3 2\omega_{X_a}}\frac{1}{2\omega_b}\int\prod_{X_b}\frac{d^3p_{X_b}}{(2\pi)^3 2\omega_{X_b}}$$
$$\times\frac{1}{2}\sum_{s_a}\sum_{s_{X_a}} <X_a|\mathcal{J}_\mu|P_a s_a><P_a s_a|\mathcal{J}_{\mu'}|X_a>(\frac{1}{q_a^2})^2$$

$$\times \frac{1}{2} \sum_{s_b} \sum_{s X_b} < X_b|\mathcal{J}_\nu|P_b s_b >< P_b s_b|\mathcal{J}_{\nu'}|X_b > (\frac{1}{q_b^2})^2$$

$$\times 4\{(q_a \cdot q_b)g^{\mu\nu} - q_a^\nu q_b^\mu)\}\{(q_a \cdot q_b)g^{\mu'\nu'} - q_a^{\nu'} q_b^{\mu'})\}(2\pi)^4 \delta^4(q_a + q_b - k), \quad (22)$$

where the sums are over all the possible hadronic final states X_a and X_b, and β is the velocity of the proton in the collider frame. We may introduce a structure function defined by

$$W_{\mu\mu'}(x,Q^2) = \frac{1}{4M_n} \int \prod_X \frac{d^3 p_X}{(2\pi)^3 2\omega_X} \frac{1}{2} \sum_s \sum_{s_X} < X|\mathcal{J}_\mu|Ps >$$
$$\times < Ps|\mathcal{J}_{\mu'}|X > (2\pi)^3 \delta^4(P_X - P + q), \quad (23)$$

where $Q^2 = -q^2$ and $x = q^2/2M_n\nu$ and M_n is the proton mass. Here ν is the time component of the momentum transfer q in the proton rest frame. It can be proven that the proton electromagnetic structure function has a general form

$$W_{\mu\nu}(x,Q^2) = W_1(x,Q^2)(-g_{\mu\nu} + \frac{q_\mu q_\nu}{q^2})$$
$$+ \frac{1}{M_n^2} W_2(x,Q^2)(P_\mu - \frac{1}{2x}q_\mu)(P_\nu - \frac{1}{2x}q_\nu). \quad (24)$$

It is usual to identify the two structure functions

$$M_n W_1(x,Q^2) = F_1(x,Q^2), \quad |\nu|W_2(x,Q^2) = F_2(x,Q^2), \quad (25)$$

which satisfies the Callen-Gross relation in the scaling limit.

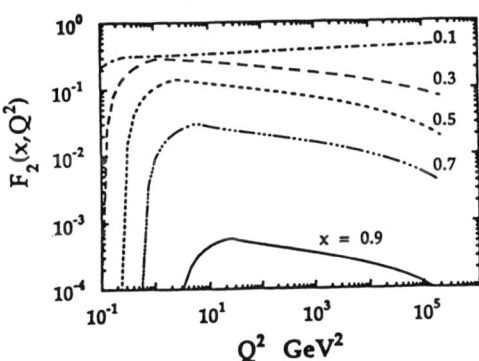

Fig. 6. The proton structure function F_2 as a function of x and Q^2.

Therefore, the total cross section is given by

$$\sigma = \frac{(4\pi\alpha)^4|I|^2}{2\beta\gamma^2} \int \frac{d^3k}{(2\pi)^3 2\omega_k} \int \frac{d^4 q_a}{(2\pi)^2} W_{\mu\mu'}^{(a)}(x_a, Q_a^2) W_{\nu\nu'}^{(b)}(x_b, Q_b^2)$$
$$\times (\frac{1}{q_a^2})^2 (\frac{1}{q_b^2})^2 4\{(q_a \cdot q_b)g^{\mu\nu} - q_a^\nu q_b^\mu)\}\{(q_a \cdot q_b)g^{\mu'\nu'} - q_a^{\nu'} q_b^{\mu'})\}. \quad (26)$$

For heavy-ion collisions we may replace the proton structure function by the nuclear structure function, which is, as a first approximation neglecting the nuclear EMC effect, the incoherent sum of the structure functions for all nucleons, i.e.

$$W_i(x,Q^2) = ZW_i^{(p)}(x,Q^2) + NW_i^{(n)}(x,Q^2), \qquad (27)$$

where $i = 1, 2$, and Z and N are the charge number and the neutron number, respectively. The neutron structure function is taken to be

$$F_2^{(n)}(x,Q^2) = (1-x)F_2^{(p)}(x,Q^2), \qquad (28)$$

which is a good approximation except for large values of x, which do not contribute much to the total cross section. In practice, the structure function is fitted to the experimental data in a limited kinematic range and extrapolated to the whole phase space consistent with energy-momentum conservation. The conservation is expressed by the delta-function appropriate to scattering from a composite system. With the inclusion of the higher-twist coefficient and target mass corrections given in ref. [14], the proton structure function for the leading twist is fitted to the EMC data [15]. Since the inelastic scattering of nucleon has a threshold energy of $t_0 = (M_n + m_\pi)c^2$, where m_π is the pion mass, the structure functions are taken to be zero below the threshold. For numerical convenience, an interpolation is introduced between t_0 and $3t_0$. Fig. 6 shows the variation of the proton structure function with x and Q^2.

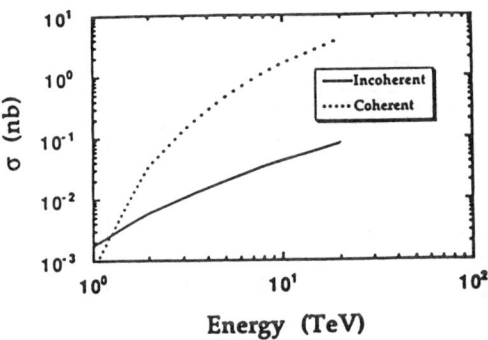

Fig. 7. The cross section for the incoherent Higgs production as a function of the beam energy per nucleon for uranium collisions. The Higgs mass and the top quark mass are both assumed to be 100 GeV.

The total cross section for Higgs production in the heavy-ion collider for uranium collisions is shown in Fig. 7 as a function of energy. The results are compared with the coherent production. The coherent production is seen to have a sharp rise in the energy from 1 to 2 TeV per nucleon, reflecting the suppression of the cross section by the nuclear form factor at lower beam energies. At higher bombarding energies, the coherent production dominates the incoherent. The ratio scales roughly as Z^4/A^2.

VI. CONCLUSION

We have calculated the coherent and incoherent production of Higgs bosons in ultrarelativistic heavy-ion collisions from the standard model. We use perturbation theory to calculate the two-photon processes and carry out the Feynman integrals numerically by Monte Carlo methods. For coherent production we use a classical field approximation to the nuclear electromagnetic fields. This method is a good approximation for coherent processes involving heavy-ions, and it is particularly suited for the description in impact parameter space when the classical trajectories of heavy ions are experimentally tagged. The results show that beyond some critical energy the coherent production with Z^4 enhancement is significant and dominates the incoherent; asymptotically, the production cross section increases with energy as $(\ln\gamma)^3$. However, the production is very sensitive to the beam energy below the critical energy. Since the coherent production provides clean signatures in peripheral collisions, the prospect of using heavy ions as a means of testing the standard model and searching for the Higgs boson is excellent at the higher energy machines.

ACKNOWLEDGMENTS

This research was sponsored by the Division of Chemical Sciences, Office of Basic Energy Sciences, and by the Division of Nuclear Physics of the U.S. Department of Energy under contract No. DE-AC05-84OR21400 with Martin Marietta Energy Systems Inc, and by the National Science Foundation and the Faculty Research Award Program of the City University of New York.

*Permanent address: Department of Physics, and Center for Nuclear Theory, Brooklyn College of the City University of New York, Brooklyn, New York 11210.

TABLE I. The table of interaction Hamiltonian used in the particle production of relativistic heavy-ion colliders, where : : indicates the normal ordered product. *In the standard model $\mathcal{K} = 1$.

with photons	$H_{int}(x)$
charged fermions	$e : \bar\psi\gamma^\mu\psi A_\mu :$
$W^\pm$ bosons	$-ie : \{\mathcal{K}(\partial_\mu A_\nu - \partial_\nu A_\mu)W^{-\mu}W^{+\nu}$ $+(\partial_\mu W_\nu^+ - \partial_\nu W_\mu^+)A^\mu W^{-\nu}$ $+(\partial_\mu W_\nu^- - \partial_\nu W_\mu^-)W^{+\mu}A^\nu\} :$ $+e^2 : (W_\mu^+ W^{-\mu}A_\nu A^\nu - W_\mu^+ W_\nu^- A^\mu A^\nu) :$
charged scalar bosons	$-ie : A_\mu(\phi^\dagger\partial^\mu\phi - \partial^\mu\phi^\dagger\phi)$ $-e^2 : A_\mu A^\mu \phi^\dagger\phi :$
neutral scalar bosons	$\frac{e^2}{2}I : F_{\mu\nu}F^{\mu\nu}\phi :$
with the Higgs boson	
fermions	$(\sqrt{2}G_F)^{1/2}m_f : H\bar{f}f :$
weak bosons	$(\sqrt{2}G_F)^{1/2} : (2M_W^2 W_\mu^+ W^{-\mu} + M_Z^2 Z_\mu Z^\mu)H :$

REFERENCES

[1] H. GOULD, in Proc. Atomic Theory Workshop on Relativistic and QED Effects in Heavy Atoms, AIP Conf. Proc., Vol. 136 (AIP, New York, 1985) p. 66.

[2] C. BOTTCHER AND M.R. STRAYER, in "Physics of Strong Fields", edited by W. Greiner, (Plenum, New York, 1987), Vol. 153, page 629; "Proceedings of the Second Workshop on Experiments and Detectors for the RHIC", Berkeley, California, May 25-29, 1987, page 279, (LBL-24604); Nucl. Inst. and Meth.

[3] C. BOTTCHER AND M.R. STRAYER, Phys. Rev. **D39**, 1330 (1989).

[4] G. BAUR AND C.A. BERTULANI, Phys. Rep. **161**, 299 (1988).

[5] M. GRABIAK, B. MÜLLER, W. GREINER, G. SOFF, AND P. KOCH, J. Phys. **G15**, L25 (1989).

[6] Elena Papageorgiu, Phys. Rev. **D40**, 92 (1989); Nucl. Phys. **A498**, 593c (1989).

[7] M. DREES, J. ELLIS AND D. ZEPPENFELD, Phys. Lett. **223B**, 454 (1989).

[8] C. BOTTCHER, A.K. KERMAN, M.R. STRAYER AND J.S. WU, Particle World, Vol. 1, No. 6, 174 (1990).

[9] G. SOFF, J. RAU, M. GRABIAK, B. MÜLLER AND W.GREINER, *Production of super-symmetric particles and Higgs bosons in ultrarelativistic heavy-ion collisions*, Preprint, 1989, (GSI-89-55).

[10] J.S. WU, C. BOTTCHER, M.R. STRAYER AND C.M. SHAKIN, Phys. Rev. **C43** 2422 (1991).

[11] J.F. GUNION, H.E. HABEN, G.L. KANE, AND S. DAWSON, "The Higgs Hunter's Guide", (Addison Wesley, New York, 1990).

[12] J.S. WU, C. BOTTCHER, M.R. STRAYER AND A.K. KERMAN, Ann. Phys. (in Press, 1991).

[13] J. ELLIS AND G.L. FOGLI, Phys. Lett. **B249**, 543 (1990).

[14] A. MILSZTAJN, Saclay Report No. DPHPE90-07, (1990).

[15] J.J. AUBERT et. al., The European Muon Collaboration, Nucl. Phys., **B259**, 189 (1985).

OPTIMAL METHODS FOR LARGE-SCALE SCIENTIFIC DATABASE AND SPARSE MATRIX APPLICATIONS

P. Rochford, S. C. Park, and J.P. Draayer*
Department of Physics and Astronomy
Louisiana State University, Baton Rouge, LA 70803

and

S.-Q. Zheng
Department of Computer Science
Louisiana State University, Baton Rouge, LA 70803

ABSTRACT

Pursuing a many-nucleon description of collective rotations of nuclei demands that very large sparse matrix representations of hamiltonians be constructed and diagonalized. Two general computational methods are presented for reducing the computer resources required for these processes. Specifically, a space and time optimal numerical database called a Weighted Search Tree that minimizes the number of redundant intermediate calculations required when calculating the matrix elements of the sparse hamiltonian matrices is given, and a matrix multiplication algorithm for a new space optimal representation for sparse matrices is given which is expected to dramatically decrease the cost of diagonalizing large sparse band matrices. Both methods can be implemented in a broad range of large-scale scientific applications.

INTRODUCTION

Because of its remarkable success at describing the various structural features observed for nuclei, the shell model is usually used when attempting to describe the bound states of a nucleus.[1] The basic tenet of the model is that the nucleus is comprised of neutrons and protons that sit within a global mean-field potential which is generated from the average effect of their interactions with each other. The harmonic oscillator is the mean-field usually employed because of the very useful algebraic properties it possesses. Within this mean field potential the nucleons are considered to interact with each other through residual forces. The mean-field plus the sum of these residual interactions yields the many-body hamiltonian for the nuclear system, which is then evaluated within the hilbert space spanned by nucleons occupying states of the oscillator.

A research area which is currently very active within the nuclear physics community is that describing the collective rotational behaviour in nuclei, i.e. motion involving the rotation of the nucleus as a whole. It has been determined through various investigations[2-4] that the residual interactions which are important for obtaining a good

---

* Supported in part by the U.S. National Science Foundation, Grant No. 89-22550.

description of the observed rotational features in nuclei are those contained in the rotationally invariant many-body hamiltonian

$$H = H' + \sum_{n=1}^{A} (Cl_n \cdot s_n + Dl_n^2) + (AJ^2 + BK_J^2) + P$$

$$H' = H_0 - \frac{1}{2}\chi\left[Q_2 \cdot Q_2 - (Q_2 \cdot Q_2)^{TE}\right]. \qquad (1)$$

The first term H_0 is the mean field harmonic oscillator hamiltonian, while the remainder are the residual interactions needed to account for deformation, single-particle energy splitting, rotational energy splitting, and nucleon pairing, respectively. The interactions appear in Eq.(1) in the order of their importance with regard to influencing collective rotational features in nuclei. Given this many-body hamiltonian, describing rotational motion now requires constructing its matrix representation for states of a fixed total angular momentum J, and diagonalizing the matrix to obtain the energy eigenvalues and eigenstates that contain the information on the state of the system..

Unfortunately, as with any many-body system, one encounters the problem that the hilbert space grows rapidly in size with increasing number of nucleons. This is true even with the usual shell model assumption that the nucleons that completely fill the lowest harmonic oscillator shells are inert and do not participate in the collective motion, while the remaining nucleons are confined to occupy only the lowest incompletely filled (valence) shell. For example, matrix dimensions for ds-shell nuclei of different total isospin T are shown in Table I,[5] where the rapid increase with number of valence nucleons n_v is evident. This proliferation is much more dramatic when valence nucleons are distributed among several shells, involving for example dimensions on the order of 10^{14} when distributing twenty nucleons among the p, ds, and fp-shells.[6] Since the interaction that generates deformation attempts to distribute nucleons among the shells, it is clear that a proper treatment of the many-body hamiltonian (1) leads to matrices which are beyond the capabilities of existing computers.

Table I
Shell model dimensions for ds-shell nuclei.

	n_v	T	J=0	J=2	J=4	J=6	J=8
^{18}O	2	1	3	5	2	0	0
^{20}Ne	4	0	21	56	44	17	3
^{22}Ne	6	1	71	307	311	169	47
^{24}Mg	8	0	325	1206	1311	835	329
^{28}Si	12	0	839	3276	3793	2667	1205

Fortunately, the deformation inducing interaction in Eq.(1) that dictates using multi-shell hilbert spaces has a group symmetry that can be exploited to reduce the number of basis states that must be considered. This group symmetry is Sp(3,R), and when basis states that possess this symmetry are employed, it results in a partitioning of the shell model space into vertical segments $N_0(\lambda_0\mu_0)$ (referred to as irreps in group theory) which span an infinite number of oscillator shells.[7] The latter are irreps of the Sp(3,R) group, and the set of states within each segment that belong to a particular oscillator shell can be partitionned into subspaces which are irreps $N(\lambda\mu)$ of the U(3) $\supset$ SU(3) subgroup of Sp(3,R). The hamiltonian H' operates only within a given Sp(3,R) irrep, and hence has a block diagonal matrix representation in this basis.

Diagonalization of these matrix blocks requires that they be truncated, and it is found that for the lowest energy eigenstates of physical interest that they converge rapidly with increasing shell number, thereby allowing truncation to a finite number of the lower oscillator shells.[2] While the single-particle splitting and pairing interactions in Eq.(1) have off-diagonal matrix elements which connect these blocks, the size of these matrix elements are much smaller than those contained in the blocks defined by H'. To a good approximation they can therefore either be ignored, or limited to those connecting only a few of the blocks, which then reduces the problem to constructing and diagonalizing a much smaller matrix. Shown in Table II are some typical dimensions for the $N_0(\lambda_0\mu_0)$ irrep that yields the lowest energy solution for H' at a given N oscillator shell truncation.

Table II
Hilbert space dimensions for truncated Sp(3,R) irreps.

	$N_0(\lambda_0,\mu_0)$	N	J=0	J=2	J=4	J=6	J=8
^{20}Ne	48.5(8,0)	20	398	1272	1851	2057	1913
^{24}Mg	62.5(8,4)	16	618	2533	4004	4763	4740

These matrix representations for the hamiltonian of Eq.(1) are also rather sparse, which is due to the symmetry basis employed being optimal with regard to describing collective rotational motion. For example a typical hamiltonian matrix is found to be only ~10% dense. This significantly reduces the amount of computer resources which are required for constructing and diagonalizing the matrices. However, the very large dimensions of these sparse matrices still result in considerable demands being made upon computer resources. If such calculations are to become feasible, then it is clear that computational techniques must be devised which more rapidly and efficiently solve this large sparse eigenmatrix problem.

In this manuscript, two techniques will be presented which help accomplish this goal. The first is a numerical database structure called a Weight Search Tree (WST) that dramatically decreases the amount of computer resources required to construct the sparse matrices. The second is a new technique for sparse matrix multiplications which can be implemented to increase the efficiency with which sparse matrix diagonalizations are performed.

WEIGHTED SEARCH TREE

When constructing the matrix representation of H in the $Sp(3,R) \supset U(3) \supset SU(3)$ hilbert space basis the hamiltonian is first expressed in terms of U(3) tensor operators $T_\alpha^{N(\lambda\mu)}$, where α denotes the remaining labels in the subgroup chain $SU(3) \supset SO(3) \supset SO(2)$. The hamiltonian matrix elements are then calculated by evaluating those for the tensors using the Wigner-Eckart theorem,[8] which states that the matrix element can be expressed as

$$\left\langle N_1(\lambda_1\mu_1)\alpha_1 \left| T_{\alpha_2}^{N_2(\lambda_2\mu_2)} \right| N_3(\lambda_3\mu_3)\alpha_3 \right\rangle = \sum_{\rho_2} \left\langle (\lambda_3\mu_3)\alpha_3; (\lambda_2\mu_2)\alpha_2 \mid (\lambda_1\mu_1)\alpha_1 \right\rangle_{\rho_2}$$
$$\times \left\langle N_1(\lambda_1\mu_1) \left\| T^{N_2(\lambda_2\mu_2)} \right\| N_3(\lambda_3\mu_3) \right\rangle_{\rho_2} \quad (2)$$

i.e. a sum over the product of a SU(3) coupling coefficient and a reduced matrix element, respectively. The coupling coefficient is determined by a well defined procedure,[9,10] and is interpreted as a geometrical factor associated with the group symmetry chain SU(3) ⊃ SO(3) ⊃ SO(2). The reduced matrix element depends upon the specific physical form of the tensor operator and is interpreted as the quantity associated with the physical properties of the system. While this factorization means the determination of matrix elements can be seperated into two parts, these separate quantities are each expensive to compute.

Since in the matrix construction of the hamiltonian (1) for any given angular momentum J the same coupling coefficients and reduced matrix elements reoccur frequently, it is desirable to save these intermediate results and thereby avoid wasteful regeneration. The most efficient algorithm for minimizing the cpu time is one which stores the new intermediate quantities as they are computed, while also retrieving previously computed quantities by a binary search. Since computer memory limitations will usually preclude one from saving all the intermediate results which would be generated when constructing the large sparse matrices considered here, the most efficient numerical database is one which is also able to retain the most valuable information when subject to imposed memory space constraints. Such a data structure has been developed by the authors, and is called a Weighted Search Tree (WST).[11, 12]

The WST is a numerical database which allows for the storage and retrieval of dynamically generated lists of length n in optimal time, $i.e.$ ~$O[\log(n)]$. When implemented, it allows one to search an already generated list for a particular element, and, if that element is not found, to also calculate the required result and add it to the list for possible reuse as necessary at a later stage of the calculation. Each list element that is stored must be uniquely identified by a key, which is usually characterized by the input variables used to determine its value, such as the quantum labels in the case of the coupling coefficients and reduced matrix elements considered here. It is the key which is actually searched for in the WST. Once a key is found the corresponding data entries associated with the key, which are also kept in the WST, provide information on the location where the previously calculated results can be found. The key and data elements do not have to be entered in any predefined order, as order is maintained by pointers associated with each element in the WST. All keys contained in the numerical database are assigned a weight, or priority, value which indicates its relative importance. These priority values provide the criteria by which it is determined whether or not new information is stored once the fixed size constraints of the database have been reached.

Successive calls to the WST package constructs a tree of the type shown in Figure 1, where the left-hand number within the circles is the key value and the right-hand number the priority. While the actual representation of the WST that has been coded as a set of FORTRAN subprograms is different from that shown in Figure 1,[12] and is employed because it requires minimal memory space, the concept is essentially the same. The construction logic for the tree involves assigning elements to nodes (represented by the circles) in a linear array and using linkage information (the pointers and superscript integers) to specify the left and right subtrees of each node. The binary search times for all insert, search, and deletion operations are achieved by insuring the tree is height-balanced. For example, a binary tree is height balanced modulo one if the difference in maximum number of steps from root to apex between left and right subtrees for every node is less than or equal to one. The WST itself is a linear integer array, ID(-10:*). The first eleven elements, ID(-10) to ID(0), are reserved for specifying the structure of the tree and other linkage and status information. The rest of the array, starting with ID(1), is for the node elements. Each node contains four sets of information: a key that serves to identify the node, its priority, the linkage instructions, and the data. The array for the WST shown in Figure 1 is given in Figure 2 as an example. The columns from left to right indicate the nodes as they occur in the WST linear array, and the LCHILD, RCHILD, and PARENT rows are the linkage instructions. The data in this example is chosen to be the actual key values. The data could specify, as in the current case of calculating matrix elements, the position in an external real array where additional information about the node can be found, such as the values for the coupling coefficients.

The deletion of a node in the WST does not occur until the numerical database is full. Any new key which is generated is only inserted in place of the existing key if that particular data item has a higher priority value than the lowest priority node in the database. When this is the case the lowest priority node is deleted, the tree is rebalanced, and a node with the new element is then inserted. The actual deletion of the node is accomplished by exploiting the fact that the WST is a linear array, and arranging the nodes in what is referred to as a heap structure [13] such that the lowest priority one starts at ID(1). This is seen to be the case for the example in Figure 2, as the leftmost column is the one with the lowest priority.

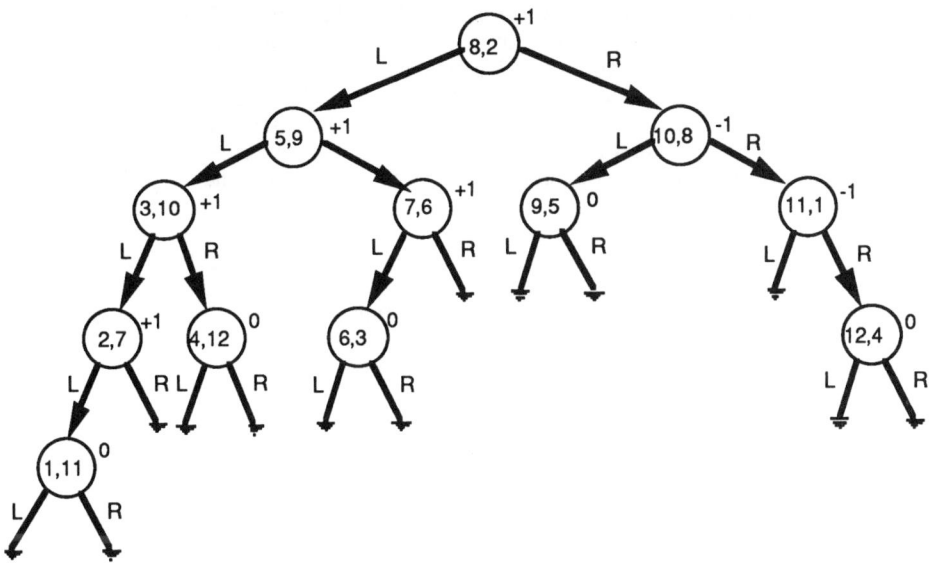

Figure 1. A weighted search tree.

	1	2	3	4	5	6	7	8	9	10	11	12
PRIORITY	1	2	5	3	4	7	10	12	6	9	8	11
KEY	11	8	9	6	12	2	3	4	7	5	10	1
LCHILD	0	10	0	0	0	12	6	0	4	7	3	0
RCHILD	5	11	0	0	0	0	8	0	0	9	1	0
PARENT	11	0	11	9	1	7	10	7	10	2	2	6

Figure 2. Linear array representation of above weighted search tree.

It is important to point out that the priority assigned to data items in the WST can incorporate the frequency of use of an item as well as its intrinsic value. The reason is that while the generation times of two data items may be quite different, this difference may be offset by use frequency. For example, if result A takes ten times longer to generate than B, but is used only once, while B is used fifty times, then B is clearly the preferred element to be retained. For the WST the priority strategy employed consists of two parts: a base priority, which reflects the computation time involved in generating the information associated with the data item, and the hit frequency, which measures the frequency of use of the data item. Specifically, when a data item is accessed its priority is updated by adding its base priority to its current priority value to get its new priority value. The base priority can be defined either internally when the element is generated or externally by the user. Internally defined priorities would normally use the cpu time involved in the calculation and/or the storage required for holding the information associated with the data item. External priorities, on the other hand, might be set by the user according to the type of computation.

When applying the WST to the matrix element calculations being considered here, *i.e.* those specified by Eq.(2), the approach involves employing two WST's, one for the SU(3) coupling coefficients and another for the reduced matrix elements. Each of these WST's will store position information where the values of these quantities are located in external real arrays. Whenever a specific coupling coefficient, etcetera, is required, the key for that quantity is calculated and a call is issued to the WST array via the fetch option to search for the occurence of the key. This requires that a unique key be defined for each distinct quantity. For all those quantities considered here this is easily accomplished using the technique of packing integer words, the details of which can be found elsewhere.[14] For example, the following packing function

$$KEY(I,J,KRO) = IOR(KRO,ISHFT(IOR(J,ISHFT(I,14)),14)) \qquad (3)$$

achieves this purpose for the reduced matrix elements of a given $T^{N_2(\lambda_2\mu_2)}$ tensor

$$\langle I || T^{(\lambda_2\mu_2)} || J \rangle_{KRO} = \langle N_1(\lambda_1\mu_1) || T^{N_2(\lambda_2\mu_2)} || N_3(\lambda_3\mu_3) \rangle_{\rho_2}, \qquad (4)$$

where I and J denote the occurence in the Sp(3,R) irrep hilbert space of the SU(3) irreps indicated.

If a node with the required key is found in the tree, the WST program increases its priority according to the user specified scheme, restructures the heap to reflect the greater priority for this node, and then executes an alternate return. On completion of a successful search, the routine provides the specific location from which the required value in the external real array can be found. This value can then be employed immediately in the calculation. When no node is found with the required key, the routine executes a normal return, thereby indicating that the desired quantity must be computed. Once the quantity and its priority value are determined, a check is made by the user to determine whether or not the WST is full. If there is room, a call is made using the insert option which adds a new node with that key, rebalances the tree, assigns the node its intrinsic priority, and reconstructs the priority heap structure as necessary. The calculated quantity is subsequently stored in the external real array within the calling program, and the location index for this quantity saved as the data of the new node.However, if the WST is found to be full, a comparison is made in the calling program of the priority value for the quantity just computed against the lowest priority value contained within the WST, which is located in the node that starts at ID(1). If the new priority is less than the lowest in the WST, then the calculated quantity is used as needed in the calculation and subsequently discarded. However, if the converse is true, then the delete option is invoked to remove the lowest priority node and rebalance the WST so that a node containing the new key can be added. The insert option is then called as indicated above for the case when the WST was not full.

This application of the WST as given up to now works well in the situation where there is one computed quantity associated with each node in the WST, such as one reduced matrix element (c.f. Eq.(4)) in an external real array for each node. However, this is not the most efficient method for using the WST, as frequently a subset of the generated list of intermediate results can be identified by a single key. For example, the reduced matrix elements occurring in the matrix element calculation of Eq.(3) may have $\rho=1, \ldots, \rho_{max}$ and can be identified by the $N(\lambda\mu)$ irrep labels. The more efficient approach in such a case is to assign only a single node in the WST for this subset, and store the starting and end positions (or the starting position and number of elements) where this information is contained in the external array (henceforth called the buffer) as its data items, thereby saving memory space and reducing the number of nodes to be searched. The WST can easily accomodate this as it has the flexibility to contain any number of data items per node as specified by the user.

While this straightforward extension is fine when constructing the WST, problems occur once either the WST or buffer reach their storage limitations and a deletion operation is required. The reason is bookkeeping information must be retained that indicates where and how much free space is available in the buffer. It must not only keep account of space which is free at several points throughout the buffer, as more than one node may have to be deleted from the WST before enough buffer space is available to accomodate the higher priority data, but also contain information as to how to retrieve it when the intermediate results are required for reuse later on. For example, a new set of reduced matrix elements having $\rho_{max}=5$ may be generated having a higher priority than two sets with $\rho_{max}=3$ and $\rho_{max}=4$ that occupy different and nonsequential locations in the buffer. While some of the elements of this new set may occupy space vacated by one set, the remainder must occupy that of the other. Since this set of five reduced matrix elements will be located in at least two different parts of the buffer, extra information must exist indicating which locations they occupy. The problem of keeping track of this information is exactly what is involved in a file directory management system. Incorporated within the WST routines is therefore a file directory system which is in the form of an array containing pointer information for the location of the data in the buffer.

With this directory system, the user stores as the data for the WST node the starting point and number of locations occupied in the index array, which in turn contains the pointers for the locations in the buffer where the intermediate results are stored. This integer array INDEX(-2:*) is dimensioned the same size as the buffer array plus three extra positions, where the first three elements INDEX(-2) to INDEX(0) contain the information on the available free space. On a successful search for a given key, the information in the buffer is retrieved by using the values of INDEX(I) as the locations in the buffer array. When inserting and deleting information the logic is almost identical to that described above for the WST. The exceptions are that an additional check must now be made to see if there is enough storage in the buffer for the new results. This information is kept in INDEX(-2:0) along with pointers that give the starting point location of the free space and number of positions available. The deletion option automatically updates INDEX for the new free space available in the buffer.

Using this system, the specific coupling coefficients and reduced matrix elements needed in the construction of the sparse matrices of interest here are calculated only once, while unneeded values are never generated. Moreover, under a fixed memory size constraint, those values which are the most expensive to generate are retained in favor of the least expensive. This means the required matrix representations can be generated such that the cpu time is minimized, as the non-zero matrix elements are computed in the most efficient manner possible. As an example, the implementation of the WST has resulted in typical gains in run times of at least a factor of two for the case of ^{20}Ne shown in Table II. This kind of reduction is indicative of the savings which can be expected by implementing a WST numerical database. They are not, however, limited to this specific problem, as the WST is a general purpose package of FORTRAN subprograms that can be implemented in any application which involves the computation and later reuse of intermediate numerical results.

SPARSE MATRIX MULTIPLICATIONS

Once the sparse matrix representations of the hamiltonian given by Eq.(1) have been constructed, one is confronted with the task of diagonalizing them in as efficient a manner as possible in order to obtain the energy eigenvalues and eigenvector wavefunctions. The best approach is clearly to exploit their sparse nature and employ a method where the memory storage required for the matrices is minimal and the calculations involving the zero matrix elements are avoided. All standard diagonalization algorithms for real symmetric matrices of a general form, be they either of modest or large size, reduce the full matrix to tridiagonal form as an intermediate step. When all eigenvalues and eigenvectors are desired the reduction from full to tridiagonal form is accomplished by a series of similarity transformations using reflector matrices, which is very costly when dealing with very large matrices.[15] When only a few of the extremal eigenvalues and their eigenvectors are desired, as in the present case, this reduction is achieved at much less cost by using the Lanczos method which constructs a guess of the tridiagonal matrix given a trial starting vector.[16] Once the tridiagonal form is achieved the QL and QR algorithms with quotient shifts and plane Jacobi rotations are employed to complete the diagonalization process. In both cases, matrix-matrix or matrix-vector multiplications are performed that involve the original matrix and its descendants that arise during execution of the algorithm. If the original matrix is sparse, then in both cases it is reasonable to expect that the successive operations will involve matrices which are sparse. A means of performing these calculations more rapidly is to therefore devise and implement a more efficient algorithm for sparse matrix-matrix and matrix-vector multiplications.

Such an algorithm has been developed by the authors and tested for matrices of a general band form, since many hamiltonians of physical interest such as that in Eq.(1) possess matrix representations of this form.[17] While there are specialized algorithms for the diagonalization of dense band matrices, these are only found to be faster than the standard algorithms for those dense matrices where the bandwidth L (maximum number of non-zero elements in a row or column of the band) is less than one third the dimension of the matrix ($L < D/3$).[15] For larger bandwidths the standard approach is found to be necessary.[15] In the results presented below it will become apparent that the sparse matrix multiplication algorithm introduced here outperforms the standard approaches even in the situation of a fully dense matrix, thereby underscoring the practicality of this algorithm when diagonalizing matrices of any bandwidth.

First, it is obvious that a two-dimensional array representation for sparse matrices not only wastes space but also prevents matrix operations from being performed in optimal time because of the additional operations that are performed on the null elements. A standard representation for a sparse matrix which enables it to be stored as a linear list, is one where each non-zero element is represented by a 3-tuple of the form (i, j, v), where i and j are the row and column numbers of the element, respectively, and v is the value of the element.[13] For example, a sparse matrix such as that shown in Figure 4 below would be stored in memory in row-major order as shown in Figure 5. That is, the nonzero elements are in increasing order of their row numbers, and all non-zero elements with the same row number are in increasing order of their columns numbers.

	col 1	col 2	col 3	col 4	col 5	col 6	col 7
row 1	2	0	0	0	0	0	0
row 2	3	4	0	0	0	0	0
row 3	0	0	0	5	0	0	0
row 4	0	0	0	6	0	0	0
row 5	0	1	0	8	4	3	0
row 6	0	0	0	0	2	2	1
row 7	0	0	0	0	5	0	0

Figure 4. Two Dimensional Representation of a 7 × 7 sparse matrix.

	0	1	2	3	4	5	6	7	8	9	10	11	12	13
1	7	1	2	2	3	4	5	5	5	5	6	6	6	7
2	7	1	1	2	4	4	2	4	5	6	5	6	7	5
3	13	2	3	4	5	6	1	8	4	3	2	2	1	5

Figure 5. Space efficient representation of a 7 × 7 sparse matrix.

Here each column contains information on a nonzero element of the matrix A. For columns with k > 0 the three rows contain the row and column index i and j of the nonzero matrix element A_{ij}, and its value, respectively. The 0 column contains the global information of the matrix A, where the first two rows contain the number of rows and columns, respectively, and the third row the number t of non-zero elements in the matrix A.

While faster matrix operations are achieved with this representation, it is easy to see that this data structure is not the most space efficient for a sparse matrix that contains adjacent non-zero elements in its rows, as most of the row and column numbers in a segment are redundant. Consider as an example the calculation of C_{ij} via C=A×B. If one of A_{ik} or B_{kj} is a zero element and the other is a non-zero element, say A_{ik}=0 and $B_{kj}\neq 0$, then B_{kj} is checked even though the multiplication operation is not performed. Clearly, it would be preferable to add additional information to the data structure for the matrix that allows for skipping those elements which do not contribute to the final result of the matrix operation. This is the idea behind the data structure designed here.

A maximal nonzero segment (or simply segment) in row i of an $m \times n$ matrix A is defined here as a closed interval $[j,k]$ such that all elements $A_{il} \neq 0$ for $j \leq l \leq k$, $A_{i,j-1}$ = 0 if $j > 1$ and $A_{i,k+1} = 0$ if $k < m$. This representation for A consists of two arrays EA(1:t_A) and SA(0:s_A, 1:3), where t_A is the number of nonzero elements in A and s_A is the number of segments in A in the row-major order. Clearly, each segment corresponds to a subarray of EA and all segments of A are in row major order in EA. SA contains s_A+1 entries, where each entry is a 3-tuple (SA(r,1),SA(r, 2),SA(r, 3)) that contains information of a segment for r > 0. SA(0,1) and SA(0, 2) are the number of rows and columns of A, respectively, whereas SA(0,3) contains the number of segments of A. If the corresponding segment $[j,k]$ is in row i, then SA(r, 1) = i, SA(r, 2) = j, and SA(r, 3) = k. Since non-zero elements and segments of A are arranged in EA in a unique linear order, the indices of non-zero elements to be used can be calculated by a linear scan of SA. Shown in Figure 6 is an example of this representation for the matrix A of Figure 4.

68 Large-Scale Scientific Database

	1	2	3	4	5	6	7	8	9	10	11	12	13
EA:	2	3	4	5	6	1	8	4	3	2	2	1	5

		0	1	2	3	4	5	6	7	8
	1	7	1	2	3	4	5	5	6	7
SA:	2	7	1	1	4	4	2	4	5	5
	3	8	1	2	4	4	2	6	7	5

Figure 6. New Representation of a sparse matrix.

It can be shown that this representation of the sparse matrix occupies less memory space when $t_A > 1.5\, s_A$, *i.e.* when each segment contains more than 1.5 elements on average.[17] This condition is one which is frequently satisfied in most scientific applications, and is found to be true for the hamiltonian matrices of interest to the authors.

Consider now the explicit multiplication of a sparse $m \times n$ matrix A with a sparse $n \times p$ matrix B to obtain a $m \times p$ matrix C. This operation involves multiplying the rows of A by the columns of B, or equivalently, by the rows of the transpose of B (B^T). In terms of the new matrix representation introduced here, matrix multiplication therefore requires first transforming the representation of B into its transpose. This can be accomplished in a standard manner by implementing the bucket sort algorithm.[18] Once this has been carried out, the problem is reduced to evaluating the dot products of the vectors corresponding to the rows of A with those of the rows of B^T, which in turn involves evaluating the overlaps of the segments contained in each of the rows. Performing the latter is relatively straightforward as there are only nine different kinds of segment overlap conditions, and these can be sorted into three groups as shown in Figure 7 below.

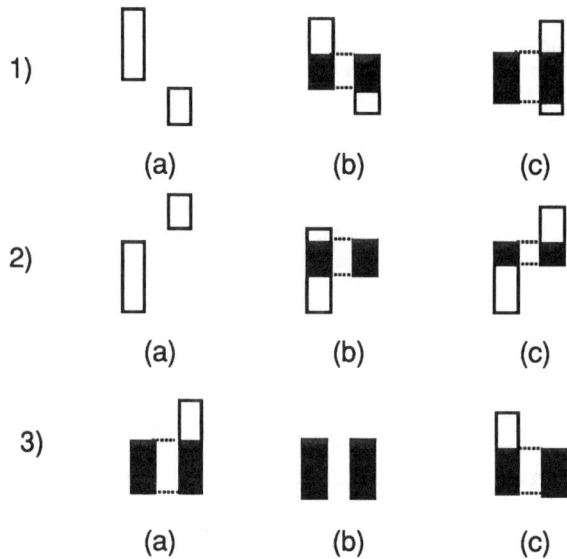

Figure 7. Overlap condition groups.

Collecting these various results, the general algorithm for the matrix multiplication of two sparse matrices in terms of the new representation introduced here is easily determined to be given by that outlined below.

Algorithm *MATRIX_MULTIPLY*(EA, SA, EB, SB, EC, SC)
   Transpose B to obtain EB^T and SB^T;
   Let $a_1, a_2, ..., a_g$ be the row numbers of non-zero rows of A such that $a_i \leq a_{i+1}$;
   Let $b_1, b_2, ..., b_h$ be the row numbers of non-zero rows of B^T such that $b_j \leq b_{j+1}$;
   For i =1 **to** g **do**
     Scan through segments $[j_a^A, k_a^A]$ in row a_i;
     For j=1 **to** h **do**
       Scan through segments $[j_b^B, k_b^B]$ in row b_j;
       if $[j_a^A, k_a^A] \cap [j_b^B, k_b^B] \neq \emptyset$ **then**
         compute the dot-product of sub-vectors of A and B defined
           by segment overlap;
       else
         null case;
       endif
       Store the result of C_{a_i, b_j} in EC and update SC;
     endfor
   endfor
end of *MATRIX_MULTIPLY*

It is clear from the above algorithm that fewer comparisons will be required when performing matrix multiplications with this representation than the standard one of Figure 5. Moreover, this algorithm will require much less time than that employing the representation of Figure 4, as a much more efficient comparison operation is performed in the former instead of a multiplication or addition of two numbers as for the latter.

To test the efficiency of this algorithm, it was coded in FORTRAN and executed for the matrix multiplication of two identical matrices A and B. Initially A and B are $D \times D$ band matrices, where D is 300 and the band width L is 31 (31 nonzero elements in a row or column), with each row containing one segment. Test data was then generated from this initial distribution iteratively by breaking a segment into two segments (*i.e.* replacing a non-zero element with 0 in a segment of at least three non-zero elements) for both A and B. The performance of the algorithm is shown in Figure 8, along with the same results obtained for the same data when using the standard multiplication algorithm and the conventional sparse matrix one.[13] It is important to note that the distribution of segments in the test data for each of the cases of different numbers of segments was closely related to a normal distribution, and was therefore representative of the situation frequently found in actual scientific applications. From the figure it is evident that when each row contains only one segment, that the new algorithm is 7-8 times faster than the conventional one[13] and 40 times faster than the standard algorithm. This performance only becomes poorer than the conventional algorithm once there are more than 12 segments in each row (except the first 15 rows and the last 15 rows), that is 4 segments with one nonzero element and 8 segments with two elements. For many sparse matrices that occur in physical problems, it is therefore highly likely that this new matrix representation will out perform the conventional method with regard to execution speed in addition to requiring less space utilization.

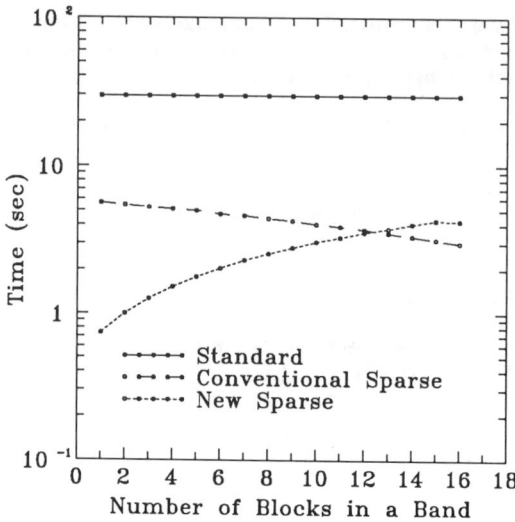

Figure 8. Performance comparison of matrix multiplication algorithms.

SUMMARY

Pursuing a many-nucleon description of collective rotations of nuclei leads to the common problem of very large matrices which must be constructed and diagonalized. By exploiting group symmetries, the dimensions of the hilbert spaces can be reduced to a level where the resulting matrices are of a size that they can be done with existing computers. Furthermore, these matrices have a sparsity on the order of 10-20%. The challenge for improving the practicality of such a description of collective nuclear rotations is to devise efficient methods by which these very large sparse matrices, which are typically of the order of 5000×5000 and greater, can be constructed and diagonalized. In response to this challenge, a numerical database called a Weighted Search Tree was developed by the authors to minimize the redundant intermediate calculations that occur in constructing the matrix representation of the many-body hamiltonian when limited by fixed memory space constraints. This numerical database can be implemented in any scientific application to minimize the use of computer resources, and has been found to yield gains of at least a factor of two in typical applications. With regard to the diagonalization of these large sparse matrices, a new storage representation has been developed, along with a matrix multiplication algorithm, which holds the potential to significantly diminish the cost of storing and reducing large sparse band matrices to tridiagonal form (an important intermediate step in all diagonalization processes). With both this matrix multiplication algorithm and the Weighted Search Tree, the cost of performing the quantum physics computations required for the many-body description of collective nuclear rotations that are of interest to the authors have become much more feasible, and thereby allows the scope of applications that can be undertaken to be much broader.

REFERENCES

1. B. A. Brown and B. H. Wildenthal, Ann. Rev. Nucl. Part. Sci. 38, 29-36 (1988).
2. O. Castaños and J. P. Draayer, Nucl. Phys. A491, 349-372 (1989).
3. H. A. Naqvi and J. P. Draayer, Nucl. Phys. A (in press).

4. P. Ring and P. Schuck, *The Nuclear Many-Body Problem* (Springer-Verlag, New York, 1980).
5. T. Sebe and M. Harvey, "Enumeration of Many Body States of the Nuclear Shell Model with Definite Angular Momentum and Isobaric Spin with Mixed Single Particle Orbits", Technical Report no. AECL-3007, Chalk River Nuclear Laboratories (1968).
6. J. P. Draayer and H. T. Valdes, Comp. Phys. Comm. 36, 313-320 (1985).
7. D. J. Rowe, Rep. Prog. Phys. 48, 1419-1480 (1985).
8. B. G. Wybourne, *Classical Groups for Physicists* (Wiley-Interscience, New York, 1974).
9. Y. Akiyama and J. P. Draayer, Comp. Phys. Comm. 5, 405-415 (1973).
10. J. P. Draayer and Y. Akiyama, J. Math. Phys. 14, 1904-1912 (1973).
11. S. C. Park, J. P. Draayer and S.-Q. Zheng, "Time-Space Optimal Numerical Database for Large-Scale Scientific Applications", in *Proceedings of the International Computer Symposium 1990* (Hsinchu, Taiwan, 1990).
12. S. C. Park, J. P. Draayer and S.-Q. Zheng, "Numerical Database as a File Directory System using a Weighted Search Tree", in *Proceedings of the* (LSU preprint, 1991).
13. E. Horowitz and S. Sahni, *Fundamentals of Data Structures* (Computer Science Press, Rockville, Maryland, 1983).
14. J. P. Draayer and P. Rochford, Notas de Fisica 13, 73-88 (1990).
15. B. N. Parlett, *The Symmetric Eigenvalue Problem* (Prentice-Hall, Englewood Cliffs, N.J., 1980).
16. K. J. Cullum and R. A. Willoughby, *Lanczos Algorithms for Large Symmetric Eigenvalue Computations* (Birkhauser, Boston, 1985).
17. S. C. Park, J. P. Draayer and S.-Q. Zheng, "An Efficient Algorithm for Sparse Matrix Computations", Technical Report no. 91-009, Department of Computer Science, Louisiana State University (1991).
18. A. V. Aho, J. E. Hopcroft and J. D. Ullman, *The Design and Analysis of Computer Algorithms* (Addison-Wesley, Reading, Massachusetts, 1974).

SIEGERT'S CURSE: TAMING AND DOMESTICATING DIVERGENT WAVE FUNCTIONS

Peter Winkler
University of Nevada, Reno, NV 89557

ABSTRACT

Computational techniques of energies and widths of metastable states are discussed when Siegert boundary conditions are applied to solutions of Schrödinger's eigenvalue equation in order to achieve the necessary analytic continuation onto the complex energy plane. Results for a number of different potentials are presented to illustrate specific technical problems.

INTRODUCTION

In his article "On the derivation of the Dispersion Formula for Nuclear Reactions" of the year 1939 Siegert[1] generalizes earlier work[2] by Kapur and Peierls evaluating the dependence of scattering cross sections on the energy of the incident particle. The chosen method of proof applies a perturbation to the boundary conditions rather than to the Hamiltonian itself. This allows for simplified model treatments of processes as complex as the decay of nuclear compound states: All the focus is on the boundary conditions and little on the actual form of the Hamiltonian. For the purpose of this lecture we will adopt this philosophy for a while: Many of the specific computational problems introduced by non-Hermitian, complex boundary conditions can be studied in the context of simple model Hamiltonians. Whenever possible we will choose problems for which very accurate solutions from other methods - preferentially analytical ones - are available for comparison. We feel, however, that more than fifty years after Siegert's landmark paper the discussion would not be complete without addressing some applications to many-body systems although here completed work for review is still rather sparse and computational difficulties are often more indicative of the complexity of such systems rather than the presence of resonance boundary conditions.

SIEGERT REVISITED

With the scattering cross section defined as the ratio S/I of the intensities of scattered and incoming waves singularities of the cross section occur when there is *only an outgoing wave*. Radioactive and autoionizing states belong into this category. Both may be caused by incident particles but this fact is of little interest for their final decay.

Mathematically, such singular behavior of the cross section is expressed by the asymptotic boundary condition

$$f'_n(r_0) / f_n(r_0) = iK_n = i(k_n + i\kappa_n) = i(2E_n)^{1/2} \qquad (1)$$

which is applied to resonance solutions of Schrödinger's eigenvalue equation

$$f''_E + 2(E - V)f_E = 0 \qquad (2)$$

We require that $f_E(r) = 0$ at the origin and $V(r) = 0$ for $r > r_0$. Asymptotically, the general wave function is

$$f_E = (1/K)\sin Kr + S\exp(iKr). \qquad (3)$$

As usual $K = (2E)^{1/2}$. If the (discrete and complex!) energies for which the denominator of the ratio S/I vanishes are denoted by the subscript n, the boundary condition (1) is obtained which aside from bound-state solutions for imaginary K_n yields also discrete eigensolutions in the complex K-plane of which the ones with small negative imaginary parts κ_n correspond to experimentally observable resonance solutions. Two prominent features of these boundary conditions warrant mentioning: (a) The wave function diverges exponentially and with it a number of relevant integrals such as the "normalization" integral. (b) The boundary condition itself depends on the eigenvalue it is supposed to determine. This connection necessitates iterative procedures.

WAYS TO SIDESTEP THE DIVERGENCIES

A number of techniques have been designed to avoid the Siegert divergences altogether. Two of the most prominent ones will be discussed here: Methods based on dilatation transformations (or complex scaling) and direct Siegert methods. A recent approach[3] that avoids divergences completely and remains real altogether belongs to the latter group and will be discussed there. In both approaches we encounter purely numerical treatments of the differential equations involved but applications in the spirit of traditional expansion approaches are more frequent. Molecular computations, in particular, seem still to favor expansion techniques in spite of considerable progress of other, e.g. algebraic techniques[4].

Figure 1: Schematic view of the effect of both uniform and exterior complex scaling (taken from ref. 5).

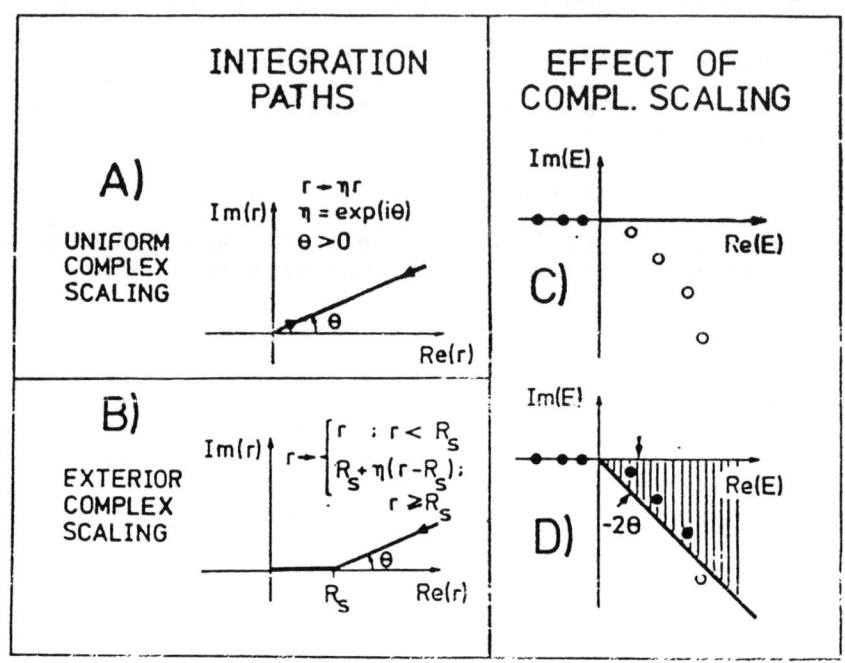

In both of these approaches the unifying idea is to change the path of integration as depicted in fig. 1 (taken from ref.5) into a region where all relevant integrals remain finite. For illustration: the biorthogonal "normalization" integral

$$\int_0^\infty e^{2iKr}\, dr = i/2K \qquad (4)$$

provided that $ImK > 0$. However, the right hand side remains finite even for $ImK < 0$ and is regarded as the unique analytic continuation of the integral. Early work by Miller and coworkers[6] utilizes this technique in the frame of the direct Siegert method for the calculation of autoionizing states of the helium atom and the negative hydrogen ion for which all integrals have an analytic representation. The application to the molecular case is aggravated by the fact that not all the integrals can be obtained in closed form[7].

In dilatation treatments the analytic continuation is achieved by replacing all radial coordinates r of the Hamiltonian by $\exp(i\theta)r$ which is often referred to as complex scaling. Part D of fig. 1 illustrates the way the continuum cut fans out into the lower complex E-plane when the dilatation parameter Θ is turned on thus uncovering resonance eigenvalues (the full circles of part D) which are not accessible as eigenvalues of undilated - or at least not sufficiently dilated - Hamiltonians as symbolized by the open circles. The asymptotic boundary condition relevant for the dilated problem is of the usual exponentially decreasing bound-state type thus facilitating expansion techniques in exponentially decreasing basis sets. The major computational obstacle is the tendency of the basis to also represent the ubiquitous continuum solutions of which there are usual several quite close to the desired metastable state. Root stabilization[8] , though very useful to obtain initial information about resonance parameters, works only up to a certain basis size. The hope in present expansion work is that, as a function of basis size, the result is sufficiently accurate before the obviously inevitable destabilization sets in.

In an effort to establish better variational criteria for resonance calculations of the expansion type we have studied various sum-rules[9] : If we start with an already dilated Hamiltonian H (corresponding to a dilatation parameter Θ) and subject it to an additional small dilatation by δ we can study stationarity and stability properties of the solutions in a perturbation framework. This leads to the perturbed Hamiltonian $\mathcal{H}$. Quantities referring to the combined dilatation (Θ + δ) will be indicated by script letters, e.g. $\mathcal{H}$, $\mathcal{f}(r)$, $\mathcal{E}$, while quantities referring to the initial dilatation by Θ only are given in straight notation (e.g. H, f(r), E). Hence, the perturbed Schrödinger equation reads as

$$\mathcal{H} |n\rangle = \mathcal{E}_n |n\rangle \qquad (5)$$

with $\mathcal{H} = T e^{-2i\delta} + \mathcal{V}$. Here T is the original unperturbed kinetic energy operator. The perturbed, general potential function $\mathcal{V}$ is obtained from the original potential function V by replacing all position vectors $\underline{r}$ by $\underline{r} e^{i\delta}$. We partition the perturbed Hamiltonian as

$$\mathcal{H} = H + \mathcal{U} = H + [(e^{-2i\delta} - 1) T + (\mathcal{V} - V)] \qquad (6)$$

and expand $\mathcal{H}$ in powers of iδ :

$$\mathcal{H} = H + \mathcal{U} = H + {}^1U\, i\delta - {}^2U\, \delta^2 + \ldots \qquad (7)$$

with ${}^1U = -2T - i\left(\dfrac{\partial \mathcal{V}}{\partial \delta}\right)_{\delta=0}$ and ${}^2U = 2T - \tfrac{1}{2}\left(\dfrac{\partial^2 \mathcal{V}}{\partial \delta^2}\right)_{\delta=0}$. (8)

Left superscripts are used to indicate the order of terms in the expansion. In the same manner we expand the eigenstates of $\mathcal{H}$ as

$$|n\rangle\!\rangle = |n\rangle + |^1n\rangle\, i\delta - |^2n\rangle\, \delta^2 + O(\delta^3) \qquad (9)$$

The complex eigenenergy of $\mathcal{H}$ can be expanded in analogy to $\mathcal{H}$ and $|n\rangle\!\rangle$:

$$\mathcal{E}_n = E_n + i\delta\, {}^1E_n - \delta^2\, {}^2E_n + \ldots \qquad (10)$$

The expectation value of H in the state $|n\rangle\!\rangle$ is, through second order, given by

$$\frac{\langle n^*|H|n\rangle\!\rangle}{\langle n^*|n\rangle\!\rangle} = \frac{E_n - \delta^2 \langle {}^1n^*|H|{}^1n\rangle + O(\delta^3)}{1 - \delta^2 \langle {}^1n^*|{}^1n\rangle + O(\delta^3)}$$

$$= E_n - \delta^2 \langle {}^1n^*|H - E_n|{}^1n\rangle + O(\delta^3). \qquad (11)$$

Here the star indicates that, except for complex conjugation of the **angular** parts, the function in the bra is identical to that in the corresponding ket as required for **biorthogonal scalar products**.

From the Taylor expansion of the perturbation $\mathcal{U}$ the complex eigenenergy of the state n has the following form

$$\mathcal{E}_n = E_n + i\delta\, [-2T_{nn} - i\left(\frac{\partial v_{nn}}{\partial \delta}\right)_{\delta=0}] -$$

$$\delta^2\, [2T_{nn} - \tfrac{1}{2}\left(\frac{\partial^2 v_{nn}}{\partial \delta^2}\right)_{\delta=0} - \langle {}^1n^*|H - E_n|{}^1n\rangle]. \qquad (12)$$

Two general relations are obtained expressing the independence of $\mathcal{E}$ from the transformation parameter δ. The first of these is the complex virial theorem

$$2T_{nn} = -i\left(\frac{\partial v_{nn}}{\partial \delta}\right)_{\delta=0}, \qquad (13)$$

while the other leads to

$$2T_{nn} = \tfrac{1}{2}\left(\frac{\partial^2 v_{nn}}{\partial \delta^2}\right)_{\delta=0} + \langle {}^1n^*|H - E_n|{}^1n\rangle. \qquad (14)$$

which is a second order sum rule. Similarly, higher order sum rules can be derived in a straightforward manner.

In our earlier work[9] these sum-rules, including third order, serve merely as diagnostic tools: The better the sum-rules are satisfied the closer the result is to the (usually unknown !) true result. For the present

illustration, however, we compare several of our
dilatation calculations of the lowest autoionizing state
of the helium atom to very accurate calculations by both
Burke and collaborators[10] and Bhatia and Temkin[11] both
employing different methods. This comparison is given in
table 1. The sum-rule indicators S_2 and S_3 in second and
third order are defined such that the smaller their
values the better the satisfaction of the sum-rules. The
virial theorem is satisfied in these calculations. The
energies and widths are given in eV.

Table 1 Sum Rules as Goodness Criteria

Previous Results		Present Results		Indicators	
Energy	Width	Energy	Width	S_2	S_3
57.84	0.124[11]	58.02	0.296	0.67	-
57.843	0.125[10]	57.80	0.089	0.178	1.69
		57.87	0.073	0.093	0.15
		57.838	0.129	0.048	0.13

We can easily pick out the best calculation just by
looking at the last two columns. In a more recent
calculation[12] we embarked on the much more challenging
task to optimize a given calculation by the use of
sum-rules. The physical problem at hand is the
calculation of low-lying s-resonances of screened Coulomb
potentials. The advantage of having to deal with just one
particle is partially offset by particularly nasty
features of the chosen potential such as a flat potential
barrier at considerable distance from the origin. This
makes it necessary to span a large region of space with a
basis here chosen as Slater s-type functions

$$\Phi_n = N_n(\alpha) \; r^{n-1} \; e^{-\alpha r} \qquad (15)$$

all with the same exponent α but different powers of r.
This basis simulates the effect of a set of scaled
Laguerre functions and is, after much of experimentation,
our favorite type of expansion functions in resonance
calculations. In order to avoid the introduction of a new
variational parameter we restrict the optimization to a
variation of the **complex** dilatation parameter θ and the
real exponent α of the simulated Laguerre set. This is
not quite sufficient to enforce the satisfaction of the
first two complex sum-rules. Hence we compromise in the
sense that, though we satisfy the complex virial theorem
strictly by varying the real and imaginary parts of θ, we
aim only at the minimization of the real quantity $|S_2|^2$

by optimizing the exponent α. As a result, the convergence with increasing basis size is greatly improved. Table 2 presents our results for a threshold resonance of the Debye-Laughton potential

$$V(r) = -\frac{A}{r} \exp\left(-\frac{r}{D}\right) + B \exp\left(-(C+1)\frac{r}{D}\right) \qquad (16)$$

for A = 1.0, B = 0.05, C = 0, and D = 20.0 for different basis sizes N.

Table 2 Basis Optimization with Sum-Rules

N	α	Θ	S_2 (times 100)	E (in 10^{-3} au)
5	0.291	0.453+i0.020	0.624+i0.004	2.6852-i0.8204
12	0.217	0.567+i0.081	0.116+i0.027	2.8146-i0.1844
20	0.209	0.590+i0.008	0.192+i0.000	2.8146-i0.1842

Without the use of the sum-rule optimization convergence is much slower. This we can tell since there is a highly accurate numerical result for this resonance at 2.814446-i0.184076 $\times$ 10^{-3} au to compare with. We obtained this result using a code which was originally developed in Stockholm by Elander and Rittby[13].

As an alternative to the dilatation approach we have studied the direct application of Siegert boundary conditions for various potentials. Following the example of Miller we expand the complex wavefunction in a set of square integrable basis functions plus one asymptotically diverging orbital which introduces the correct boundary condition. For a short range potential a reasonable choice is

$$u_K(r) = \left[1-\exp(-\eta r)\right]^{l+1} \exp(iKr - i\pi l/2) \qquad (17)$$

where the cut-off factor in square brackets is needed for regularization at the origin. The divergent integrals are avoided by not diagonalizing the Hamiltonian itself but rather the operator

$$A = H - K^2/2 \qquad (18)$$

and search iteratively for zero eigenvalues. It has been shown[14], that all divergent integrals can be cancelled exactly before the matrix is set up, even though we do not know the correct value of K until the self-consistency condition $E = K^2/2$ is satisfied. As a point of technical interest it may be mentioned that during the iterative search for zero eigenvalues of the matrix **A** it is not necessary to diagonalize the complex symmetric matrix each time the input value of K is changed. We rather prediagonalize the Hamiltonian first in the space of the real and square integrablefunctions and then solve the following algebraic equation iteratively until self-consistency is achieved:

$$A_{KK} + \sum_{n=1}^{N} \frac{2 A_{Kn} A_{nK}}{K_n^2 - K^2} = 0 \qquad (19)$$

Here the matrix elements are taken between the eigenvectors from the real prediagonalization and the Siegert orbital (eq. 17). The quantities $K_n^2/2$ are the eigenvalues from the prediagonalization. Frequently this formula produces quite acceptable results with only a few terms of the sum included.

The iterative search for zero eigenvalues has been found to be the bottleneck of the computation. Our first attempt is usually to start a two-dimensional Newton - Raphson procedure since both the real and imaginary parts must vanish simultaneously. In the majority of cases a few iterations is all that is needed to pin down the desired root with arbitrary accuracy (within the model space defined by the finite basis). In other, less favorable cases, however, two (or more) of the surfaces that represent the eigenvalues of **A** in a complex K-space come so close to each other in the region of interest that surface jumping occurs during the zero search. We have not found a save remedy yet for these situations. Moreover, the occurance of close surfaces tends to increase with basis size. Changing the non-linear parameters of the finite L^2 basis is usually the best alternative. So far, however, no established ways have been worked out to direct such changes away from the trial and error stage. The following observation, however, may provide some guidance to that effect: In obtaining the results reported in table 2 we had to perform an exponent optimization in the framework of the dilatation transformation calculation. If we use this optimal value of α in the direct Siegert approach (after a meaningful combination with the imaginary part of θ) we obtain results of the same high quality as before which is much better than the quality obtainable with arbitrary values of α. This result is somewhat surprising because the optimization in the first approach is performed by enforcing the virial theorem (aside from minimizing $|S_2|^2$). The virial relation is usually interpreted in the context of spatially confined classical motion which relates to localized wavefunctions in quantum mechanics. Exactly this is not the case for Siegert solutions!

So far we have not yet studied whether the optimal exponents from the dilatation approach are also optimal in some sense in the direct Siegert method or just very good. But we have picked up the hint that the divergent tail of the wave function is possibly not all that interesting. We know it anyhow once we know the correct energy! In an attempt[3] to reformulate the problem from the start without divergent functions we first rewrite the eigenvalue equation with a real potential function but an explicitly complex wave function as

$$(T+V)(\Psi_R + i\Psi_I) = (E_R + iE_I)(\Psi_R + \Psi_I) \tag{20}$$

with $E_R = +(K_R^2 - K_I^2)$ and $E_I = K_R K_I$. We then introduce functions $\Psi_R = \Phi_R \exp(-K_I r)$ and $\Psi_I = \Phi_I \exp(-K_I r)$. This leads to the following system of differential equations without divergent functions:

$$\Phi_R'' - 2K_I \Phi_R = +\left[K_R^2 - \frac{l(l+1)}{r^2} - 2V\right]\Phi_R = 2K_R K_I \Phi_I \tag{21}$$

$$\Phi_I'' - 2K_I \Phi_I = +\left[K_R^2 - \frac{l(l+1)}{r^2} - 2V\right]\Phi_I = -2K_R K_I \Phi_R . \tag{22}$$

The asypmtotic boundary conditions are $\Phi_R \propto \cos K_R r$, and $\Phi_I \propto \sin K_R r$. As usual, we convert the system to four equations of first order plus two additional simple differential equations to account for the eigenvalue character of the problem. Again the real and imaginary parts of the eigenvalue are introduced as input guesses and modified during the course of the solution until self-consistency is achieved. This novel method proved very efficient in some resonance calculations which so far resisted every solution attempt by other methods. In table 3 we present results for the lowest s-wave resonance of the Woods - Saxon potential

$$V(r) = -V_0/\{1 + \exp[(r-r_0)/a\} . \tag{23}$$

The particular interest in these results is both the occurance of resonances in the s-channel without any potential barrier and with smooth walls (as opposed to the abrupt step in the finite square well potential) and the feasibilty of the calculation of resonances with very large widths. The actual solution was achieved by a real finite difference scheme with a modest 500 grid points using relaxation techniques.

Table 3 Virtual state results for the Woods-Saxon potential with strength $V_0 = 0.5$ a.u. and $r_0 = 1.0$.

a	Re E in a.u.	Im E in a.u.
10^{-8}	4.8771	-10.889
0.01	4.83585	-10.8719
0.05	3.96125	-10.5594
0.08	2.84015	-10.0928
0.10	2.04277	- 9.6763

The results in the first line can be meaningfully compared to the well known resonance[15] of the finite square well potential at 4.8763 and -10.885 a.u. Accompanying expansion results of ours[16] corroborate the results for larger values of thickness parameter a for which no other data seems to be known.

DISCUSSION

Although the technique of Siegert resonance computations is far from fully explored we can formulate a few general observations:
- There seems not to be one method that fits all cases equally well. It is therefore warranted to doublecheck all results by verifying them by other approaches. all results reported by us here and elsewhere have been corroborated this way.
- Since the complex calculations are far less stable than their real counterparts one should work in the highest accuracy option available on the compiler. Most of our computations were done in extended (quadruple) precision. in particular the variational calculations invoking sum-rules did converge only in extended precision.
- It is expected that in the future the trend goes more toward purely numerical computations as opposed to expansion methods. Expansion methods, however, will remain very valuable because they can more easily yield approximate starting values without which all schemes tend to fail.

ACKNOWLEDGEMENTS

Longlasting collaborations with J. H. Brink, L. S. Cederbaum, N. Elander, B. T. Pickup and Z. Wang have made this work not only possible but also enjoyable. Financial support by the US Department of Energy in the form of a research grant (DE-FG08-90ER14160) is gratefully acknowledged.

REFERENCES

1. A. J. F. Siegert, Phys. Rev. 56, 750-52 (1939).
2. P. L. Kapur and R. Peierls, Proc. Roy. Soc. A166, 27-95 (1938).
3. H. Brink and P. Winkler, Int. J. Quantum Chem. S24, 321 (1990).
4. B. I. Schneider and L. A. Collins in Resonances in Electron-Molecule Scattering, van der Waals Complexes, and Reactive Chemical Dynamics (ed. D. G. Truhlar, American Chemical Society, Washington 1984) p.65-88.
5. P. Krylstedt, Thesis, Stockholm (1986).
6. A. D. Isaacson, C. W. McCurdy and W. H. Miller Chem. Phys. 34, 311 (1978).
7. A. D. Isaacson and W. H. Miller, Chem. Phys. Lett. 62, 374 (1979).
8. A. U. Hazi and H. S. Taylor, Phys. Rev. A1, 1109 (1970).
9. P. Winkler, Int. Jour. Quant. Chem. S19, 201 (1986).
10. P. G. Burke, D. D. McVicar and K. Smith, Phys. Rev. Lett. 11, 559 (1963)
11. A. K. Bhatia and A. Temkin, Phys. Rev. A11, 2010 (1975).
12. Z. Wang, P. Winkler, B. T. Pickup and N. Elander, Chem. Phys. 135, 247-253 (1989).
13. M. Rittby, N. Elander and E. Brandas, Chem. Phys. 87, 55, (1984).
14. R. Yaris, R. Lovett, and P. Winkler, Chem. Phys. 43, 29 (1979).
15. D. L. Huestis, J. Math. Phys. 16, 2148 (1975).
16. B. T. Pickup and P. Winkler, Nucl. Phys. A, July 1990.
17. P. Winkler, Nucl. Instr. Methods B42, 563-565 (1989).

EXOTIC GAMOW STATES OF ATOMS, NUCLEI, AND PARTICLES*

Rubin H Landau
Physics Department, Oregon State University, Corvallis, Oregon 97331

Abstract

Results are presented showing recent progress in the momentum space computation of atomic and nuclear Gamow states. Coupled bound and reaction channels and an exact treatment of the Coulomb plus nuclear potentials are included. Three–dimensional visualizations help uncover some of the physics present on the complex energy sheets.

1 EXOTIC PHYSICAL SYSTEMS

The purpose of this talk is to describe an application and reinterpretation in quantum mechanics needed to undertake first principle computations describing quasi–bound systems. These systems are somewhat of a computational physics analog of the omnibus with something of interest for the atomic, nuclear, and particle physicists. The systems are exotic, hydrogen–like bound states in which a negatively–charged hadron, such as an antikaon K^- or an antiproton $\bar{p}$, orbits a nucleus (or proton)— as sketched in Fig. 1. Since the Bohr radii for these atoms are about 1000 times smaller than in electronic atoms, the orbiting hadron lies deep within the electronic cloud and is close enough to the nucleus to experience its force, possibly to react with it, and possibly to get captured by it into a nuclear bound state (which would be some 30 times *smaller yet* in size).

To be specific, let us consider *antikaonic hydrogen* K^-p (while similar physics is present for both K^-'s and $\bar{p}$'s we concentrate on just kaons). It has a Bohr radius $R_B = 84fm$ and a ground state energy $E_{1S} \simeq -8613eV$. The physics of this type of atom is very rich since near threshold there are several strongly–interacting *channels* which couple in

$$K^-p \rightarrow \begin{cases} K^-p & +8.6KeV & \text{atom; bound} \\ \Sigma^{\pm,0}\pi^{\mp 0} & +100MeV & \text{annihilation; open} \\ \Lambda^0\pi^0 & +180MeV & \text{annihilation; open} \\ \overline{K}^0 n & -5MeV & \text{charge exchange; closed} \end{cases} \qquad (1)$$

Thus an atom bound essentially by the attractive and well understood Coulomb force also experiences a very attractive, very absorptive, highly *nonlocal*, and rather poorly understood strong interaction which induces spontaneous transitions to different pairs of particles.

As interesting as this system may seem because of its complexity, just what is happening in the coupled channels is also of basic interest. We would like to learn about the strong scattering at the low energies in these channels, and particularly so when there are elementary states, such as the $\Lambda^*(1405)$, formed in these channels. In the $\Lambda^*(1405)$'s case it would be nice to know if it is a $\Sigma\pi$ resonance, an unstable $\overline{K}N$ bound state (a quark molecule), an elementary three-quark state, or some combination of the above[1].

---

*Work done in collaboration with Paul Fink, Guangliang He, and Jeffrey Schnick, and supported in part by the US Department of Energy.

84 Exotic Gamow States

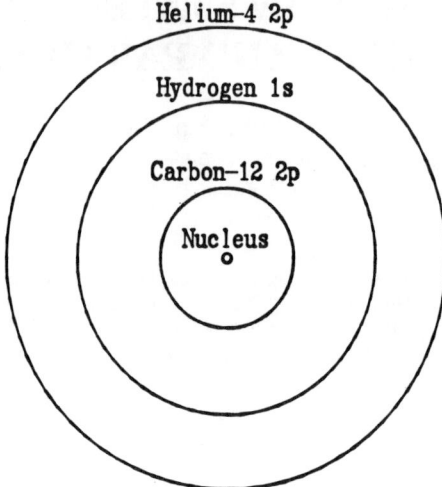

Figure 1: Sizes of K^- Bohr orbits relative to actual size of carbon nucleus.

Since the hadron in an exotic atom reacts with the nucleus, its energy E_{atom} is shifted from the Bohr value E_n and the level acquires a width Γ. This physics is experimentally accessible by measuring the level shift and width

$$\epsilon + i\Gamma/2 \equiv -(E_{atom} - E_n) \tag{2}$$
$$\Rightarrow Im E_{atom} = = -\Gamma/2 = -(2\tau)^{-1} \tag{3}$$

(more about the imaginary part latter). Typical values for Γ are $\approx (1000, 100)eV$ for $(K^- p, \bar{p}p)$ with larger values for heavier nuclei. The level shifts ϵ are $\approx -(100 - 1000)eV$'s and are uniformly *negative*, that is, to the *less bound*. While conventional wisdom leads us to believe that a nuclear force whose addition diminishes the binding must be repulsive, this is not true for systems with very strong absorption (the annihilation channels in (1)). This reversal arises from the strong absorption effectively repelling the wave function away from the origin, or equivalently, from the nuclear interaction being strong enough to form an odd number of nuclear or "inner" bound states confined within the atomic or "outer" ones. In either case this is not a problem to be treated in perturbation theory — as indicated by the occurrence of oscillations in the predicted shifts and widths as the nuclear interaction's strength is increased monotonically.

2 QUANTUM MECHANICAL PROBLEMS

The conventional and simple interpretation of the strong interaction shift (2) follows from the quasipotential or Trueman formula

$$\frac{E_{atom} - E_n}{E_n} \simeq -4 \frac{f(E_n)}{n R_B} \tag{4}$$

in which the shift effectively measures the nuclear scattering amplitude $f(E_n)$ of the orbiting system at a slightly negative energy. Yet since the use of this formula to interpret the kaonic hydrogen shifts leads to an apparent conflict with the low energy K^-p scattering data[2] — or the prediction of some too – unusual nuclear states, we will discuss here our effort to construct a computable quantum mechanical description of this system. This is a nice story for this conference since these computational advances have led to surprising new physics and — not too surprisingly — to the realization that the conventional, analytic formula (4) is inaccurate for amplitudes with strong energy dependence near threshold (as arises from strongly coupled channels).

There are four essentials to include in converting this problem to computable form

1. Relativity, field theory, and many–body effects leading to an effective, short–range potential which is *nonlocal*. Thus even if the problem were single channel, the appropriate Schrödinger equation would be

$$\sqrt{(\vec{\nabla}/i)^2 + \mu^2}\,\psi(\vec{r}) + \int d\vec{r'} V(\vec{r},\vec{r'})\psi(\vec{r'}) = E\psi(\vec{r}) \qquad (5)$$

This tricky integro-differential equation becomes a straight–forward integral equation in momentum space. (While (5) is not a covariant equation, it is equivalent to the Blankenbecler–Sugar equation without crossing, or a 3-D reduction of the Bethe–Salpeter equation.)

2. The direct inclusion of the coupled channels nature of a system such as (1) by use of coupled dynamical equations rather than an effective or "optical" potential.

3. The combination of open and closed channels requires care in defining the meaning of a "bound" state — especially in momentum space where there are no boundary conditions to apply to wave functions.

4. The need to accurately include both the short–range nuclear force as well as a long–range Coulomb force which is singular in momentum space.

2.1 GAMOW STATES IN COUPLED MOMENTUM SPACES

A quantum state which decays exponentially in time,

$$|\psi|^2 \propto e^{-\Gamma t},\ \psi \propto e^{-iE_R t}e^{E_I t} \qquad (6)$$

is not a conventional eigenstate of a one particle Hamiltonian since the decreasing probability indicates a coupling to another system which removes flux from the original system. This type of state may be called "Gamow", "resonant," or "quasi–bound," and its place in quantum mechanics is open to interpretation[3, 2, 4]. In some sense an exponentially decaying state or a resonance is just a quantum model for a complicated natural event since a state which "decays" must really lie in some continuum and accordingly is just a (possibly long–lived) scattering event. Yet since continuum states are not normalizable in the usual sense, Gamow states are also not normalizable in the usual sense.

In our definition of the shift formula (2) we already alluded to a "bound state" with a negative imaginary part E_I of its energy, a part related to its lifetime τ. The connection to coupled–channels physical systems such as (1) is made by considering the *complex momentum* corresponding to a complex energy with negative imaginary part

$$E = E_R + iE_I = k^2 = k_R^2 - k_I^2 + 2ik_Rk_I \qquad (7)$$

There are two possible solutions here

$$E_I < 0 \ \Rightarrow 2k_Rk_I < 0\ \Rightarrow \left\{\begin{array}{l} k_R < 0,\ k_I > 0,\ \text{case A} \\ k_R > 0,\ k_I < 0,\ \text{case B} \end{array}\right. \qquad (8)$$

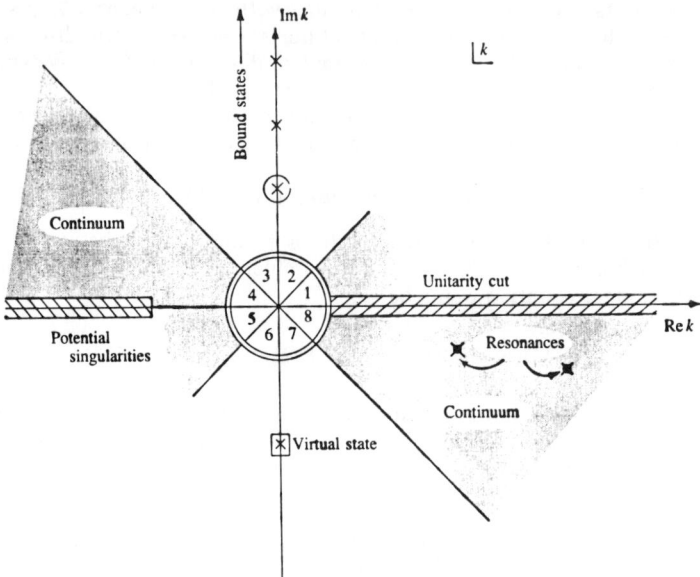

Figure 2: Complex k plane showing analytic properties of T matrix. The shaded area corresponds to $Re\,E > 0$, and the eight octants correspond to the two sets of quadrants in the two energy sheets. The continuum has $E_R > 0$.

This means that for a state with negative imaginary energy (needed for exponential decay in time), the coordinate space wave function at asymptotically large separations must have one of two possible plane wave forms

$$\psi(r) \sim e^{\pm ikr} \Rightarrow \begin{cases} \psi_A \sim & e^{-|k_I|r}e^{-i|k_R|r} \\ \psi_B \sim & e^{+|k_I|r}e^{+i|k_R|r} \end{cases} \quad (9)$$

Since ψ_A decays exponentially in space, it represents a normalizable bound state appropriate to a closed channel such as k^-p (its oscillatory *incoming* wave part is required to "feed" the state at the origin so that it can maintain the exponential decay in time). In Fig. 2 we look at the complex k plane. We see ψ_A as a bound state with momentum on the *positive imaginary* axis — if the energy was real negative— but now migrating to the third octant since there is a negative imaginary part to the energy. Since ψ_B contains an outgoing wave part, it represents the wave which "leaks" into the open $\Sigma\pi$ channel and escapes. The wave function ψ_B is non–normalizable and represents a resonance (its exponentially growing space part indicates a state decaying with time which was stronger at an earlier time — as seen at large r). In the complex k plane of Fig. 2, ψ_B's momentum is on the *positive real* axis — when the energy is real positive— but migrates to the eighth octant when there is a negative imaginary part to the energy.

Although the asymptotic forms ψ_A and ψ_B may appear unusual, they are forced upon us by the imaginary part to the energy of a Gamow state. This is relevant to our coupled channels problem in that it means we cannot find these energies by solving the conventional eigenvalue problem

$$H\psi = (E_R + iE_I)\psi \quad (10)$$

when open channels are included since these procedures yield normalizable solutions such as ψ_A, whereas we need to include non–normalizable states such as ψ_B in open channels. Further, the non–normalizability means we cannot Fourier transform them from coordinate space to momentum space and therefore must find another way to build boundary conditions into the problem.

We attempt to include all the essential dynamics of this type of physical systems by using the integral (Lippmann-Schwinger) form of the Schrödinger equation expressed in terms of the transition matrix T

$$T(k',k;E) = V(k',k) + \frac{2}{\pi}\int_0^\infty dp\, p^2\, V(k',p)\, G(p)\, T(p,k;E) \tag{11}$$

The coupled-channels physics is included by T, V and G being matrices, for example in the simplest 2–channel case

$$V \equiv \begin{bmatrix} V_{11} & V_{12} \\ V_{21} & V_{22} \end{bmatrix},\ T \equiv \begin{bmatrix} T_{11} & T_{12} \\ T_{21} & T_{22} \end{bmatrix},\ G = \begin{bmatrix} \frac{1}{E-E_1(p)} & 0 \\ 0 & \frac{1}{E+i\epsilon+\Delta M - E_2(p)} \end{bmatrix} \tag{12}$$

Since in solving integral equations it is impossible to build appropriate boundary conditions into a momentum space wave functions, we place them in the Green's function. The $i\epsilon$ in G is used for open channels to guarantee the outgoing-wave boundary conditions, its principal value and delta function part are separated, and then the analytic continuation to complex energies is made.

The connection between these integral equations and exotic atom states follows from first examining the operator form and corresponding formal solution of the Lippmann Schwinger equation (10)

$$T = V + VG_E T \quad \Rightarrow \quad T = (1-VG_E)^{-1} V \tag{13}$$

Likewise, the operator bound-state Schrödinger equation can be written in the the equivalent forms

$$<\psi|(E-H_0) = <\psi|V \quad \Rightarrow \quad <\psi|(1-VG_E) = 0 \tag{14}$$

If (12) and (13) are compared, we see that *the condition (i.e. energy) for bound-state solutions of the Schrödinger equation to occur is the same as that for the T-matrix to have a pole, namely*

$$\det(1-VG_E) = 0 \tag{15}$$

Our computational problem is to search for solutions of (14) for large, complex, momentum space matrices V. When these poles are on the positive imaginary k axis they are conventional bound states, otherwise they have complex energies which we view as the energies of Gamow states with widths Γ given by (2). Whether these poles are "bound states" or "resonances" depends on its location on the complex energy sheets — and the path they took to get there as channel coupling is introduced. For states with negative imaginary energies (positive lifetimes), they will always be in the lower half of the energy planes, and will be resonances if their home is on the second sheet, and bound states if their home is on the first.

3 COMPUTATIONAL ASPECTS

To handle the nonlocal nature of the nuclear potential we solve the dynamical equations in momentum space. Yet handling the Coulomb potential there is a problem since the long range of the potential in coordinate space produces a momentum space singularity in both the 3-D and partial wave potentials

$$< \vec{k}|V^C|\vec{k'}> = \frac{-Ze^2}{2\pi^2(\vec{k}-\vec{k'})^2} \qquad (16)$$

$$V^C_{l=0}(k',k) = -\frac{Ze^2}{2kk'}\ln\left|\frac{k+k'}{k-k'}\right| \qquad (17)$$

Fock[5] way back in 1935 was able to solve the momentum space Schrödinger equation directly for the hydrogen states, for example,

$$\psi_{1S}(p) = \frac{1}{(p^2 R_B^2 + 1)^2} \Rightarrow \text{pole at } p^2 = -\frac{1}{R_B^2} \qquad (18)$$

Yet we need do it numerically since we want more than just a *point* charge and want to also add in a rather complicated nuclear potential. To avoid the logarithmetic singularity in V^C, we modify the Landé[6] – Kwon – Tabakin[7] technique to subtract off the singularity in the Coulomb potential while simultaneously solving the integral equation

$$T(k',k) = V^C(k',k) + \frac{2}{\pi}\int dp V^C(k',p)\left[p^2 G(p)T(p,k) - \frac{k'^2 G(k')T(k',k;E)}{P_l(z_{k'p})}\right]$$

$$+\frac{2}{\pi}k'^2 G(k')T(k',k)S(k') \qquad (19)$$

$$S(k') = \int_0^\infty \frac{V^C(k',p)}{P_l(z_{k'p})}dp, \quad z_{k'p} = (k'^2 + p^2)/2k'p \qquad (20)$$

The integrand is now nonsingular at $k' = p$ and the integral of the subtracted term exactly cancels the S_l term. With 40 grid points we obtain six–place accuracy for the analytically–known pure Coulomb E_{1S} for K^-p.

The Green's function G for an annihilation channel is singular since the channel is open. We make the singularity numerically acceptable by separating the integral of VGT into delta function and principal value parts, and then evaluating the principal value part with a Haftel-Tabakin[8] subtraction

$$\int dp p^2 \frac{V(k',p)T(p,k)}{E + i\epsilon - E(p)} = P\int dp p^2 \frac{V(k',p)T(p,k)}{E - E(p)} - i\pi\mu k_0 V(k',k_0)T(k_0,k) \qquad (21)$$

$$= \int dp \left[p^2\frac{V(k',p)T(p,k)}{E - E(p)} - k_0^2\frac{V(k',k_0)T(k_0,k)}{\frac{k_0^2}{2\mu} - \frac{p^2}{2\mu}}\right]$$

$$- i\pi\mu k_0 V(k',k_0)T(k_0,k) \qquad (22)$$

Here the effective mass μ arises from the expansion of the denominator around the pole at k_0 and is present even in this relativistic calculation. Now that the outgoing wave boundary conditions are built into the solution, we make the analytic continuation to complex energies.

4 VISUALIZATION OF RESULTS

We have searched for and found atomic and nuclear "bound states" for the kaonic and antiprotonic atoms represented in Fig.1. A typical comparison with experiment for the $\bar{p}p$ atom is given in Fig. 3. This figure shows a range of predictions which demonstrates how these calculations test different models for the nuclear force. The comparison of solid and dotted levels shows a comparison between the exact calculation and the analytic but

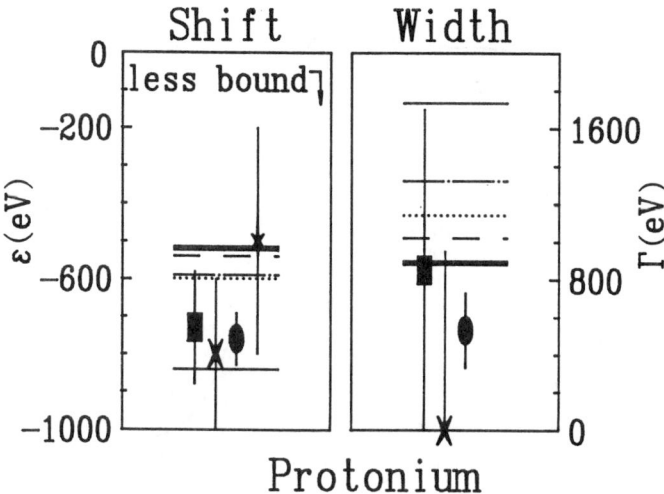

Figure 3: Shift and width of the $n = 1\,^1S_0$ protonium level as measured experimentally and computed with different models for the nuclear force. The solid line is the exact computation while the dotted line is the approximate Trueman formula for the same potential.

approximate Trueman formula (4); quantitative comparison clearly demands the computed result.

Considering the nature of this conference, I refrain from displaying tables of energy level and instead concentrate on some basic physics in the results and how computation has helped uncover that physics. In Fig. 4 we see the results of a computer experiment in which we took a model for the strong interactions in the coupled K^-p system (1) and varied the strength Λ with which the nuclear channels couple to the atom. We see that for weak coupling, $\Lambda \simeq 0$, the nuclear force increases the binding of the atomic state by about 8% and hardly broadens the level at all (as expected for an attractive potential which cannot absorb flux). However, as the coupling increases to $\Lambda \simeq 0.65$, the shift passes through 0 as the width simultaneously passes through a large maximum. Contrary to intuition, further increases in the "attractive" channel coupling only serve to make the atomic level less bound and narrower. We understand this behavior by noting that below the crossover point our pole search finds no strongly bound (Λ^*) state, yet at the crossover point a nuclear state first appears, and then gets bound even more deeply (from $14 - 50 MeV$) as the coupling increases. Accordingly, Fig. 4 is showing a level–crossing effect between a kaonic hydrogen $1S$ state and a Λ^* $1S$ nuclear resonance in a coupled channel. (This means that what we call the $1S$ state in kaonic hydrogen is *really* a $2S$ state, the *real* $1S$ being the inner nuclear state.)

While the strong interaction model used in Fig. 4 did not originally agree with the sign of the level shift, we have developed a $\overline{K}N$ interaction ($r2$) which agrees with the kaonic hydrogen level shift as well as the low–energy scattering data. As a further test of this (and other) interaction models, we have used them[10] to construct a theoretical optical potential describing the $K^- - C$ interaction, and then to search for atomic and strong bound state poles for this heavier atom. The computation of this optical potential was

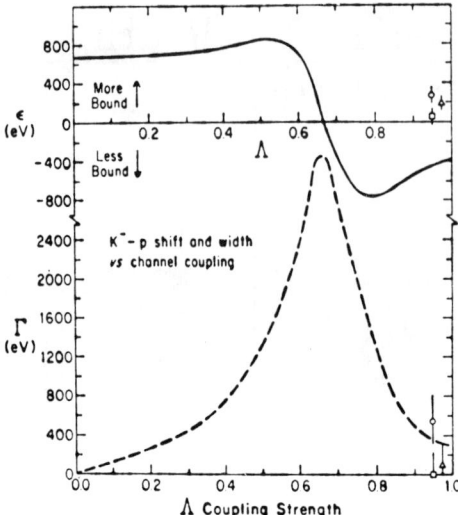

Figure 4: Kaonic hydrogen 1S level shift as function of channel coupling strength. The data points are for kaonic hydrogen. At $\Lambda \approx 0.65$ a nuclear state is first formed.

itself a supercomputer calculation

$$V(\mathbf{k'}, \mathbf{k}, E) = A \int d^3p <\mathbf{p}-\mathbf{q}, \mathbf{k'}|t^{\overline{K}N}(\omega)|\mathbf{p}, \mathbf{k}> \mathcal{F}(\mathbf{p}-\mathbf{q}, \mathbf{p}) \qquad (23)$$

$$\mathcal{F}(\mathbf{p'}, \mathbf{p}) = \int d^3p_2 d^3p_3 \ldots d^3p_A \, \Psi^*(\mathbf{p'}, \mathbf{p}_2, \ldots, \mathbf{p}_A) \Psi(\mathbf{p}, \mathbf{p}_2, \ldots, \mathbf{p}_A) \delta(\mathbf{P}_{tot}) \qquad (24)$$

$$\Psi(p_1, p_2, \ldots, p_Z) = \frac{1}{\sqrt{Z!}} \begin{vmatrix} \phi_1(p_1) & \phi_2(p_1) & \cdots & \phi_Z(p_1) \\ \phi_1(p_2) & \phi_2(p_2) & \cdots & \phi_Z(p_2) \\ \vdots & & & \vdots \\ \phi_1(p_Z) & \phi_2(p_Z) & \cdots & \phi_Z(p_Z) \end{vmatrix} \qquad (25)$$

The determinantal wave functions for the nucleons within the nucleus had to be convoluted 1000's of times with off–energy–shell, coupled–channels, $\overline{K}N$ T matrices (which themselves are solutions of integral equations) — and all within a Pauli principle constraint. While the details are inappropriate here, it is interesting to note that the convoluted nature of the logic leads to a significantly *smaller speedup* after vectorization (40%) than for other computational science problems (80 – 300%).

A visualization of the space we have traveled in our search for kaonic carbon bound state poles is given in Fig. 5. This shows the absolute value of the Fredholm determinant $|det[1-GV]|$ as a function of complex energy (recall poles of T correspond to zeros of this determinant). The atomic states are all squeezed into the the steep hole near the origin. As Fig. 1 shows, the kaon is closer to the nucleus in the 2p state of carbon than in the 1s state of hydrogen, and accordingly the levels are shifted further in carbon than in hydrogen. The atomic shift to the less bound indicates an odd number of nuclear bound states.

The nuclear or hypernucler bound states in Fig. 5 correspond to a $\overline{K}$ or Σ entering a nucleus and converting a nucleon to a Λ^* — with the entire nucleus still remaining bound. It is seen as the broad dip near $(-50 - 10i)MeV$. The spikes (and the zeros between them)

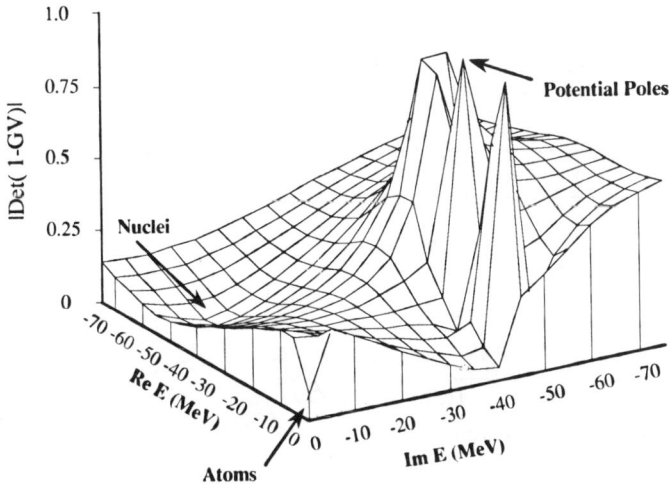

Figure 5: Fredholm determinant as a function of complex energy for $K^- - C$. The zeros are nuclear and atomic bound states, the spikes, optical potential poles.

are poles in the determinant arising from *potential poles* in the optical potential (which in turn arise from poles in the elementary $\overline{K}N$ T matrices). While of little dynamic interest, these spikes do block some search routes.

Finally, in Fig. 6 (A) and (B) we explore the classic question[9] of the relation between the properties of a resonance (the Λ^*) and the poles of the corresponding T matrix. In this case the interaction model has a bag of quarks surrounded by cloud of mesons[1]. Physical experiments are conducted along the real energy axis where this bag model predicts a characteristic resonance signal in T — as shown in Fig. 6(A). Yet as we see by examining, in Fig. 6(B), the scattering amplitude throughout the entire energies sheets, the interaction actually produces two poles, with the apparent resonance signal *below* threshold arising from a pole right *above threshold*! A significant error is thereby made if the behavior along the real axis is used to deduce the resonance's properties. Furthermore, identifying the nature of this state can be tricky without a dynamical study; in the $\Sigma\pi$ channel these poles are on the second (nonphysical) sheet, but on the first (physical) sheet for the $\overline{K}N$ channel. Correspondingly, in the $\overline{K}N$ channel they are unstable bound states (one of which can decay into the $\overline{K}N$ continuum, the other of which decays into the $\Sigma\pi$ continuum), while in the $\Sigma\pi$ channel channel they are resonances. Yet in all cases they are Gamow states.

In conclusion, we have presented results which demonstrate the feasibility of computing coupled–channels, atomic and nuclear Gamow states in momentum space. With these techniques we have been able to connect the elementary particle two–body scattering data to exotic hydrogen and exotic heavy atom data, and beyond that to the prediction of hypernuclear states. In the process, models for elementary $\overline{K}$ and $\overline{p}$ interactions have been compared to atomic data and improved ones developed.

While the coupled channels, Coulomb plus nuclear problem is a challenge, its solution shows the way for doing more quantum mechanics in momentum space and also shows that not all advances in computational physics arises from using big computers; some require new formulations of quantum mechanics which in turn provide further insight into the

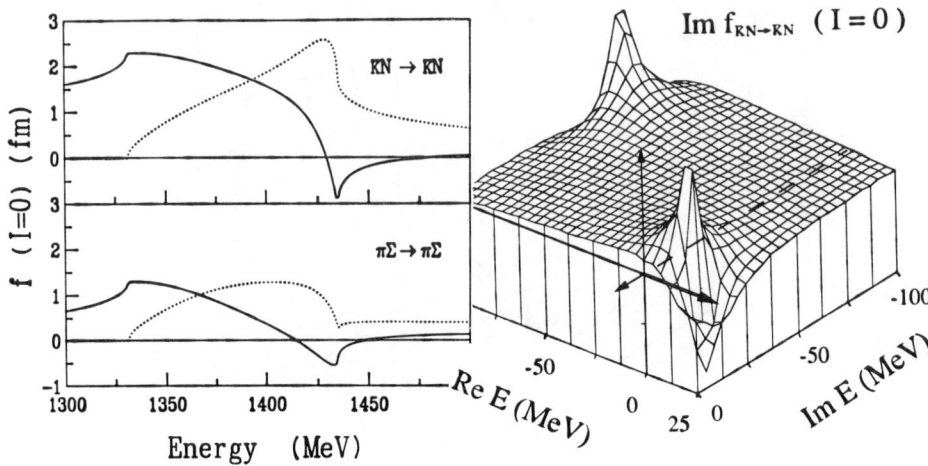

Figure 6: (A) $\overline{K}N$ and $\Sigma\pi$ scattering amplitudes calculated with a cloudy, quark bag model for isospin 0. The solid curves are real parts, the dotted imaginary parts as a function of the real part of the energy. (B) The imaginary part of the same $\overline{K}N$ amplitude as a function of complex energy.

physics.

References

[1] E.A. Veit, B.K. Jennings, A.W. Thomas, and R.C. Barrett, Phys. Rev. D **31**, 1033 (1985); B.K. Jennings, Phys. Lett. **176B**, 229 (1986).

[2] R.H. Landau, Phys. Rev. C **28**, 1324 (1983); R.H. Landau and B. Cheng, Phys. Rev. C**33**, 734 (1986); J. Schnick and R.H. Landau, Phys. Rev. Letters **58**, 1719 (1987).

[3] G. Garcia–Calderon and R. Peierls, Nucl. Phys. **A265**, 443 (1976); P.L. Kapur and R. Peierls, Proc. R. Soc. London Ser. A **166**, 277 (1938).

[4] E. Hernàndez and A. Mondragon, Phys. Rev C **29**, 722 (1984).

[5] V. Fock, Z. Physik **98**, 145 (1935).

[6] A. Landé, personal communication

[7] Y.R. Kwon and F. Tabakin, Phys. Rev C **18**, 932 (1978); D.P. Heddle, Y.R. Kwon and F. Tabakin, Comput. Phys. Commun. **38**, 71 (1985); R.J. Luce and F. Tabakin, Comput. Phys. Commun. **46**, 193 (1987).

[8] M.I. Haftel and F. Tabakin, Nucl. Phys. **158**, 1 (1970).

[9] R.G. Newton, *Scattering Theory of Waves and Particles*, McGraw-Hill, New York (1966); J.R. Taylor, *Scattering Theory*, John Wiley & Sons, New York (1972); D. Park, *Introduction to Strong Interactions*, W.A. Benjamin, New York, (1966).

[10] P.J. Fink Jr., J.W. Schnick, and R.H. Landau, Phys. Rev. C **42**, 232 (1990).

ATOMIC NEGATIVE IONS

Tomas Brage
Vanderbilt University, Nashville, Tn. 37235

Abstract

We review some of the recent progress in the studies of alkaline-earth, negative ions. Computations of autodetachment rates, electron affinities and transition wavelengths are discussed and some new and improved results are given.

INTRODUCTION

In this talk we will discuss and review some of the latest progress in the computational studies of negative ions. These quite exotic systems are interesting for a number of reasons, and they constitute a challenge for atomic structure calculations. Let us clarify this by some examples.

In negative ions, the *correlation* energy is often larger than the binding energy for the "outermost" electron. Since correlation is defined as[1],

$$E_{corr} = E_{NR} - E_{HF} \qquad (1)$$

where E_{NR} and E_{HF} are the exact, nonrelativistic and Hartree-Fock energies respectively, the correction to our first approximation is then larger than one of the properties we are computing. As a matter of fact, the bound Hartree-Fock solution does often not exist, and correlated models are required as a starting point.

The negative ions are also unique in atomic physics since they have only a few bound states. We here define a bound state to be an atomic state that in a nonrelativistic approach can be represented by a square-integrable wavefunction. The reason for this definition will be apparent below. As an example we show in Fig. 1 the structure of one of the simplest negative ions, He$^-$. The only predicted and observed[2] bound states in this ion belong to one of the two terms $1s2s2p\,^4P^o$ and $2p^3\,^4S^e$. From this figure we can also see that a requirement for a bound state is that it exists below its nonrelativistic limit, $1s2s\,^3S^e$ and $2p^2\,^3P^e$ in this case. Equivalently, the two states in neutral helium have positive electron affinity.

Finally, the negative ions are difficult to observe, so accurate calculations of their different properties are needed to guide the analyses of experimental data.

In this talk, we discuss recent progress in the calculation of autodetachment rates, electron affinities and transition energies for alkaline-earth, negative ions.

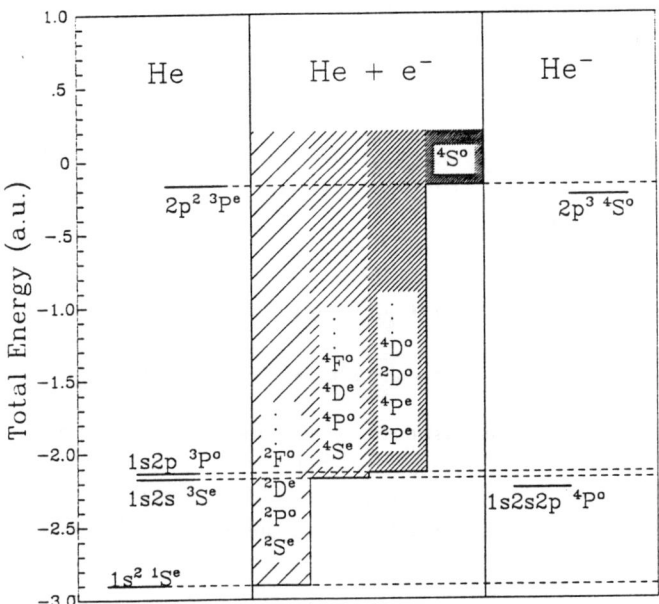

Figure 1: The predicted terms of He⁻ (right), together with their limits in neutral helium (left) and the onset of different continua (middle).

METHOD OF CALCULATION

The negative ion in our approach is represented by an *Atomic State Function* (ASF);

$$\Psi_{ASF} = \Psi_D + \Psi_C \quad (2)$$

where Ψ_D is the square-integrable, discrete part and Ψ_C represents an open channel. Our computations now consist of two different stages. First an optimization stage, where we compute the radial wavefunctions, and second a perturbation stage, where some specific properties can be obtained.

In the *optimization* stage, we use a nonrelativistic Hamiltonian for the N electron atomic system;

$$H_{NR} = -\frac{1}{2}\sum_{i=1}^{N}\left(\Delta_i + \frac{2Z}{r_i}\right) + \sum_{i>j}\frac{1}{r_{ij}} \quad (3)$$

The discrete part of the ASF has the form:

$$\Psi_D \equiv \Psi(\gamma LS) = \sum_i c_i \Phi(\alpha_i LS) \quad (4)$$

where $\Phi(\alpha_i LS)$ is a *Configuration State Function* (CSF). Each of these CSFs are constructed as coupled (to total L and S, as specified by α_i), antisymmetric sums of products of *spin-orbitals*;

$$\phi_{nlm_lm_s}(\vec{r}) = \frac{1}{r}P_{nl}(r)Y^l_{m_l}(\theta,\phi)\chi_{m_s} \tag{5}$$

where $Y^l_{m_l}(\theta,\phi)$ and χ_{m_s} are spherical harmonics and spinfunctions, respectively. The stationary condition, with respect to all variations in $P_{nl}(r)$'s and c_i's gives the Multiconfiguration Hartree-Fock (MCHF) equations[3], together with a secular equation. The solution is obtained by self-consistent field, numerical calculations, using the MCHF_ASP atomic structure package[4].

In the second, *perturbative* stage we use a Breit-Pauli Hamiltonian[5];

$$H_{BP} = H_{NR} + H_{RS} + H_J(1) + H_J(2) \tag{6}$$

where H_{RS} consists of relativistic shift operators (mass-correction, spin-spin contact and Darwin terms), $H_J(1)$ of J-dependent, one-body operators (spin-orbit interaction) and $H_J(2)$ of J-dependent, two-body operators (spin-other-orbit and spin-spin interaction). The discrete part of the ASF now has the form

$$\Psi_D \equiv \Psi(\gamma J) = \sum_i c_i \Phi(\alpha_i L_i S_i J) \tag{7}$$

where we now include mixing between different LS-terms.

As an example, we can use Be$^-$. Its structure, known and predicted, is shown in Fig. 2. For the lowest bound state, we use a discrete representation

$$|2s2p^2\, {}^4P^e{}_J\rangle = \sum_i c_i|\alpha_i\, {}^4P_J\rangle + \sum_k c'_k|\alpha'_k\, {}^2D_J\rangle \tag{8}$$

It turns out that our approach, as described above, is well justified since both $\frac{\max_k |c'_k|}{\max_i |c_i|}$ and $\left|\frac{\langle H_J(1)+H_J(2)\rangle}{\langle H_{NR}\rangle}\right|$ are of the order of 10^{-6}.

DECAY MODES OF NEGATIVE IONS

An accurate calculation of different properties for decay modes of negative ions is one of our most important goals, since they each represent a way of observing the system. Our first example is a *radiative transition* between an excited state and lower excited or ground state. This decay can be represented as

$$X^{-*} \to X^- + \gamma \tag{9}$$

and the rate can be written

$$A(D \to D') \propto |\langle \Psi_D|e\vec{r}|\Psi_{D'}\rangle|^2 \tag{10}$$

While this type of decay is the most important source of information on neutral atoms and positive ions, it represents quite an exotic process for negative ions

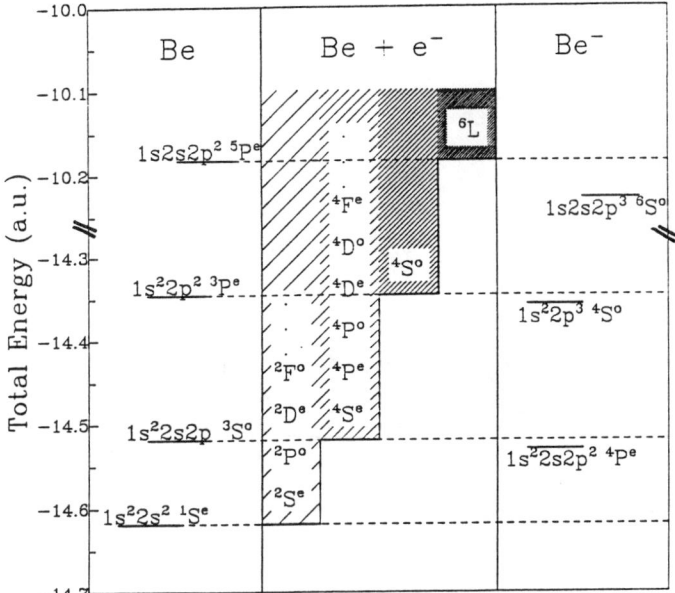

Figure 2: The predicted termsystem of Be⁻ (right).

and only a few transitions have been observed. Each ion might only have one or a few possible transitions, due to the small number of bound states. The accurate calculation of the corresponding wavelengths is therefore of utmost importance to make an identification possible. We will return to this in a later section.

The inclusion of the J-dependent operators in our Hamiltonian, opens up a different type of decay. The bound state can interact with a continuum, which results in emission of an electron, i.e. *autodetachment*. This process can be represented as

$$X^{-*} \to X + e^- \qquad (11)$$

and the rate, in our calculations, is given by the "Golden rule" formula

$$A(D \to C) \propto |\langle \Psi_D | H_{BP} | \Psi_{C'} \rangle|^2 \qquad (12)$$

This process is the subject of our next section.

We should also mention the possibility of *radiative autodetachment*, which represents a simultaneous emission of an electron and a photon;

$$X^{-*} \to X + e^- + \gamma \qquad (13)$$

Its rate is given by

$$A(D \to C') \propto |\langle \Psi_D | e\vec{r} | \Psi_{C'} \rangle|^2 \qquad (14)$$

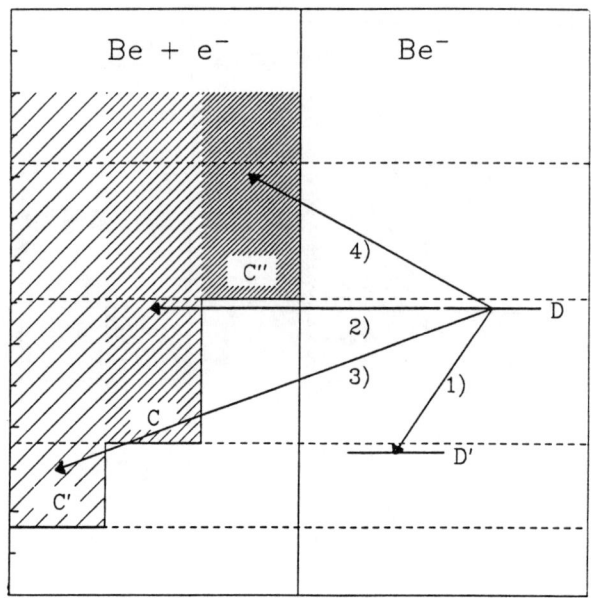

Figure 3: Decay modes for "bound" states of Be$^-$: 1) optical transition, 2) autodetachment, 3) radiative autodetachment and 4) photodetachment.

This process has been treated by Nicolaides and coworkers[6], among others. Since its rate is much slower than that for autodetachment, we ignore it in the rest of this report.

Finally, we have the process of *photodetachment*;

$$\gamma + X^- \rightarrow X + e^- \quad (15)$$

Even though it only corresponds to a decay mode in an external electromagnetic field, it represents a very important possibility for observing negative ions. The rate is in this case

$$A(D \rightarrow C''') \propto |\langle \Psi_D | e\vec{r} | \Psi_{C''} \rangle|^2 \quad (16)$$

All these different processes are illustrated in Fig. 3.

AUTODETACHMENT

Negative ions are some of the least relativistic atomic systems, but in many cases relativistically induced autodetachment is the main decay channel. In this section we discuss recent calculations[7,8] of the decay of $nsnp^2\ ^4P^e$ in Be$^-$ ($n=2$), Mg$^-$ ($n=3$) and Ca$^-$ ($n=4$). We refer to Fig. 2, for Be$^-$, and Fig. 4, for Ca$^-$.

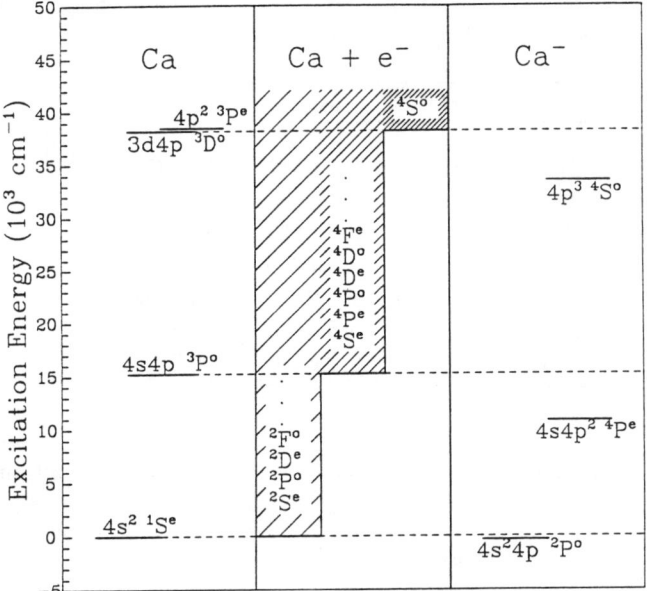

Figure 4: Structure of the negative calcium ion.

If we use Be⁻ as an example, we can list the possible decay channels for the $^4P^e$ bound states;

$$Be^-(1s^22s2p^2\,^4P_{5/2}) \rightarrow Be(1s^22s^2kd\,^2D_{5/2}) + e^-$$
$$Be^-(1s^22s2p^2\,^4P_{3/2}) \rightarrow Be(1s^22s^2kd\,^2D_{3/2}) + e^-$$
$$Be^-(1s^22s2p^2\,^4P_{1/2}) \rightarrow Be(1s^22s^2ks\,^2S_{1/2}) + e^-$$

The expansion in Eq. 7 is for $J = 3/2$ and $5/2$

$$\Psi_D = |2s2p^2\,^4P_J\rangle = \sum_i c_i |\alpha_i\,^4P_J\rangle + \sum_k c'_k |\alpha'_k\,^2D_J\rangle \tag{17}$$

The rate of Eq. 12, is proportional to the square of the *interaction element*:

$$M = \langle \Psi_D | H_{BP} | \Psi_C \rangle \tag{18}$$

where H_{BP} is given by Eq. 6. From this it is obvious that, to order α^2, there are two different contributions to this element. First the *Direct Relativistic*

$$M_{DR} = \sum_i c_i \langle \alpha_i\,^4P_J | H_J(1) + H_J(2) | 2s^2kd\,^2D_J \rangle \tag{19}$$

and second, the *Induced Coulomb* contribution:

$$M_{IC} = \sum_k c'_k \langle \alpha'_k\,^2D_J | H_{NR} + H_{RS} | 2s^2kd\,^2D_J \rangle \tag{20}$$

Table I: Contribution to interaction element (in 10^{-7} au), for the $nsnp^2\,^4P_J$ ($J = 3/2, 5/2$) states of Be⁻, Mg⁻ and Ca⁻.

	Be⁻	Mg⁻	Ca⁻
		$J = 3/2$	
$M_J(1)$	8.7301	199.0542	689.3833
$M_J(2)$	−8.1828	−7.9863	−4.9216
Total	0.5473	191.0679	684.4617
		$J = 5/2$	
$M_J(1)$	24.7287	576.0132	2112.5822
$M_J(2)$	−4.5254	−5.3496	−3.3210
Total	20.2033	570.6636	2109.3500

It is also possible to divide the different contributions to the interaction element into one- and two-body effects;

$$\langle \Psi_D | H_{BP} | \Psi_C \rangle \approx M_J(1) + M_J(2) \tag{21}$$

In Table I we list the contributions from one- and two-body terms to the interaction element, for the $nsnp^2\,^4P_{3/2}$ and $^4P_{5/2}$ states of Be⁻, Mg⁻ and Ca⁻. The one-body term, which is a correction to the nuclear potential (screened by the core), increases drastically from the light Be⁻ ion to the heavy Ca⁻. The two-body, which is a correction to the interaction between the electrons, is almost constant. For Be⁻$(2s2p^2\,^4P^e{}_{1/2})$ these two terms cancel, which leads to a small interaction element and thereby a low autodetachment rate and a long lifetime.

If we look at the $J = 1/2$ state, which decays to the $2s^2ks\,^2S_{1/2}$ continuum, the picture is even more complex. To fully understand its "decay dynamics" we list in Table II not only one- and two-body contributions, but also the direct relativistic and induced Coulomb terms. As expected, the absolute value of both $M_{IC}(1)$ and $M_{DR}(1)$ increases when moving from Be⁻ to Ca⁻, and are of the same order-of-magnitude. The fact that these two contributions have different signs for the $J = 1/2$ states leads to a cancellation within the one-body contribution, most prominent for Ca⁻. The total one-body part is therefore of the same order of magnitude as the two-body part in this case, which opens up the possibility for a cancellation in the interaction element.

As a conclusion, the long lifetime of the Ca⁻$(4s4p^2\,^4P_{1/2})$ state, where the difference between the M_{DR} and M_{IC} elements is about 1 % of each of them, is extremely sensitive to the accuracy of our calculations. In the quite simple approach described here, we can probably just expect qualitative results. For the Be⁻$(2s2p^2\,^4P_{3/2})$ state the difference is about 10 % of the one- and two-body terms, and the result is more reliable. The calculated lifetimes are given in Table III, together with experimental values and some other calculations. We

Table II: Contributions to the interaction element, in 10^{-7} a.u., for the $nsnp^2\,{}^4P^e_{1/2}$ states in Be⁻, Mg⁻ and Ca⁻. DR = direct relativistic, IC = induced Coulomb.

	X = DR	X = IC	DR + IC
	Be⁻		
$M_X(1)$	−0.9612	−0.2819	−1.2431
$M_X(2)$	−1.0870	0.2152	−0.8718
M_X	−2.0482	−0.0667	−2.1149
	Mg⁻		
$M_X(1)$	−13.8981	12.0314	−1.8667
$M_X(2)$	−0.9429	0.0894	−0.8535
M_X	−14.8410	12.1207	−2.7203
	Ca⁻		
$M_X(1)$	−49.6949	50.8720	1.1771
$M_X(2)$	−0.6144	0.0743	−0.5401
M_X	−50.3093	50.9462	0.6370

would like to point out, that the goal of the experiments reported is not to determine the *J*-value of the most long-lived state, but to detect the existance of it. The designations given earlier relied on comparison with known systems, such as He⁻. This sometimes leads to the wrong conclusions. We have therefore, in Table III, changed the experimental designations for Be⁻ and Ca⁻ in light of our new calculations.

ELECTRON AFFINITY OF CALCIUM

Until recently it was believed that none of the alkaline-earths formed a *stable* negative ion. The electron affinity for the ground state was in all calculations predicted to be negative. In 1987, however, Froese Fischer[12] predicted that calcium does form a stable negative ion, in the $4s^24p\,^2P$ state. This was confirmed experimentally by Pegg and coworkers[13], and a number of later calculations (see Table VI). The first calculations[14-20] only included correlation between the outer, valence electrons, that is the two 4s electrons in neutral calcium, and between the two 4s and the 4p electrons in the negative ion. A systematic trend towards larger theoretical than experimental affinities were observed from these studies. In later computations[21,22] the *core-valence correlation*, or core polarization, has also been included, the most important contribution being the correlation between the 3p-subshell and the valence electrons. Due to the diffuse nature of the outer electrons in the negative ion, the core-valence correlation is larger in the atom and its inclusion was expected to decrease the electron affinity. It was surprising to find, that according to the most recent calculation[22] the contribution from the

Table III: Lifetimes for states of Be⁻, Mg⁻ and Ca⁻. $x[-n]$ implies $x \cdot 10^{-n}$ s.

Element	Term	J	This work	Experiment	Other Theories
Be⁻	$2s2p^2\,^4P^e$	5/2	9.43[−7]	∼ 1[−6][9]	1.0[−6][10]
		3/2	1.29[−4]	∼ 1[−4][9]	2.0[−3][10]
		1/2	8.61[−7]	∼ 1[−5][9]	8.0[−8][10]
Mg⁻	$3s3p^2\,^4P^e$	5/2	1.18[−9]		
		3/2	1.05[−8]		
		1/2	5.20[−7]		
Ca⁻	$4s4p^2\,^4P^e$	5/2	8.65[−11]		
		3/2	8.22[−10]		
		1/2	9.49[−6]	$2.9 \pm 1.0[-4]$[11]	

core-valence correlation completely cancels the outer correlation, questioning the stability of the negative calcium ion. This has lead us to return to the earlier MCHF calculations, to investigate the effect of the core-valence correlation.

Table IV: Total energies for $4s^2\,^1S$ in calcium and $4s^24p\,^2P$ of Ca⁻, and electron affinty of calcium, from calculation with different orbitals (in a.u.).

Step	$4s^2\,^1S$		$4s^24p\,^2P$		E_{ea}
	# CSF	E_{Tot}	# CSF	E_{Tot}	
$4s, 3d, 4p$	3	−676.7848025	7	−676.7756043	−0.00920
$+5s, 5p, 4d, 4f$	7	−676.7857616	83	−676.7875719	0.00181
$+6s, 6p, 5d, 5f, 5g$	12	−676.7858291	407	−676.7886240	0.00279
$+7s, 7p, 6s$	34	−676.7858337	804	−676.7887374	0.00290
$+8s, 8p, 7d$	49	−676.7858350	1388	−676.7887631	0.00293

Our first goal is to test the convergence of the calculation of outer correlation. To do that we need a systematic approach to select CSFs in the expansion 4. The method we use is based on two concepts, an active set of orbitals and the Generalized Brillouins Theorem.

From an *active set* of orbitals, we generate all possible CSFs with a given number of electrons, parity and total L and S. In such an approach, the wave-function, and therefore also the energy expression, is invariant under rotations of two orbitals with the same quantum number l. This leads to instability in our variational, self-consistent field approach. This "degree of freedom" in the variations of the radial functions $P_{nl}(r)$ will disappear if we exclude some CSFs according to a *Generalized Brillouins Theorem* (GBT). This theorem states that, when increasing our active set, we should exclude CSFs generated from a single-electron, coupling preserving substitution $(P_{nl}(r) \rightarrow P_{n'l}(r))$ from a major contributor to expansion 4. As an example, we perform computations for the ground state $(4s^2\,^1S)$ of neutral calcium, and our old active set is $\{4s, 4p, 4d, 4f\}$ which is extended to include $\{5s, 5p, 5d, 5f, 5g\}$. The GBT then states that we should exclude $4s5s\,^1S$, $4p5p\,^1S$, $4d5d\,^1S$ and $4f5f\,^1S$ since they are obtained by the $4s \rightarrow 5s$, $4p \rightarrow 5p$, $4d \rightarrow 5d$ and $4f \rightarrow 5f$ substitutions, respectively, from the major contributors $4s^2, 4p^2, 4d^2$ and $4f^2\,^1S$. The inclusion of $5g$ does not require any exclusion, since it is the first orbital of its symmetry.

Table V: Total Energies and Electron Affinity, in a.u., with common orbitals.

Step	$4s^2\,^1S$	$4s^24p\,^2P$	E_{ea}
$+4s, 3d, 4p$	-676.7683672	-676.7756043	0.00724
$+5s, 5p, 4d, 4f$	-676.7845758	-676.7875719	0.00300
$+6s, 6p, 5d, 5f, 5g$	-676.7856809	-676.7886240	0.00294
$+7s, 7p, 6d$	-676.7857936	-676.7887374	0.00294
$+8s, 8p, 7d$	-676.7858083	-676.7887631	0.00295
$+$ CV1	-676.8058413	-676.8091358	0.00329
$+$ CV2	-676.8021445	-676.8041956	0.00205
$+$ RS	-679.5826663	-679.5843782	0.00171

In Table IV we report on active set, GBT calculations of the electron affinity of calcium. The core orbitals, $1s - 3p$, are common for the neutral atom and the negative ion, but the valence orbitals are optimized separately. In Table V, we report on the same type of calculations, but with all orbitals in common. They are in this case optimized on the negative ion state. It is interesting to observe that the convergence is much faster in this calculation, as if the larger complexity of the $4s^24p\,^2P$ is compensated by the less accuracy in the calculation for the $4s^2\,^1S$. The last calculation, which includes 1388 CSFs for the negative ion seems to show good convergence. It probably represents the electron affinity with only the outer correlation included, in an accurate way (0.082 eV).

There are two corrections to this approach, core-valence correlation and relativistic effects. The core-valence correlation, can be estimated by using modelpotentials. We choose a form, introduced by Baylis and coworkers[23], with a one-body part;

$$V_{cv1}(r) = -\frac{1}{2}\alpha_d \frac{r^2}{(r^2+r_c^2)^3} \qquad (22)$$

and a two-body part;

$$V_{cv2}(\vec{r_1}, \vec{r_2}) = -\alpha_d \frac{\vec{r_1} \cdot \vec{r_2}}{[(r_1^2 + r_c^2)(r_2^2 + r_c^2)]^{3/2}} \tag{23}$$

where α_d is the dipole polarizability of the $3p^6$ core, and r_c is a suitable cut-off radius. The former parameter is quite well-determined (3.254 au[24]), while the cut-off radius is an adjustable parameter. We here use the expectation value of r for the outermost core subshell, 1.26432 au, to represent r_c.

The second important correction is the relativistic shift. Both these two corrections are introduced in the perturbation stage of our calculations. When we change the Hamiltonian we have to reintroduce the deleted CSFs, since they can contribute through the new interactions.

The results are given in Table V for the case of common orbitals. As expected, the electron affinity is reduced and if we look at Table VI the result seems to be in excellent agreement with experiment. However the dependence of our results on the actual choice of r_c has to be investigated further. It is also desirable to treat the core-valence correlation with ab initio methods, in the form of expansion of CSFs with one hole in the core. The main problem is that the size of the expansion increases quickly with the size of the active set, and has quite a slow convergence.

Table VI: The recent history of the electron affinity (EA) of calcium.

Reference	Method	EA (eV)
Theoretical Prediction:		
Froese Fischer 1987[12]	MCHF	0.045
Experimental Discovery:		
Pegg et al 1987[13]	EXP	0.043 ± 0.007
Later Theory, with only outer correlation:		
Froese Fischer 1989[14]	MCHF	0.062
Vosko et al 1989[15]	Density-functional	0.132
Kim and Greene 1989[16]	R-matrix	0.071
Bauschlicher et al 1989[18]	SOCI	0.022
Gribakin et al 1990[19]	Dyson equation	0.058
Cowan and Wilson 1991[20]	HFR+Modelpotential	0.082
Later theory with core-valence correlation:		
Johnson et al 1989[21]	MBPT	0.058
Fuentealba et al 1990[22]	CI	0.0
This work 1991	MCHF+CV	0.047

OPTICAL TRANSITIONS

The last property we will discuss in this talk is optical transitions in negative ions. In the identification of these, the experimentalist cannot take advantage of the very powerful techniques used for atoms and positive ions, since there are no closed loops of transition. Therefore very accurate predictions of wavelengths are needed to find these evasive features in a spectrum that often contains many other ionization stages. In spite of a careful search[25] in Mg^-, the only observed optical transitions are $1s2s2p^2\,{}^5P^e - 1s2p^3\,{}^5S^o$ of negative lithium[26] (3489.7 Å) and $1s^22s2p^2\,{}^4P^e - 1s^22p^3\,{}^4S^o$ of negative beryllium[27] (2654.01 Å).

Recently, Froese Fischer[28] proposed a method to deduce the wavelengths of transitions in negative ions, by using the well known transition between the limits in the neutral atom, together with accurately calculated electron affinities. If we look at Be^-, the transition energy for the $2s2p\,{}^3P - 2p^2\,{}^3P$ is 37715.64 cm^{-1}. By using the active set, GBT technique we can get quite accurate values for the electron affinities of these two states, from just the outer correlation calculations. As corrections, we include relativistic shifts, mass-polarization and core-valence interference[28]. From this we can deduce the wavelength of the ${}^4P^e - {}^4S^o$ transition. Since the table in ref. 28 contains a misprint, we give the correct contributions to the transition wavelength in Table VII.

Table VII: Electron affinities (in a.u.) of the $2s2p\,{}^3P$ and $2p^2\,{}^3P$ of beryllium, and wave length (in Å) for the $2s2p^2\,{}^4P^e - 2p^3\,{}^4S^o$ Transition in Be^-

Step	$EA(2s2p\,{}^3P)$	$EA(2p^2\,{}^3P)$	ΔEA	λ_{air}
$\{2s, 2p, 3s, 3p, 3d\}^a$	0.0087357	0.0076737	0.0010620	2634.36
$+\{4s, 4p, 4d, 4f\}^a$	0.0103993	0.0107950	-0.0003957	2656.75
$+\{5s, 5p, 5d, 5f, 5g\}^a$	0.0106480	0.0110156	-0.0003676	2656.31
+ RS and MP^b	0.0106380	0.0110078	-0.0003698	2656.35
+ CVI^c	0.0106380	0.0108516	-0.0002136	2653.93
Bunge[29]				2645.0±3.0
B&N[30]				2654.0±9.0
Exp.[27]				2653.01±0.05

^{a} The active set used
^{b} RS = Relativistic shifts, MP = masspolarization
^{c} CVI = Core-valence interference

CONCLUSIONS AND FUTURE GOALS

We have described calculations of different properties of negative ions. To predict and identify long-lived components in a beam of negative ions, fairly accurate computations are needed of autodetachment rates. The relativistic effects, that causes the detachment, are in these cases about 10^{-6} the nonrelativistic correlation energies, and the representation of the interplay and cancellation of the different contributions raises high demands on our computations.

High accuracy is also required to predict stable negative ions, through the calculations of electron affinities. As Bates pointed out in a recent review article[30]: "since 0.043 eV is only about 2×10^{-6} the total energy of $(4s^2\,^1S)$Ca or $(4s^2 4p\,^2P)$Ca$^-$, it is evident that the calculation of EA(Ca) is a formidable task".

Finally, to make it possible to identify optical transitions in negative ions, unambiguously, a prediction of the wavelength within 0.1 %, or better, is required. We have shown that ab initio electron affinities can be used, together with experimentally known transition energies for the neutral atom.

The computations described here are all part of an ongoing project. There are a number of different improvement and extensions that are in progress. Among them we can mention ab initio treatment of core-valence interaction in Ca$^-$ and studies of optical transitions in Mg$^-$ and Ca$^-$. Recently some theoretical results on photodetachment cross section of Ca$^-$ have appeared[19,32]. The treatment of correlation and resonances in this type of calculations is one of the most interesting challenges for the future.

ACKNOWLEDGEMENT

This work was supported by the U.S. Department of Energy, Office of Basic Energy Sciences.

REFERENCES

1. P.-O. Löwdin, Phys. Rev. **97**, 1509 (1955).

2. C.A. Nicolaides and G. Aspromallis, J. of Molecular Structure **199**, 283 (1989) and references therein.

3. C. Froese Fischer, The Hartree-Fock Method for Atoms (John Wiley & Sons, N.Y., 1977).

4. C. Froese Fischer, Comput. Phys. Commun **64**, in press.

5. R. Glass and A. Hibbert, Comput. Phys. Commun. **16**, 19 (1978).

6. G. Aspromallis, C.A. Nicolaides and D.R. Beck, Phys. Rev. A **28**, 1879 (1983), and references therein.

7. T. Brage and C. Froese Fischer, Phys. Rev. A**44**, in press.

8. T. Brage, G. Miecznik and C. Froese Fischer, submitted to J. Phys. B:At. Mol. Opt. Phys.

9. Y.K. Bae and J.R. Peterson, Phys. Rev. A**30**, 2145 (1984).

10. G. Aspromallis, C.A. Nicolaides and D.R. Beck, J. Phys. B:At. Mo. Opt. Phys. **19**, 1713 (1986).

11. D. Hanstorp. P. Devynck, W.G. Graham, and J.R. Peterson, Phys. Rev. Lett. **63**, 368 (1989).

12. C. Froese Fischer, J.B. Lagowski and S.H. Vosko, Phys. Rev. Lett. —bf 59, 2263 (1987).

13. D.J. Pegg, J.S. Thompson, R.N. Compton and G.D. Alton, Phys. Rev. Lett. **59**, 2267 (1987).

14. C. Froese Fischer, Phys. Rev. A **39**, 963 (1989).

15. S.H. Vosko, J.B. Lagowski and I.L. Mayer, Phys. Rev. A **39**, 446 (1989).

16. L. Kim and C.H. Greene, J. Phys. B:At. Mol. Opt. Phys. **22**, L175 (1989).

17. Y. Guo and M.A. Whitehead, Phys. Rev. A **40**, 28 (1989).

18. C.W. Bauschlicher, S.R. Langhoff and P.R. Taylor, Chem. Phys. Lett. **158**, 245 (1989).

19. G.F. Gribakin, B.V. Gul'tsev, V.K. Ivanov and M. Yu Kuchiev, J. Phys. B:At. Mol. Opt. Phys. **23**, 4505 (1990).

20. R.D. Cowan and M. Wilson, Physica Scripta **43**, 244 (1991).

21. W.R. Johnson, J. Sapirstein and S.A. Blundell, J. Phys. B:At. Mol. Opt. Phys. **22**, 2341 (1989).

22. P. Fuentealba, A. Savin, H. Stoll and H. Preuss, Phys. Rev. A **41**, 1238 (1990).

23. W.E. Baylis, J. Phys. B:At. Mol. Opt. Phys. **10**, L583 (1977).

24. W.R. Johnson, D. Kolb and K.-N. Huang, At. Data Nucl. Data Tables **28**, 333 (1983).

25. T. Andersen, J.O. Gaardsted, L.Eg Sørensen and T. Brage, Phys. Rev. A **42**, 2728 (1990).

26. J. Bromander, S. Hultberg, B. Jelenkovic, L. Liljeby and S. Mannervik, J. Phys. (Paris) Colloq. Suppl. **40**, C1-10 (1979).

27. J.O. Gaardsted and T. Andersen, J. Phys. B:At. Mol. Phys. **22**, L51 (1989).

28. C. Froese Fischer, Phys. Rev. A **41**, 3481 (1990).

29. A.V. Bunge, Phys. Rev. A **33**, 82 (1986).

30. D.R. Beck and C.A. Nicolaides, Int. J. Quantum Chem. Symp. **18**, 467 (1984).

31. D.R. Bates, Adv. Atom. Mol. Opt. Phys. **27**, 1 (1991).

32. C. Froese Fischer and J.E. Hansen, Phys. Rev. A **44**, in press.

LORENTZ COVARIANT MULTIPLE SCATTERING WITH APPLICATION TO KAON–NUCLEUS SCATTERING

C. M. Chen and D. J. Ernst
Department of Physics and Center for Theoretical Physics
Texas A&M University
College Station, Texas 77843

Abstract

The construction of Lorentz covariant three–body states and their angular momentum decomposition is reviewed. It is shown how they can be used to construct the impulse approximation to the optical potential in momentum space. The potential thus constructed incorporates covariant kinematics, relativistic normalizations and phase space factors, a covariant treatment of the recoil of the target nucleus, utilizes invariant amplitudes which are free of kinematic singularities, and allows an exact performance of the fermi integration over the momentum of the struck nucleon. The interaction of the K^+ meson with ^{12}C and ^{40}Ca is examined and it is found that the present data indicate an enhanced, in–medium, two–body interaction. We find that all the data is consistent with a angle and energy independent 25% enhancement of the theoretical calculations.

I. Introduction

Modern accelerators enable the nuclear physicist to probe the nucleus with projectiles that are moving with relativistic velocities. This requires a generalization of conventional multiple scattering theory in which the kinematics is treated covariantly. We review a procedure to construct many–body free particle wave functions[1] and angular momentum decompose them in such a way that they may be used in a kinematically covariant multiple scattering theory. We explicitly construct three–body states. These states are then used to construct in momentum space a covariant impulse approximation to the optical potential. This is possible because the impulse approximation has the structure of a three–body problem with the three bodies being the incident projectile, the struck nucleon, and the $A - 1$ residual nucleons. The calculation requires the unitary transformation, called the relativistic three–body recoupling coefficients, between two different covariant three–particle states. As the wave function for a relativistic projectile will contain many partial waves, such calculations become computer intensive. Typically, several hundred thousand recoupling coefficients are required. This approach not only allows the incorporation of covariant kinematics, it also incorporates invariant normalizations and phase space factors and naturally utilizes invariant amplitudes which are free of kinematical singularities.

We utilize the covariant optical potential to calculate elastic K^+ scattering from ^{12}C and ^{40}Ca at a laboratory momentum of 800 MeV/c. The K^+ is the weakest of the strongly interacting particles. This gives it a long mean–free–path[2] in the nucleus which allows it to interact with all of the nucleons. It has been suggested[3-6] that nucleons in a nucleus are "fatter" than free nucleons. The K^+ is the projectile of choice to investigate this phenomenon. Because it scatters from all the nucleons, its nuclear cross section is roughly A times the two–body cross section. Thus, if the in medium two–body cross section increases by a certain percent, the K^+–nucleus cross section will be enhanced by an equal percent. For other strongly interacting probes, the mean–free–path is less than the nuclear radius. The interaction is thus surface dominated and is diffractive in character. The nuclear cross section is roughly proportional to the sum of the squares of the nuclear radius and the radius of the projectile–nucleon interaction. An increase in the radius of the projectile-nucleon interaction thus will only produce a small percentage change in the projectile–nucleus cross section. Siegel et al.[7] calculated K^+ elastic scattering from C^{12} and Ca^{40} to compare with the data of Ref. 8. They found cross sections which were significantly smaller than the experimental data. They found that an increase of 15% in the S_{11} phase shift (an increase of 26% in the two-body cross section) raised the calculated cross section to agree with the data. Recent measurements[9] of the ratio of the total cross section of K^+ scattering from ^{12}C to six times the scattering from the deuteron also indicate an enhanced cross section for ^{12}C.

The covariant optical potential approach reviewed here is ideal for this problem. We first review the construction of the covariant free particle states, their angular momentum decomposition, the three–body recoupling coefficient and their use in constructing the optical potential, and then present results for K^+ scattering from ^{12}C and ^{40}Ca.

II. Lorentz covariant three–body states and recoupling coefficients

We begin by constructing the single particle states, then proceed to the two–body and three–body states. For a single particle with mass m, spin s, and rest-frame spin projection λ, the state for a particle at rest can be written as $|0\lambda\rangle$. A state with momentum $\vec{k}$ is defined by boosting this state,

$$|\vec{k}\lambda\rangle = U(L_{\vec{k}})|0\lambda\rangle , \qquad (1)$$

where $L_{\vec{k}}$ is the Lorentz boost operator. For spin 1/2 nucleons, it is a 2×2 representation of the Lorentz boost. The approach has been extended to particles of arbitrary spin in Ref. 10. We utilize invariant normalizations,

$$\langle \vec{k}'\lambda' | \vec{k}\lambda \rangle = 2\bar{E}\,\delta^3(\vec{k}' - \vec{k})\delta_{\lambda'\lambda} , \qquad (2)$$

where $\bar{E} = (k^2 + m^2)^{1/2}$. A Lorentz transformation from one state to another is given by

$$U(L)|\vec{k}\lambda\rangle = \sum_{\lambda'} D^s_{\lambda'\lambda}(R_W)|\vec{k}'\lambda'\rangle , \qquad (3)$$

where $D^s_{\lambda'\lambda}$ is a rotation matrix with R_W the Wigner rotation defined by $R_W = L_{\vec{k}'}^{-1} L L_{\vec{k}}$.

The two–body states are constructed[1] from the direct product of two single particle states. The construction starts with the two–body states constructed in the center–of–momentum frame

$$|0(\vec{p}\lambda_1\lambda_2)\rangle = |\vec{p}\lambda_1\rangle \otimes |-\vec{p}\lambda_2\rangle . \qquad (4)$$

The states in all other frames are defined as

$$|\vec{K}(\vec{p}\lambda_1\lambda_2)\rangle = U(L_{\vec{K}})|0(\vec{p}\lambda_1\lambda_2)\rangle . \qquad (5)$$

For this two-particle system, the total momentum and total energy are $\vec{K} = \vec{K}_1 + \vec{K}_2$ and $\bar{W} = \bar{W}_1 + \bar{W}_2$ and the center-of-momentum energies are $\bar{\omega}_i =$

$(p^2 + m_i^2)^{1/2}$ and $\bar{\omega} = \bar{\omega}_1 + \bar{\omega}_2$. An important point to realize is that the total energy $\bar{W}$ is related to the center-of-momentum energy and the total momentum by $\bar{W} = (\bar{\omega}^2 + K^2)^{1/2}$, i.e. the center-of-momentum energy plays the role of a mass in this relationship. Also note the the center-of-momentum momentum $\vec{p}$ serves as a covariant definition of a relative momentum.

We may angular momentum decompose the relative momentum $\vec{p}$,

$$|\vec{K}(plm_l\lambda_1\lambda_2)\rangle = \left[\frac{2l+1}{4\pi}\right]^{1/2} \int d^2\hat{p} D^l_{m_l 0}(R(\hat{p}))^* |\vec{K}(\vec{p}\lambda_1\lambda_2)\rangle \ . \tag{6}$$

Coupling the two spins s_i to a total spin s and then coupling s to l to give a total momentum j gives the two-body states in a total angular momentum basis

$$|\vec{K}(pjmls)\rangle = \sum_{\lambda_1\lambda_2 m_l m_s} C^{s_1\ s_2\ s}_{\lambda_1\ \lambda_2\ m_s} C^{l\ s\ j}_{m_l\ m_s\ m} |\vec{K}(plm_l\lambda_1\lambda_2)\rangle \ , \tag{7}$$

where C is a Clebsch-Gordon coefficient. The state $|0(pjmls)\rangle$ rotates as a state of angular momentum j and projection m and, as already noted, behaves as a state of mass $\bar{\omega}$ under a boost. Thus the total angular momentum basis state in Eq. 7 behaves like a particle of total momentum $\vec{K}$, mass $\bar{\omega}$, spin j, spin projection m, and internal quantum numbers l, s, s_1 and s_2.

For three-body states we may repeat the construction by combining the third body with the two-body states which we have just constructed. This works because the two-body states transform as a single body state. The kinematics for these states goes as follows. The total momentum and total energy are $\vec{P} = \vec{P}_1 + \vec{P}_2 + \vec{P}_3$ and $\bar{E} = \bar{E}_1 + \bar{E}_2 + \bar{E}_3$. We use the three-body notation and label pair quantities by the 'odd man out'; the total momentum and energy for the (jk) pair of particles are thus $\vec{P}_{(i)} = \vec{P}_j + \vec{P}_k$ and $E_{(i)} = \bar{E}_j + \bar{E}_k$. The momentum of particle j (or k) in the center-of-momentum frame of the (jk) pair is written as $\vec{p}_i$ (or $-\vec{p}_i$). Thus the energies of particles j and k in the center-of-momentum frame of the (jk) pair are $\bar{\omega}_{j(i)} = (p_i^2 + m_j^2)^{1/2}$ and $\bar{\omega}_{k(i)} = (p_i^2 + m_k^2)^{1/2}$ with the total given by $\bar{\omega}_{(i)} = \bar{\omega}_{j(i)} + \bar{\omega}_{k(i)}$. The relative momentum of the (jk) pair with respect to particle i in the three-body center-of-momentum frame is defined as $\vec{q}_i$. The energy of particle i in the three-body center-of-momentum frame is $W_i = (q_i^2 + m_i^2)^{1/2}$ and the energy of the (jk) pair in this same frame is $\bar{W}_{(i)} = (q_i^2 + \bar{\omega}_{(i)}^2)^{1/2}$ with the total energy in the three–body center–of–momentum frame given by $\bar{W} = \bar{W}_{(i)} + \bar{W}_i$. Finally, the total energy $\bar{E}$ in any frame is $\bar{E} = (\bar{W}^2 + P^2)^{1/2}$.

With this notation, we form a two–body state from the (jk) pair with a relative momentum $\vec{p}_i$ and a total momentum $\vec{q}_i$, $|\vec{q}_i(\vec{p}_i\lambda_j\lambda_k)\rangle$, and then form the three–body state by combining it with particle i with spin projection λ_i and momentum $-\vec{q}_i$ and then boosting to a frame with total momentum $\vec{P}$. We note this state by $|\vec{P}[\vec{q}_i(\vec{p}_i\lambda_j\lambda_k)\lambda_i]\rangle$. We also could perform the angular momentum decomposition as in Eq. 7. We start with the state $|\vec{q}_i(p_ij_il_is_i)\rangle$ and combine it with particle i and then perform the angular decomposition of the $\vec{q}_i$ as in Eq. 6. Then couple the spins (note that this is coupling the spin of particle i with j_i which transforms like an internal angular momentum) to give a total spin S_i and coupling S_i with the orbital angular momentum of q_i which we call L_i to produce a total angular momentum J and projection M. In analogy to Eq. 7, we write this state as $|\vec{P}[q_iJML_iS_i(p_ij_il_is_i)_{jk}]\rangle$.

Rather than coupling the (jk) pair first and then coupling this two–body state to i, we could have coupled (ki) first and then coupled to j. The unitary transformation between these two basis is the relativistic three–body recoupling coefficient,[1] $C_J\{(ki)j;(jk)i\}$. These are defined by

$$\langle \vec{P}'(q_jJ'M'L_jS_j(p_jj_js_j)_{ki} | \vec{P}(q_iJML_iS_i(p_ij_il_is_i)_{jk}\rangle =$$
$$8\frac{\bar{\omega}_{(i)}\bar{\omega}_{(j)}}{q_ip_iq_jp_j}\delta(\bar{E}'-\bar{E})\delta^3(\vec{P}'-\vec{P})\delta_{J'J}\delta_{M'M}$$
$$\times C_J\{q_jL_jS_j(p_jj_jl_js_j)_{ki}; q_iL_iS_i(p_ij_il_is_i)_{jk}\} \quad , \quad (8)$$

and an explicit expression is

$$C_J\{q_jL_jS_j(p_jj_jl_js_j)_{ki}; q_iL_iS_i(p_ij_il_is_i)_{jk}\} =$$
$$\tfrac{1}{2}[(2L_j+1)(2l_j+1)(2L_i+1)(2l_i+1)]^{1/2}$$
$$\times \sum (-1)^{s_j+s_i-S}\frac{1}{(2L+1)}\begin{bmatrix}s_j & s_k & s_i \\ s_i & S & s_j\end{bmatrix}\begin{bmatrix}l_j & s_j & j_j \\ s_j & S_j & S\end{bmatrix}\begin{bmatrix}L_j & l_j & L \\ S & J & S_j\end{bmatrix}$$
$$\times \begin{bmatrix}l_i & s_i & j_i \\ s_i & S_i & S\end{bmatrix}\begin{bmatrix}L_i & l_i & L \\ S & J & S_i\end{bmatrix}C^{L_j\ l_j\ L}_{M_{L_j}\ m_{l_j}\ M_L}C^{L_i\ l_i\ L}_{0\ m_{l_i}\ M_L}$$
$$\times d^{L_j}_{M_{L_j},0}(\xi_{ji}-\pi)d^{l_j}_{m_{l_j},0}(\theta_j+\xi_{ji}-\pi)d^{l_i}_{m_{l_i},0}(\theta_i) \quad (9)$$

where the 6–j's and Clebsch–Gordon coefficients are defined as in Ref. 11. The angles ξ_{ji}, θ_j and θ_i are depicted in Fig. 1.

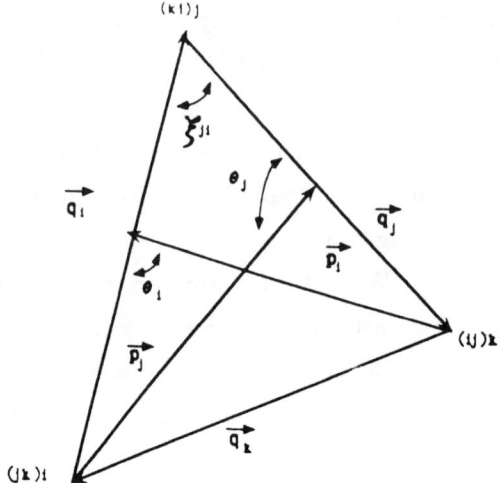

Fig. 1 Relative momenta relations and the angles in Eq. 9.

III. Relativistic optical potential

A useful application of the relativistic three–body recoupling coefficients is in the calculation[12] of the impulse approximation to the optical potential. The first order optical potential is illustrated in Fig. 2. Here, particle 1 is the projectile, particle 2 is a nucleon in an A–nucleon target nucleus, and particle 3 is the $(A-1)$–nucleon residual nucleus. The four–point interaction labeled T is the projectile–nucleon two–body T–matrix, and the vertices labeled ψ are bound state wave functions of the nucleons. The equation for the optical potential which corresponds to this diagram is

$$\langle \vec{P}_1' \vec{P}_A' | U(E) | \vec{P}_1 \vec{P}_A \rangle =$$
$$\sum_{\alpha_3} \int \frac{d^3 \vec{P}_3}{2\bar{E}} \langle \psi'(\vec{P}_A') | \vec{P}_2' \vec{P}_3 \alpha_3 \rangle \langle \vec{P}_1' \vec{P}_2' | T(E) | \vec{P}_1 \vec{P}_2 \rangle \langle \vec{P}_2 \vec{P}_3 \alpha_3 | \psi(\vec{P}_A) \rangle \ , \quad (10)$$

where E is the asymptotic incident energy of the projectile–nucleus system, and α_3 is a set of quantum numbers that specifically label the nuclear bound state.

The two–body T–matrix is a function of the relative momenta between the projectile and the nucleon $(\vec{P}_1, \vec{P}_2)$ and $(\vec{P}_1', \vec{P}_2')$. However, the bound state nuclear wave functions is a function of the relative momenta between the nucleon and the $A-1$ residual nucleons, $(\vec{P}_2, \vec{P}_3)$ and $(\vec{P}_2', \vec{P}_3')$. We thus insert two-complete sets of free three–body angular momentum decomposed states between

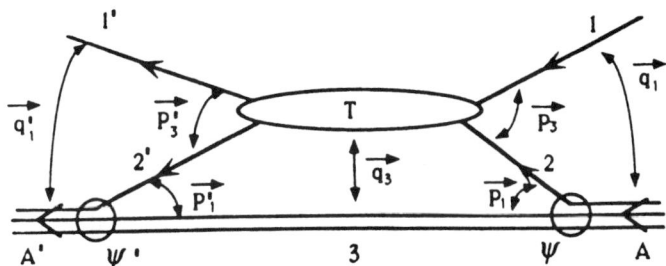

Fig. 2. The first order optical potential and the definitions of the momenta.

the T-matrix and the nuclear wave function. The first set couples $|3(1,2)\rangle$ and the second $|1(2,3)\rangle$. Inserting the same set, but in the reverse order, to the left of the T-matrix gives, after some algebra,

$$\langle \vec{P}'(q_1' J' M' L_1' S_1') | U(E) | \vec{P}(q_1 J M L_1 S_1) \rangle =$$
$$\delta^3(\vec{P}' - \vec{P}) \delta_{J'J} \delta_{M'M} U_J(W; q_1' L_1' S_1'; q_1 L_1 S_1) =$$
$$\delta^3(\vec{P}' - \vec{P}) \delta_{J'J} \delta_{M'M} \sum \int dq_3 d\cos\xi_{13}' d\cos\xi_{13} \frac{q_3^2}{8\overline{W}_2' \overline{W}_2 \overline{W}_3}$$
$$\times \psi_{s'_A}'(p_1' l_1' s_1' s_3 \alpha_3)^* T_{j_3}(\omega_{(3)}; p_3' l_3' s_3'; p_3 l_3 s_3) \psi_{s_A}(p_1 l_1 s_1 s_3 \alpha_3)$$
$$\times C_J \{q_3 L_3 S_3 (p_3' j_3' l_3' s_3')_{12}; q_1' L_1' S_1'(p_1' s_A' l_1' s_1')_{23}\}$$
$$\times C_J \{q_3 L_3 S_3 (p_3 j_3 l_3 s_3)_{12}; q_1 L_1 S_1(p_1 s_A l_1 s_1)_{23}\} , \quad (11)$$

where the sum extends over all the angular-momentum variables that do not appear on the left-hand side of the equation.

IV. K^+ nucleus scattering

As noted in the introduction, the K^+-nucleon interaction is the weakest of the strong interactions. It is a meson with isospin 1/2, is much more massive than the pion, but has a comparable lifetime. Its quark structure is $u\bar{s}$. The lack of a $\bar{u}$ or $\bar{d}$ quark is the origin of its weak interaction with the nucleon. This is in striking contrast to the K^- whose structure is $\bar{u}s$ and has a large resonant interaction with the nucleon. This is illustrated in Fig. 3. The difference in the strength of the two-body interactions produces a difference in the penetrability of the two kaons into a nucleus. The mean-free-paths of the K^+ and the K^- are pictured in Fig. 4.

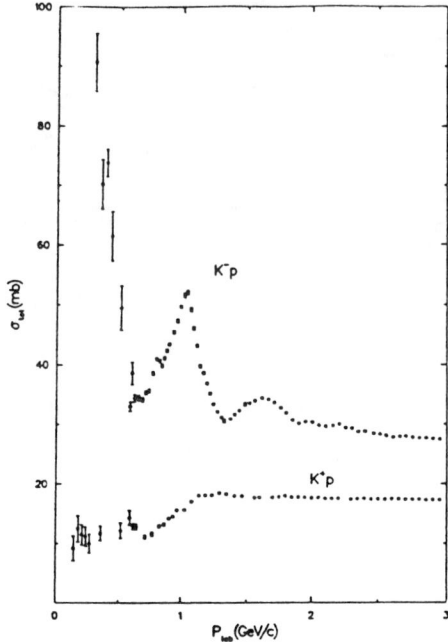

Fig. 3. $K^\pm$ total cross sections as a function of the kaon laboratory momentum, P'_{lab}, (from Arndt et al.[13]).

We here examine elastic scattering of the K^+ from ^{12}C and ^{40}Ca at a laboratory momentum of 800 MeV/c. We require an off-shell model for the two-body T-matrix. We will use a simple separable form

$$\langle q' | T_\alpha(\omega) | q \rangle = v_\alpha(q') \lambda_\alpha(\omega) v_\alpha(q) , \qquad (12)$$

where $v_\alpha(q)$ determines the off-shell behavior of the amplitude. It is important to realize that, whatever the underlying dynamics of the K^+-nucleon interaction, it could be approximated[14] over a reasonable on-shell and off-shell range by a separable form. The subscript α labels a particular spin and isospin channel. The function $\lambda_\alpha(\omega)$ is chosen to reproduce the measured on-shell amplitude. We used both Martin's[15] and Arndt's[13] amplitudes and include partial waves through the D-wave. We use a simple Gaussian function for the form factors, $v_\alpha(q) = \exp(-q^2/\beta_\alpha^2)$, and the target wave functions were obtained from Hartree-Fock calculations, Ref. 16 for ^{12}C and Ref. 17 for ^{40}Ca. The predicted elastic differential cross section are pictured in Figs. 5 and 6. The results are

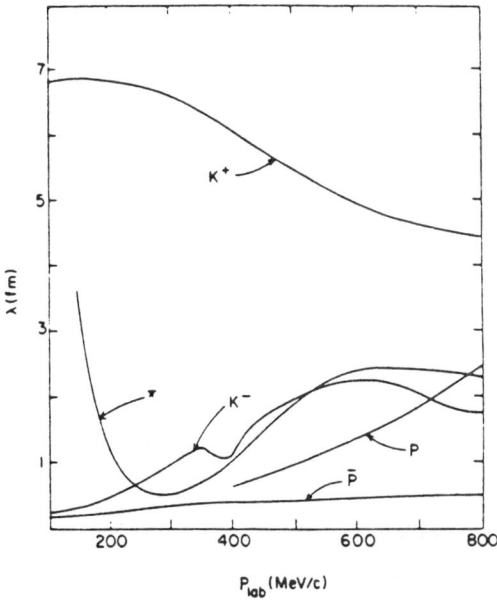

Fig. 4. The mean–free–path in nuclear matter of stongly interacting particles as a function of P_{lab} (from Dover et al.[2]).

in agreement with those of Siegel et al., Ref. 7. We too find that the theoretical cross sections are consistently and evenly below the data. We find that the results do not depend significantly on which K^+-nucleon amplitude we use or on the range β_α which we use. The K^+-nucleon D-wave amplitudes do not contribute at this energy as expected.

A simple renormalization of the data by 25% for both ^{12}C and ^{40}Ca produces a remarkable agreement with the theoretical curve as the difference between the theory and the data is, in percent, angle independent. This is beyond the systematic error of 18% quoted for the data. We also follow the suggestion of Ref. 7 and increase the strength of the two–body K^+-nucleon interaction. We describe no particular model[18,19] to the origin of this medium modification of the amplitude. We find that for ^{12}C the phase shifts must be increased by 25% (an increase in the two-body total cross section of 36%) while for ^{40}Ca the phase shifts must be increased by 42% (an increase in the two–body total cross section of 64%). Note that increasing the two–body cross section does not

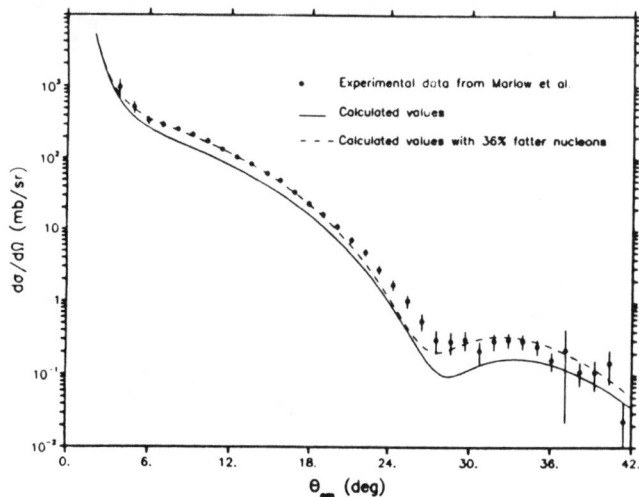

Fig. 5. Differential cross section as a function of angle for K^+ ^{12}C elastic scattering at 800 MeV/c. The data are from Ref. 8. The solid curve is our result using the free K^+-nucleon amplitude; the dashed curve is our result with a 36% increase in the two-body total cross section.

produce the same quality of fit as does renormalizing the data. This is because the discrepancy is angle independent. Even though the two–body cross section is dominated by the S–wave, the transformation from the two–body center–of–momentum to the lab frame produces a forward angle enhancement. The larger number for ^{40}Ca is easily understood. Although the kaon can penetrate deeply into ^{40}Ca, a mean-free-path of 5 fermi means that a fraction of the nucleons are 'hidden'. The ^{40}Ca cross section is thus not proportional to A times the two–body cross section but to a somewhat lower power.

The results we find are rather model independent. The Born approximation that results from setting the many–body T–matrix equal to the optical potential agrees with the full calculation to a few percent. This is a strong indication that second order multiple scattering corrections to the optical potential (Pauli or correlation effects) are quite small as they should be roughly of the same order of magnitude at the second iterate of the optical potential. They were also estimated in Ref. 7 and found to be quite small.

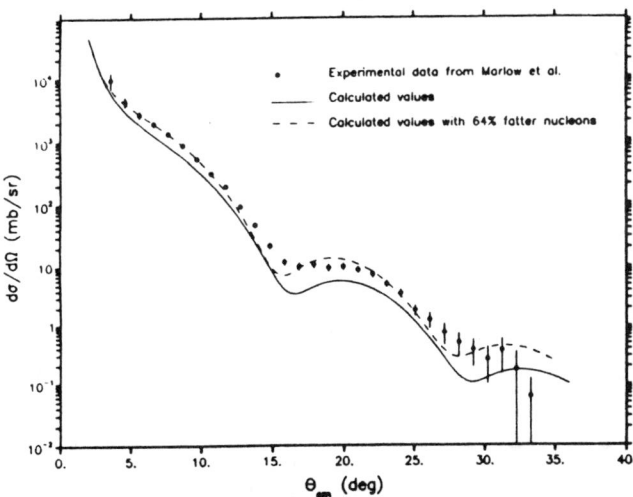

Fig. 6. The same as Fig. 5 except the target is ^{40}Ca and the dashed curve is our result with a 64% increase in the two-body total cross section.

The ratio[9] of the total cross section for K^+ scattering on ^{12}C to the total cross section on the deuteron has recently been measured. We show these measurements together with our calculations in Fig. 7. The measured ratio of the ^{12}C cross section to six times the measured deuteron cross section is greater than one at the low energies and falls smoothly toward one as the energy increases. On the other hand, our calculation, like those given in Ref. 8, are everywhere less than one. This again indicates an enhanced in medium K^+ cross section of about 25%. Results for a 25% enhanced two-body cross section are also pictured in Fig. 7.

Conclusion

We have reviewed the construction of covariant many-body free-particle states and their angular momentum decomposition. Their use in the construction of the impulse approximation to the optical potential has been presented and the scattering of K^+ from nuclei has been calculated. The calculations are numerically intensive as they require several hundred thousand recoupling coefficients. We find that the existing data indicate a medium enhancement of the K^+-nucleon amplitude. An enhancement, which is independent of energy and

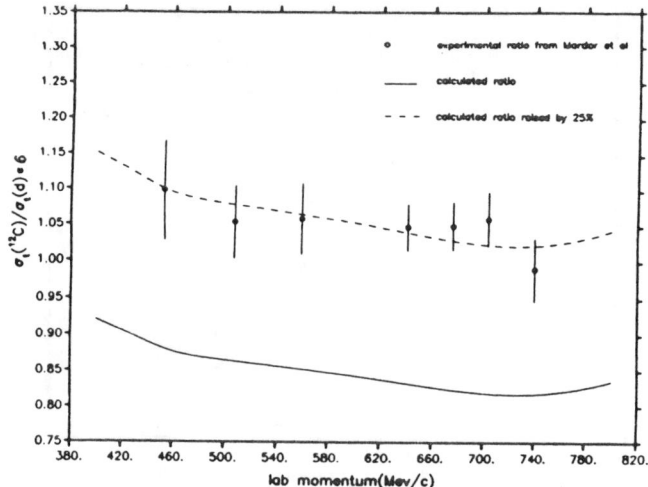

Fig. 7. The ratio of the total cross section for K$^+$ scattering from ^{12}C to six time the total cross section for the deuteron. The data are from Ref. 9. The solid curve is our theoretical cross section divided by six times a smooth fit to the experimental cross section for the deuteron. The dashed curve is the solid curve enhanced by 25%.

angle, of the theoretically predicted cross section by 25% is consistent with all the existing data. Additional data could substantiate this effect. In the absence of the total[9] cross section data, a renormalization of the data of Ref. 8 might well have been the most likely explanation of the discrepancy. Differential cross sections with careful attention paid to the normalization for ^{4}He would be desirable. Even though ^{4}He is spatially very small, it has an exceptionally large central density. One would expect to see a similar enhancement. Differential cross sections at lower energies where the two-body amplitude is even weaker would be useful. Finally, in order to understand the dynamical origin of this medium enhancement, the systematic energy and target dependence of the effect should be measured.

Reference

1. D. R. Giebink, Phys. Rev. C **32**, 502 (1985).
2. C. B. Dover and G. E. Walker, Phys. Rep. **89**, 1 (1982).

3. J.V. Noble, Phys. Rev. Lett. **46**, 412 (1981).

4. T. Goldman and G. J. Stephenson, Jr., Phys. Lett. **146B**, 143 (1984).

5. F. Close, R. Roberts, and G. Ross, Rutherford Appleton Laboratory Report 84-029,(1984).

6. L. Celenza, A. Rosenthal, and C. Shakin, Brooklyn College of CUNY, Report 84/041/123 (1984).

7. P. B. Siegel, W. B. Kaufmann, and W. R. Gibbs, Phys. Rev. C **30**, 1256 (1984); P. B. Siegel, W. B. Kaufmann, and W. R. Gibbs, Phys. Rev. C **31**, 2184 (1985).

8. D. Marlow, P. D. Barnes, N. J. Colella, S. A. Dytman, R .A. Eisenstein. R. Grace, F. Takeutchi, and W.R. Wharton; S. Bart, D. Hancock, R. Hackenberg, E. Hungerford, W. Mayes, L. Pinsky, and T. Williams; R. Chrien, H. Palevsky, and R. Sutter, Phys. Rev. C **25**, 2619 (1982).

9. Y. Mardor, E. Piasetsky, J. Alster, D. Ashery, M. A. Moinester, and A. I. Yavin; S. Bart, R. E. Chrien, P. H. Pile, and R. J. Sutter; R. A. Krauss and J. C. Hiebert; R. L. Stearns; T. Kishimoto; R. R. Johnson and R. Olshevsky, Phys. Rev. Lett. **65**, 2110 (1990).

10. D. V. Ahluwalia and D. J. Ernst, talk presented at this conference; D. V. Ahluwalia and D. J. Ernst, to be submitted to Phys. Rev. D.

11. L.C. Biedenharn and J.D. Louck, *Angular Momentum in Quantum Physics. Encyclopedia of Mathematics and Its Applications* (Addison-Wesley, Reading, Mass., 1981), Vol. 8.

12. D. R. Giebink and D. J. Ernst, Comp. Phys. Comm. **48**, 407 (1988).

13. R. A. Arndt, L. D. Roper and P. II. Steinberg, Phys. Rev. D **18**, 3278 (1978); R. A. Arndt and L. D. Roper, Phys. Rev. D **31**, 2230 (1985).

14. D. J. Ernst, C. M. Shakin, and R. M. Thaler, Phys. Rev. C **8**, 46 (1973).

15. B. R. Martin, Nucl. Phys. **B94**, 413 (1975).

16. M. Beiner, H. Flocard, N. Van Gai and P. Quentin, Nucl. Phys. **A238**, 29 (1975).

17. J. Negele, Phys. Rev. C **1**, 1260 (1970).

18. F. E. Close, R. L. Jaffe, R. G. Roberts, and G. G. Ross, Phys. Rev. D **31**, 1004 (1985).

19. G. E. Brown, C. B. Dover, P. B. Siegel, and W. Weise, Phys. Rev. Lett. **60**, 2723 (1988).

EXACT MONTE CARLO FOR FEW-ELECTRON SYSTEMS

Shiwei Zhang and M.H. Kalos

Laboratory of Atomic and Solid State Physics
and
Center for Theory and Simulation in Science and Engineering

Cornell University
Ithaca, New York 14853, USA

ABSTRACT

We have constructed an exact and stable algorithm for few-electron systems, with which we have determined the energies of an excited state of the He atom and of the ground state of the Li atom. Significant features of the new method include the following: There are equal populations of random walkers that carry positive and negative weights. The positions of walkers are selected from a distribution that couples all walkers using a Green's function with a known form. The correct importance functions that take into account the global interactions of the populations are different for positive and negative walkers. This symmetry breaking is an essential element for long-term stability. A novel transformation of the potential energy facilitates the exact evaluation of the energy E.

PACS numbers: 02.50.+s, 31.15.+q, 05.30.Fk, 02.70.+d

Despite the existence of various methods that yield accurate and useful results[1], the fundamental sign problem in Quantum Monte Carlo for fermion systems remains unsolved. The difficulty arises from the antisymmetry of the solutions to the Schroedinger equation for such systems and from the nature of Monte Carlo (MC) methods. To represent a wavefunction that is antisymmetric under interchange of like spins, an equal number of positive and negative walkers are needed. The usual algorithm, with the two populations asymptotically identical, does not provide a long-term stable solution.

Building on earlier work[2],[3], we propose here an algorithm which implements several new ideas to attack this difficulty. We use Green's function Monte Carlo (GFMC) with a Green's function whose analytical form is known. Also, rather than using independent walkers, we use equal numbers of interacting positive and negative walkers. This interaction, in the dynamical process of sampling, has the effect of a repulsion between the two types of walkers in configuration space. We use different importance functions for positive and negative walkers. This difference breaks the symmetry that otherwise would exist between an ensemble of walkers and an ensemble derived by interchanging positive and negative walkers, and is a key requirement for long-term stability. Finally, we exploit the antisymmetry under interchange of like spins, antithetically smoothing the population. The repulsion, the breaking of the symmetry between positive and negative walkers, and the smoothing permit an asymptotically stable solution. At the end, we give preliminary results of this new approach applied to the He (1s2s) excited state and to the Li ground state.

In 1962, Kalos introduced a way of carrying out GFMC for many particle systems using a Green's function of known analytical form[4]. In atomic units, the Schroedinger equation for a system of N electrons $\mathbf{x}_1, \mathbf{x}_2, ..., \mathbf{x}_N$ is:

$$[-\sum_{i=1}^{N}\frac{1}{2}\nabla_i^2 + V(\mathbf{x}_1, \mathbf{x}_2, ..., \mathbf{x}_N)]\Phi(\mathbf{x}_1, \mathbf{x}_2, ..., \mathbf{x}_N) = E\Phi(\mathbf{x}_1, \mathbf{x}_2, ..., \mathbf{x}_N). \quad (1)$$

Each $\mathbf{x}_i$ denotes the position of an electron in 3D real space. We assume the energy of the system, E, to be negative, as is true in most cases, and rescale as

follows:
$$\mathbf{r}_i = \sqrt{2|E|}\mathbf{x}_i. \qquad (2)$$

For the Coulombic systems we are dealing with, V scales as $1/|\mathbf{x}|$ and we have:
$$V(\mathbf{x}_1, \mathbf{x}_2, ..., \mathbf{x}_N) = \sqrt{2|E|}V(\mathbf{r}_1, \mathbf{r}_2, ..., \mathbf{r}_N). \qquad (3)$$

Now (1) reduces to
$$(-\nabla_\mathbf{R}^2 + 1)\psi(\mathbf{R}) = -\sqrt{\frac{2}{|E|}}V(\mathbf{R})\psi(\mathbf{R}). \qquad (4)$$

Here $\mathbf{R}$ is understood to denote the $3N$ dimensional vector $(\mathbf{r}_1, \mathbf{r}_2, ..., \mathbf{r}_N)$ and $\psi(\mathbf{R})$ stands for the original wavefunction $\Phi(\mathbf{x}_1, \mathbf{x}_2, ..., \mathbf{x}_N)$ expressed in terms of $\mathbf{R}$.

With the knowledge of $\psi(\mathbf{R})$ the eigenvalue of the system can be obtained from:
$$\sqrt{\frac{2}{|E|}} = \frac{\int d\mathbf{R}\,\psi_T(\mathbf{R})\left[-\nabla_\mathbf{R}^2 + 1\right]\psi(\mathbf{R})}{\int d\mathbf{R}\,\psi_T(\mathbf{R})\left[-V(\mathbf{R})\right]\psi(\mathbf{R})}, \qquad (5)$$

where $\psi_T(\mathbf{R})$ is a trial wavefunction for the system. In the random walk process, an ensemble of configurations (i.e., random walkers) that represent $\psi(\mathbf{R})$ will be sampled. Therefore, the novel transformation of the potential energy described above permits the exact evaluation of the energy E without knowing the length scaling factor in (2). The integrals in (5) are evaluated as sums over the positions of the walkers.

Let $\lambda = \sqrt{\frac{2}{|E|}}$. By using Green's function for the operator $(-\nabla_\mathbf{R}^2 + 1)$, we can transform (4) into the following integral equation:
$$\psi(\mathbf{R}) = \lambda \int d\mathbf{R}' G(\mathbf{R}, \mathbf{R}')[-V(\mathbf{R}')]\psi(\mathbf{R}'). \qquad (6)$$

Explicitly,
$$G(\mathbf{R}, \mathbf{R}') = (2\pi)^{-3N/2} K_{3N/2-1}(|\mathbf{R} - \mathbf{R}'|)|\mathbf{R} - \mathbf{R}'|^{1-3N/2}, \qquad (7)$$

where K_ν is the Bessel function of imaginary argument. In principle, equation (6) can be solved by iteration. The solution ψ corresponding to the lowest λ, i.e., the state with the lowest E, will eventually be projected out.

The sign problem will arise for two reasons: (i) the requirement of antisymmetry and, (ii) negative values of the potential term $-V$. We now describe our modification to the conventional MC procedure to solve the integral equation for the lowest energy state with antisymmetry.

Write
$$\psi(\mathbf{R}) = \psi^+(\mathbf{R}) - \psi^-(\mathbf{R}), \qquad (8)$$
where $\psi^\pm(\mathbf{R})$ map into each other by any odd permutation of like spins. In the usual MC approach, $\psi^\pm(\mathbf{R})$ separately satisfy (6), and when found by solving that equation, their asymptotic forms are the same. This leads to the decay of signal to noise characteristic of the sign problem. We use ψ^+ to denote the positive part of the antisymmetric wavefunction and ψ^- the absolute value of the negative part and rewrite (6) in a coupled way:

$$\psi^\pm(\mathbf{R}) = \lambda \max\left\{0, \pm \int d\mathbf{R}' G(\mathbf{R}, \mathbf{R}')[-V(\mathbf{R}')]\left(\psi^+(\mathbf{R}') - \psi^-(\mathbf{R}')\right)\right\}. \qquad (9)$$

The difference $(\psi^+ - \psi^-)$ satisfies Eq (6), as expected. We now introduce a transformed wavefunction

$$\tilde{\psi}^\pm(\mathbf{R}) = V'(\mathbf{R}) i^\pm(\mathbf{R}) \psi^\pm(\mathbf{R}), \qquad (10)$$

where $V'(\mathbf{R})$ is positive and symmetric and $i^\pm(\mathbf{R})$ map into each other like $\psi^\pm(\mathbf{R})$. Letting $\tilde{G}(\mathbf{R},\mathbf{R}') = V'(\mathbf{R})G(\mathbf{R},\mathbf{R}')$ and $W(\mathbf{R}) = -V(\mathbf{R})/V'(\mathbf{R})$, we can write (9) in terms of the transformed wavefunction:

$$\tilde{\psi}^\pm(\mathbf{R}) = \lambda i^\pm(\mathbf{R}) \max\left\{0, \pm \int d\mathbf{R}' \tilde{G}(\mathbf{R},\mathbf{R}') W(\mathbf{R}') \left[\frac{\tilde{\psi}^+(\mathbf{R}')}{i^+(\mathbf{R}')} - \frac{\tilde{\psi}^-(\mathbf{R}')}{i^-(\mathbf{R}')}\right]\right\}. \qquad (11)$$

In carrying out the MC sampling, we will need to treat $W(\mathbf{R}')$ in (11) as a weight factor for walkers. But the Coulomb potential is singular at certain points of the configuration space. We can regulate the behavior of the weights by choosing proper forms for V'. To illustrate this better, we consider an atom whose nucleus is fixed at the origin with charge Z. Let $\mathbf{r}_i$ ($i = 1, 2, ..., Z$) be the positions of the Z electrons. The potential energy for the system can be written as:

$$V(\mathbf{R}) = -\sum_i \frac{Z}{|\mathbf{r}_i|} + \sum_{i<j} \frac{1}{|\mathbf{r}_i - \mathbf{r}_j|}. \qquad (12)$$

Choose, for example,

$$V'(\mathbf{R}) = \sum_i \frac{Z}{|\mathbf{r}_i|} + \sum_{i<j} \frac{1}{|\mathbf{r}_i - \mathbf{r}_j|}. \tag{13}$$

Now the weight W in (11) satisfies $|W(\mathbf{R}')| < 1$, and the kernel of the form $\tilde{G}_i(\mathbf{R},\mathbf{R}') = \frac{1}{|\mathbf{r}_i|}G(\mathbf{R},\mathbf{R}')$ can be sampled directly[5]. (The other kernel $\tilde{G}_{ij} = \frac{1}{|\mathbf{r}_i - \mathbf{r}_j|}G$ can be transformed to the same form by a spatial rotation in $3N$ dimensions.)

In the MC process, the solution to the integral equation, i.e., the wavefunction $\tilde{\psi} = \tilde{\psi}^+ - \tilde{\psi}^-$, is represented by a collection of equal populations of positive and negative walkers, with each walker being a certain configuration of all electrons in the system. The function i has the effect of a one-body force which acts differently on positive and negative walkers. This serves to break the symmetry that exists between them; otherwise, positive and negative walkers can be exchanged yielding a distribution that satisfies the same integral equation. Asymptotically the two populations would be equal and they would cancel. The importance function $i(\mathbf{R})$ gives the overlap with an antisymmetric test function of the distribution deriving from a walker starting at $\mathbf{R}$. Qualitatively, the importance function $i^+(\mathbf{R})$ (> 0) for positive walkers will have the effect of biasing the walker toward regions in configuration space where the wavefunction is positive. Where the wavefunction is likely to be negative, it falls off because of the repulsion by negative walkers. The importance function for negative walkers, according to (10), is also positive and is related to $i^+(\mathbf{R})$ by $i^-(\mathbf{R}) = i^+(P\mathbf{R})$, where P denotes an odd permutation of like spins. In our He excited state and Li ground state simulations, the importance function was assumed to take a simple form:

$$i^+(\mathbf{R}) \propto \frac{1 + ar_2 e^{-\alpha r_2}}{1 + be^{-\beta(r_2 - r_1)}}, \tag{14}$$

where r_1 and r_2 are the distances between the nucleus and the two like electrons respectively, while a, b, α and β are adjustable parameters. We emphasize that the choice of the actual functional form of i is somewhat arbitrary. The importance function is used primarily to filter out preferred positions for walkers. It is desirable that we build into the importance function information from an antisymmetric trial wavefunction. But its quantitative properties are not essential to the stability nor to the expected eigenvalue of the algorithm.

The idea of correlating walkers of opposite sign was first introduced by Arnow,et al.[6], who used an interaction between pairs of positive and negative walkers. For problems in higher dimensions, the "repulsion" needs to be long-ranged and universal. Therefore in our algorithm, every walker interacts with the rest of the ensemble of walkers. In addition, we maintain constant and equal numbers of positive and negative walkers. Using the MC representation of the antisymmetric solution, we have:

$$\tilde{\psi}^{\pm}(\mathbf{R}) = \sum_{k=1}^{M} \delta(\mathbf{R} - \mathbf{R}_k^{\pm}), \tag{15}$$

where $\mathbf{R}_k^+$ and $\mathbf{R}_k^-$ are the positions of positive and negative walkers in configuration space. Let

$$K(\mathbf{R}, \{\mathbf{R}'\}) = \sum_{k=1}^{M} \left[\tilde{G}(\mathbf{R}, \mathbf{R}_k^+) \frac{W(\mathbf{R}_k^+)}{i^+(\mathbf{R}_k^+)} - \tilde{G}(\mathbf{R}, \mathbf{R}_k^-) \frac{W(\mathbf{R}_k^-)}{i^-(\mathbf{R}_k^-)} \right], \tag{16}$$

where $\{\mathbf{R}'\}$ denotes the current population, i.e., the collection of M positive walkers $\mathbf{R}_k^+$ and M negative walkers $\mathbf{R}_k^-$ ($k = 1, 2, ...M$). Combining (11) and (15) with (16), we now rewrite the integral equation so that it resembles more closely the actual MC process:

$$\tilde{\psi}(\mathbf{R}) = \begin{cases} \tilde{\psi}^+(\mathbf{R}) \propto i^+(\mathbf{R}) K(\mathbf{R}, \{\mathbf{R}'\}), & \text{if } K > 0; \\ -\tilde{\psi}^-(\mathbf{R}) \propto i^-(\mathbf{R}) K(\mathbf{R}, \{\mathbf{R}'\}), & \text{if } K < 0. \end{cases} \tag{17}$$

Here $\tilde{\psi}(\mathbf{R})$ represents the distribution of the next generation. Therefore, samples for $\tilde{\psi}^+(\mathbf{R})$, i.e., positive walkers for the next generation, are produced when $K(\mathbf{R}, \{\mathbf{R}'\})$ is positive. The actual sampling of the kernel K, i.e., the selection of positive and negative walkers from the current population according to the function $K(\mathbf{R}, \{\mathbf{R}'\})$, can be accomplished by a rejection technique: First sample new walkers from a probability density distribution,

$$f(\mathbf{R}, \{\mathbf{R}'\}) \propto K'(\mathbf{R}, \{\mathbf{R}'\}) \equiv \sum_{k=1}^{M} \left[\left| \tilde{G}(\mathbf{R}, \mathbf{R}_k^+) \frac{W(\mathbf{R}_k^+)}{i^+(\mathbf{R}_k^+)} \right| + \left| \tilde{G}(\mathbf{R}, \mathbf{R}_k^-) \frac{W(\mathbf{R}_k^-)}{i^-(\mathbf{R}_k^-)} \right| \right]; \tag{18}$$

Then use $p = |K(\mathbf{R}, \{\mathbf{R}'\})|/K'(\mathbf{R}, \{\mathbf{R}'\})$ to accept. We have $p \leq 1$, as required. The sign of K determines the sign of the new walker. The importance functions $i^{\pm}$ in (17) can be dealt with by another rejection using an upper bound to them.

The next key ingredient of the new algorithm is to smooth the population by permutation to help stabilization. A positive walker becomes a negative one under an odd permutation of like spins, and vice versa. This smoothing, when done randomly, prevents a local build up in the population and therefore maintains the stability. In our sampling of the He excited state and the Li ground state, we did the following: Each MC step produces lists of positive and negative walkers in random order as described by (18). After the step, corresponding entries in the lists are interchanged with probability 1/2 followed by the permutation of positions of the two electrons with like spins in each walker.

Computer programs were written to apply this algorithm to the He $1s2s$ triplet state and to the Li ground state. Needless to say, our interest was not in competing for numerical precision, but in testing and understanding the algorithm. In both cases, it remained stable throughout the simulations. In fact, for Li, we experimented with pure noise as an initial distribution and only 40 walkers for each population. The two populations were quickly separated by the algorithm and converged to a satisfactory pair of distributions.

We used the functional form described in equation (13) as our V'. For the importance function (14), several sets of parameters were tried. For the He excited state, we used $a = 8$, $\alpha = 0.15$, $b = 2$ and $\beta = 2$. We scaled i such that its maximum value approached 1 and therefore it could be used directly in a rejection method. The population was mixed at random at the end of each MC step, as described above. With an ensemble of walkers $\{\mathbf{R}_k\}$ that samples $\tilde{\psi}$ and a trial wavefunction $\psi_T(\mathbf{R})$, the energy eigenvalue E for the system can be easily calculated from (5):

$$\sqrt{\frac{2}{|E|}} = \frac{\sum_{k=1}^{M}(-\nabla^2+1)\psi_T/V'i^+|_{\mathbf{R}_k^+} - \sum_{k=1}^{M}(-\nabla^2+1)\psi_T/V'i^-|_{\mathbf{R}_k^-}}{\sum_{k=1}^{M}\psi_T W/i^+|_{\mathbf{R}_k^+} - \sum_{k=1}^{M}\psi_T W/i^-|_{\mathbf{R}_k^-}} \quad (19)$$

In testing the method, we first deliberately used very poor trial wavefunctions. In both cases, the results agreed well with the experimental values. For example, with a trial wavefunction $\psi_T(\mathbf{R}) = (r_2 - r_1)e^{-\lambda(r_2-r_1)^2}$, we computed the eigenvalue for the He 1s2s triplet and after extrapolating to infinite population size to account for MC bias, obtained $E = -2.1752 \pm 0.0013$H, compared with the actual energy $E = -2.175229$H[7]. We then used as trial wavefunction a product of a simple determinant and a simple Jastrow factor[8]. For the same state of the He atom, the computed energy eigenvalue was -2.175243 ± 0.000066H. For the ground state of Li, with a similar type of trial wavefunction, we obtained an energy of -7.47760 ± 0.00060H, in good agreement with the estimate by Veillard and Clementi of -7.4781H[9].

In summary, we have developed an algorithm for exact fermion GFMC calculations that works well for few-electron systems. Its scaling behavior remains to be elucidated. We comment here that the scaling can be improved by using convolutions of Green's functions to couple the walkers. Since $G(\mathbf{R}, \mathbf{R}')$ in Eq. (7) can be written as a superposition of gaussians[10], this will be technically possible to do.

We thank C.J. Umrigar, K.J. Runge, and S.A. Vitiello for helpful discussions. One of us (MHK) wishes to express his gratitude to Michel Moreau and Claire Lhuillier for their hospitality and support during a visit to the Laboratoire de Physique Théorique des Liquides of the Université Pierre et Marie Curie (Jussieu), where some of the work was undertaken. The Cornell Theory Center is funded by the U.S. National Science Foundation, by New York State, by IBM, and by Cornell University.

REFERENCES

1. K.E. Schmidt and M.H. Kalos, in *Applications of the Monte Carlo Method in Statistical Physics*, ed. by K. Binder (Springer Verlag 1984).
2. M.H. Kalos, in *Computational Atomic and Nuclear Physics*, ed. by C. Bottcher, M.R. Strayer and J.B. McGrory (World Scientific 1989).
3. M.H. Kalos, in press, J. Stat. Phys.; Cornell Theory Center Technical Note (CTC91TR50).
4. M.H. Kalos, Phys. Rev. $\underline{128}$, 1791, (1962).
5. M.H. Kalos, J. Comp. Phys. $\underline{2}$, 257, (1967).
6. D.M. Arnow, M.H. Kalos, M.A. Lee and K.E. Schmidt, J. Chem. Phys. $\underline{77}$, 5562 (1982).
7. C.J. Umrigar, K.G. Wilson and J.W. Wilkins in *Computer Simulation Studies in Condensed Matter Physics: Recent Developments*, ed. by D.P. Landau, K.K. Mon and H.B. Schuttler (Springer Verlag 1988).
8. C.J. Umrigar, private communication.
9. K.E. Schmidt and J.W. Moskowitz, J. Chem. Phys. $\underline{93}$, 4172 (1990).
10. G.N. Watson, "Treatise on the Theory of Bessel Functions", 2nd edition, p. 183. Cambridge Univ. Press, New York (1945).

MANY-BODY CALCULATIONS OF PHOTOIONIZATION CROSS SECTIONS

Hugh P. Kelly
J.W. Beams Laboratory of Physics, University of Virginia
Charlottesville, VA 22901

ABSTRACT

The calculation of photoionization cross sections of atoms is discussed in the framework of many-body perturbation theory. The use of appropriate coupled equations allows the summation to infinite order of certain classes of diagrams and can account for resonance structure in the cross sections. The calculations are carried out using complete sets of single-particle states with discretization of the continuum. Examples are given for the process of photoionization with excitation for helium and argon atoms. The cross section of atomic tungsten is calculated as an example of a complex, open shell atom. As an example of an alternative method, an extensive R-matrix calculation for the 1s photoionization cross section of beryllium is presented.

INTRODUCTION

There is considerable experimental and theoretical interest[1] in photoionization cross sections in which a photon is absorbed by a quantum system and an electron is ejected. In fact, two or more electrons may be ejected, and the quantum system can also experience excitation of electrons to excited bound levels in addition to the ejection of one or more electrons. The calculation of photoionization cross sections presents a challenge, since the effects of electron correlations must in general be included to obtain good agreement with experiment. An excellent review of theoretical methods has been given by Starace.[2] Methods which include effects of electron correlations are the random phase approximation with exchange (RPAE)[3,4] and the relativistic random phase approximation (RRPA);[5] the time-dependent local density approximation (TDLDA)[6] which solves the RPA equations, with orbitals obtained from a local density approximation (LDA) and calculated in a potential which is asymptotically zero rather than coulombic; the R-matrix method[7,8] which has been used both for photoionization and electron scattering cross sections and has been very effective for open shell atoms. Recently, successful calculations have been carried out using the multiconfiguration Hartree-Fock method (MCHF)[9,10] and also using configuration interaction (CI) with relativistic orbitals.[11] In this paper the focus is mainly on the use of many-body perturbation theory (MBPT)[12]. An example will also be given of the use of the R-matrix method to calculate the 1s cross sections for the beryllium atom.

In employing MBPT we employ the relation between the photoionization cross section $\sigma(\omega)$ and the frequency-dependent polarizability $\alpha(\omega)$[13]

$$\sigma(\omega) = 4\pi \frac{\omega}{c} \alpha(\omega) \quad , \tag{1}$$

where

$$\alpha(\omega) = -\sum_k |\langle \Psi_k | \sum_{i=1}^{N} z_i | \Psi_0 \rangle|^2 \times \left\{ \frac{1}{E_0 - E_k + \omega} + \frac{1}{E_0 - E_k - \omega} \right\} \quad , \tag{2}$$

where $|\Psi_0\rangle$ is the exact initial many particle state and $|\Psi_k\rangle$ is an exact many-particle excited state. Atomic units are used throughout. A small positive imaginary part $i\eta$ is added to the second denominator which can vanish.

Using the relation

$$\lim_{\eta \to 0^+} (D+i\eta)^{-1} - PD^{-1} - i\pi\delta(D) \quad , \tag{3}$$

where P represents principal value integration, and replacing Σ_k by $\int dk$, $\sigma(\omega)$ is proportional to $|\langle \Psi_k | \sum_i z_i | \Psi_0 \rangle|^2$. We then use the standard many-body perturbation expansion for $\alpha(\omega)$ or for $\langle \Psi_k | \Sigma z_i | \Psi_0 \rangle$. There are terms for $\alpha(\omega)$ representing normalization correction[12] which are not included in the perturbation expansion for $\langle \Psi_k | \Sigma z_i | \Psi_0 \rangle$.

The perturbation calculations are carried out by choosing a single-particle potential $V(r)$ and calculating a "complete set" of single-particle states which satisfy

$$(T+V)\phi_n = e_n \phi_n \quad . \tag{4}$$

In practice, the number of orbital angular momentum values is limited (usually approximately to 4) and each channel contains a finite number of bound states (approximately 10) and a finite number of continuum states (approximately 40) since the continuum is discretized. The "length" form of the dipole matrix element $\langle \Psi_k | \Sigma z_i | \Psi_0 \rangle$ is related to the "velocity" form $\langle \Psi_k | \Sigma d/dz_i | \Psi_0 \rangle$, by

$$\langle \Psi_k | \Sigma z_i | \Psi_0 \rangle - (E_0 - E_k)^{-1} \langle \Psi_k | \Sigma d/dZ_i | \Psi_0 \rangle \quad , \tag{5}$$

where Ψ_k and Ψ_0 are eigenstates of H with eigenvalues E_k and E_0.

In calculating $\mathrm{Im}\,\alpha(\omega)$ or $\langle \Psi_k | \Sigma z_i | \Psi_0 \rangle$ starting from single-particle states, the perturbation is

$$H' - \sum_{i<j} v_{ij} - \sum_i V_i \quad , \tag{6}$$

where $v_{ij} = e^2/r_{ij}$ and V_i is the single-particle potential of Eq. (4).

It is customary to choose V to be the usual Hartree-Fock potential R_{HF} for orbitals ϕ_n in Φ_0, the unperturbed many-particle state. The general potential can be written[14]

$$V - R_{HF} + (1-P)\Omega(1-P) \quad , \tag{7}$$

where Ω is an arbitrary Hermitian operator which is chosen so that excited orbitals are calculated in an appropriate potential,[12] and P in Eq. (7) is $\sum_i |n_i\rangle\langle n_i|$ where n_i are the orbitals occupied in Φ_0.

Low-order diagrams in the perturbation expansion of $\langle \Psi_k | \Sigma z_i | \Psi_0 \rangle$ are shown in Fig. 1 for a transition in which an electron in orbital p is excited to orbital k. The diagram of Fig. 1(a) is the lowest-order contribution in perturbation theory and is given by $\langle k|z|p \rangle$. In the next order are the ground state correlation diagram (b) and the final state correlation diagram (c). The dashed line ending in a dot represents the dipole operator. The other dashed lines represent coulomb interactions from H'. Diagrams (d) and (e) are some of the diagrams second-order in H' and are included in the random phase approximation formalisms. Diagrams are read from bottom to top corresponding to right to left in $\langle \Psi_k | \Sigma z_i | \Psi_0 \rangle$. Exchange diagrams also included. The expression for diagram (c) is

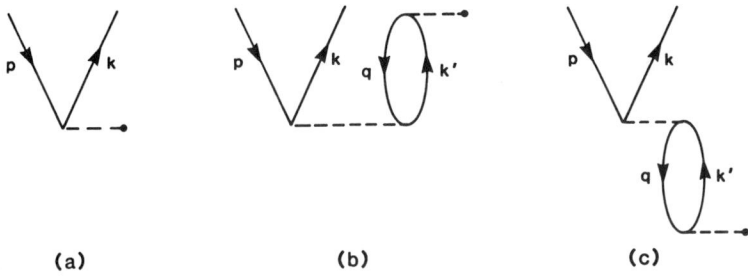

(a)   (b)   (c)

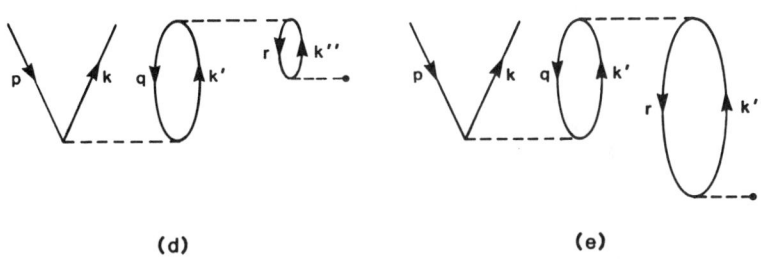

(d)   (e)

Fig. 1. Low-order diagrams for $<\Psi_k|\Sigma z_i|\Psi_0>$. Electron in orbital p is excited to orbital k. Dashed line ending in dot represents dipole operator. Other dashed lines represent coulomb interaction with H'.

$$\lim_{\eta \to 0^+} \sum_{k'} \frac{<kq|v|pk'><k'|z|q>}{e_q - e_{k'} + \omega + i\eta} , \quad (8)$$

where ω is the photon energy present in the initial state.

Excitations from different channels may be coupled by use of the coupled-equations method[15] which is a K-matrix evaluation[2] and is related to the RPAE methods. In this way certain diagrams are summed to all orders in perturbation theory. The coupled equations are indicated in Fig. 2, where the double line represents a correlated dipole matrix element. In the next section two examples are given of the use of this method.

PHOTOIONIZATION WITH EXCITATION

In this section two examples of the process of photoionization with excitation are discussed. The first of these is photoionization of $He(1s)^2$ with the resulting ion left in the $He^+(n=2)$ level or a $He^+(n=3)$ level. An interesting feature of the $He^+(n=2)$ cross section is the large resonance structure measured by Woodruff and Samson[16] and by Lindle et al.[17] Calculations were made using MBPT by Salomonson et al[18] who used the coupled equations method[15] for final state correlations and pair functions[19] to account for initial state correlations.

Nine final state channels were included: 1sεp, 2sεp, 2pεd, 2pεs, 3sεp, 3pεs, 3pεd, 3dεp, and 3dεf. In each channel ε was represented by eight bound orbitals and forty-two continuum orbitals with wave vectors in the range 0.05-14.0 a.u.

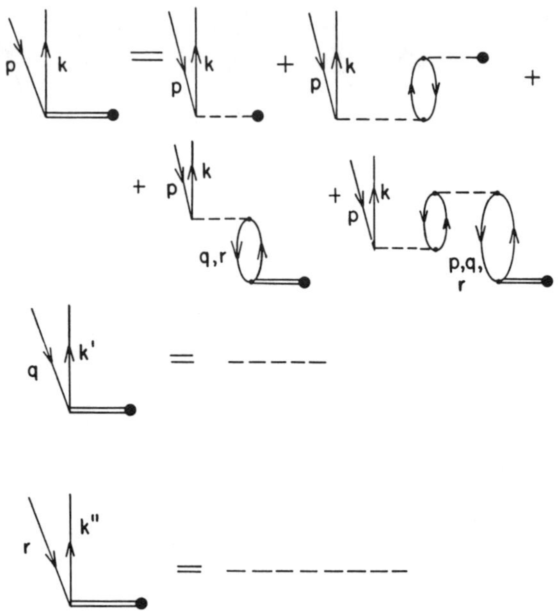

Fig. 2. Diagrammatic representation of the coupled equations methods. The double line represents a correlated dipole matrix element.

Correlations in the initial state were accounted for by use of pair functions[19] which provide an essentially exact description of a bound two-electron system, depending on the number of partial waves included (up to $\ell=6$ in this calculation). In each of the nine channels there are fifty unknown correlated dipole matrix elements (corresponding to the 8 bound states and 42 continuum states) making a total of 450 complex correlated dipole matrix elements to be determined. The coupled equations may be written as a matrix equation with a 450 x 450 matrix to be inverted for each value of photon energy ω (approximately 1000 values of ω).

Results for the He^+ (n=2) cross section are shown in Fig. 3 and are compared with experiment. The striking effects of resonance structure are reproduced. For example, the large resonance near 70eV is due to the excited configuration 3s3p which is degenerate with the configuration He^+ (n=2) and an electron in the continuum.

The second example of photoionization with excitation is for the argon atom $1s^22s^22p^63s^23p^6(^1S)$. This calculation[20] included single electron excitations leaving the ion in the levels $3s^23p^5$ and $3s3p^6$. The excited ionic levels were $3s^23p^4(^1D)md(^2S)$ with m=3, 4, 5, and $3s^23p^4(^3P, {}^1D, {}^1S)4p(^2P)$. Low-order diagrams for photoionization with excitation are shown in Fig. 4 with r representing a bound excited state and k a continuum orbital. Sums over intermediate states k' and k" include all excited states, bound and continuum. Diagrams (a)-(c) are the lowest order final state correlation diagrams and (d) and (e) represent correlations in the initial state. Diagrams (f) and (g) are higher-order diagrams which are second-order in H'. Interactions among all final state channels were included by the coupled equations methods.[15] Diagram (c) with the dipole 3s → k corresponds to mixing in the final ionic state between the configurations $3s3p^6(^2S)$ and $3s^23p^4(^1D)md(^2S)$. This mixing was accounted for by a configuration interaction (CI) calculation using 10 md orbitals (m = 3-12).

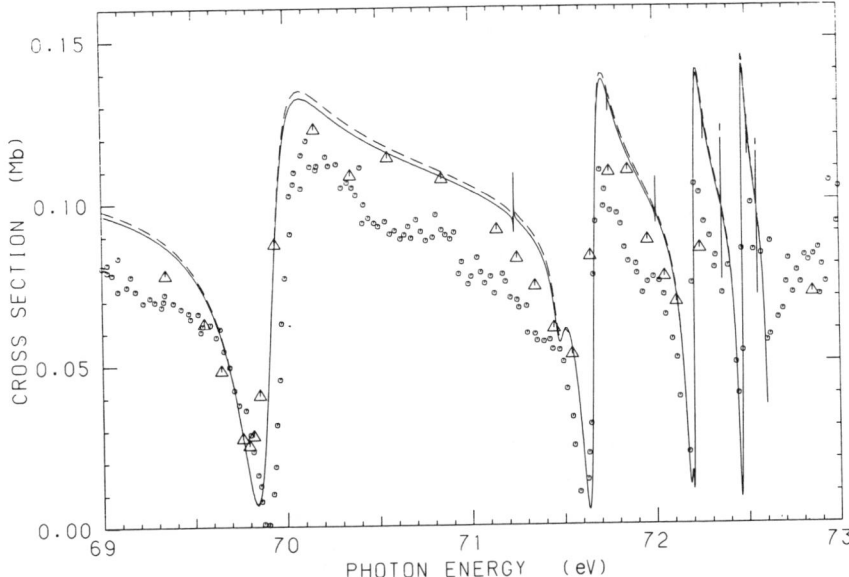

Fig. 3. The total photoionization cross section for leaving He$^+$ in the n=2 level. the full (broken) curve is the calculation by Salomonson et al, reference 18, using the length (velocity) form of the dipole operator. The open circles are experimental results by Woodruff and Samson, reference 16. Triangles are experimental points by Lindle et al, reference 17.

The results for the 3s → ϵp cross section are shown in Fig. 5 and are compared with experimental results.[21-23] The cross section differs markedly from the lowest-order Hartree-Fock results which rise monotonically from threshold. The structure near 35 eV is due to resonances with double-electron excitations of the configurations $3s^23p^4npn's$ or $3s^23p^4ndn'p$. This cross section including correlations was first calculated by Amusia et al[24] by the RPAE method and is in generally good agreement with the calculation of Fig. 5. Other methods such as the R-matrix approach have also given good agreement for the 3s cross section[8]. The calculation by Wijesundera and Kelly[20] is the only one thus far to include effects of double electron resonances on the 3s cross section of argon.

Calculated results for the $3s^23p^4(^1D)md(^2S)\epsilon p$(m=3,4,5) cross sections are presented in Fig. 6 and compared with experimental results measured by Becker et al,[25] Kossmann et al,[26] and Samson et al.[27] The solid dot at 77eV for the 4d cross section was obtained by Samson et al[27] who showed that the previous measurement[26] of the $3p^4(^1D)4d(^2S)$ lines in the photoelectron spectrum contains a contribution of approximately 25% from the $3p^4(^3P)5d(^2D,^2P)$ lines. The shape of the 3d(^2S) cross section is similar to that of the $3s3p^6(^2S)$ cross section of Figure 5 and is interpreted as due to the fact that the coupling with the $3s3p^6\epsilon p$ channel due to the ionic configuration mixing is driving the 3d(^2S)ϵp channel.[28] The 4d and 5d cross sections are expected to show the shape of the $3s3p^6$ cross section if the oscillator strengths are extended into the region of bound Rydberg states.

There is a need to carry out much more extensive calculations, coupling more channels in order to untangle the cross sections corresponding to the many possible ionic final states. If, for example, the calculation included 100 channels, we would be

136 Photoionization Cross Sections

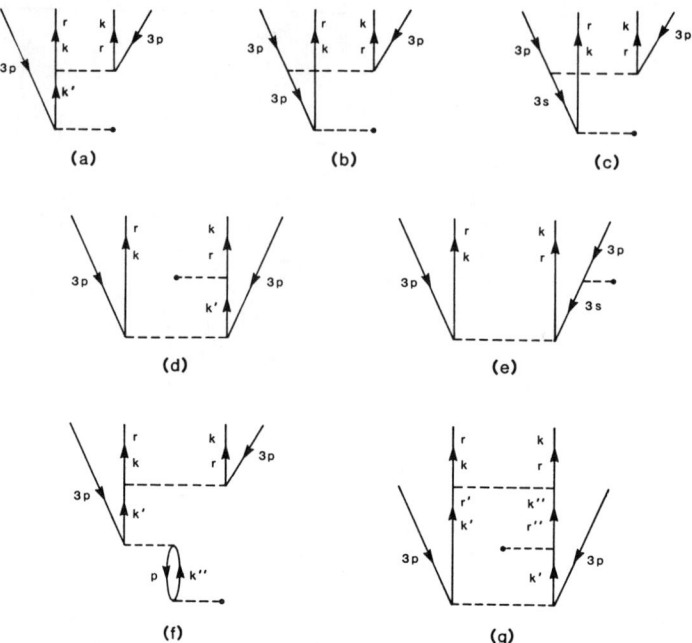

Fig. 4. Low-order diagrams contributing to photoionization with excitation for the argon atom.

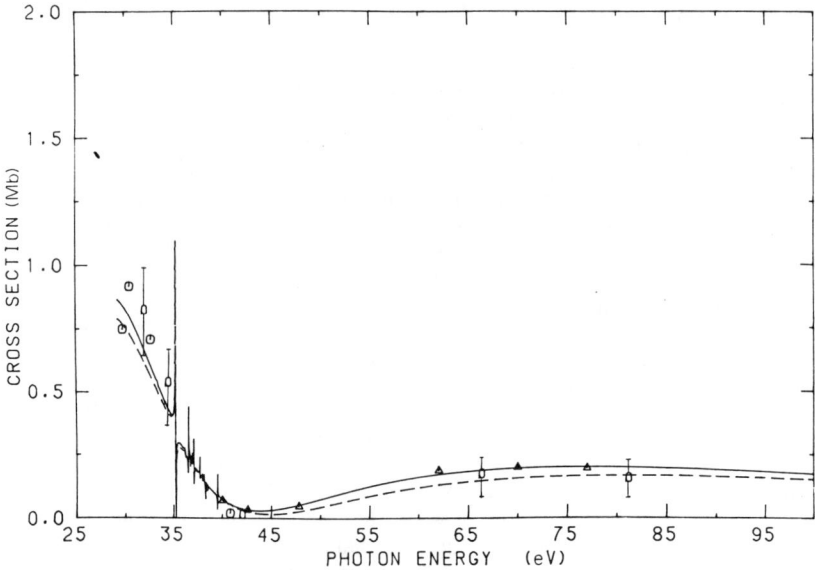

Fig. 5. Argon 3s photoionization cross section. Calculations by Wijesundera and Kelly, reference 20, with solid (dashed) line representing length (velocity) formalism. Experimental results: ▲, Adam et al, ref. 21; ⊙, Samson and Gardner, ref. 22; 0, Houlgate et al, ref. 23.

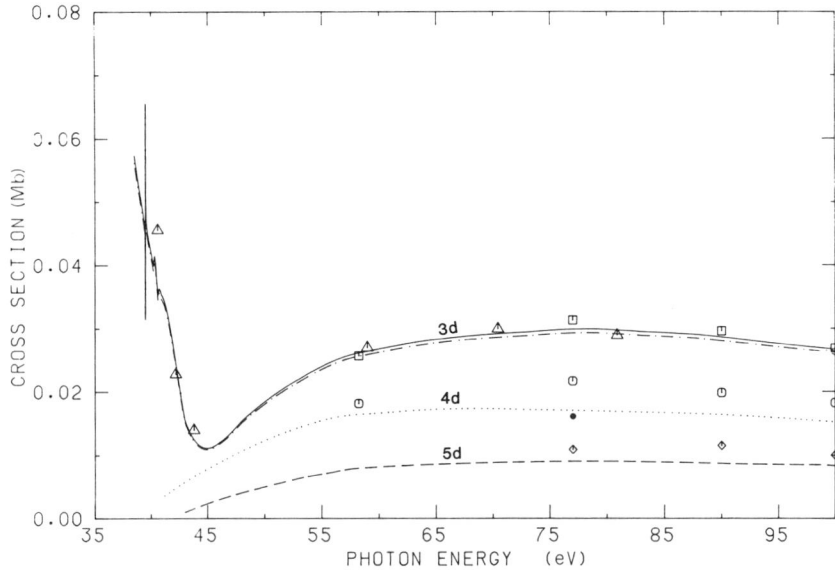

Fig. 6. Argon photoionization cross sections of $3s^2 3p^4(^1D)md(^2S)$ satellites. —(—·—) 3d calculations of reference 20 in length (velocity) formalism. ··· and --- are 4d and 5d calculations of ref. 20 in the length formalism. Experimental results: △, 3d cross section, ref. 25; □, ⊙, ◊ are 3d, 4d, 5d cross sections, ref. 26. ●, 4d cross section measured in ref. 27.

inverting 5,000 by 5,000 matrices and this will require extensive computing power. Future calculations will also use relativistic orbitals and jj-coupling, resulting in a significant further increase in the number of channels.

PHOTOIONIZATION OF TUNGSTEN

In order to test the applicability of MBPT to complicated open-shell systems, calculations have recently been carried out by Boyle et al[29] for the photoionization cross section of atomic tungsten for which recent experimental data were obtained by Kennedy et al.[30] The ground state of atomic tungsten is $5p^6 4f^{14} 5d^4 6s^2 (^5D_0)$. In the photon range 15 to 100 eV, the photoionization cross section is due mainly to the 5d subshell cross section, and large resonance structure is expected due to 5p → 5d excitations. In LS-coupling, there are ionic cores $5d^3(^4P)$ and $5d^3(^4F)$. Interaction among final-state channels based on these ionic states was accounted for by use of the effective potential of Qian et al.[31]

The resonance states of the configuration $5p^5 5d^5$ were obtained by diagonalizing the spin-orbit Hamiltonian using thirty-two LSJ-coupled states. The spin-orbit zeta parameters were calculated with the relativistic code of Cowan.[32] Resonance states due to 4f → 5d excitations were also included, but these proved to be less important than the 5p → 5d excitations. Ground state and final-state correlations were included by calculating many-body diagrams. Calculations were first carried out for the $5d^4(^5D_{J=0})$ state omitting spin-orbit effects for $5d^4$. A second calculation used a ground state for $5d^4$ in which the spin-orbit Hamiltonian had been diagonalized among the $5d^4$

138 Photoionization Cross Sections

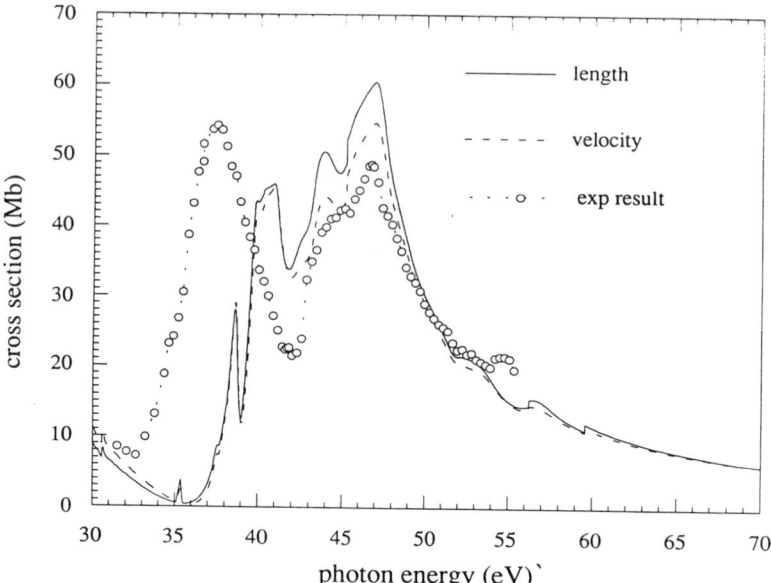

Fig. 7. Photoionization cross section of the tungsten atom in the region of 5p→5d resonances. Solid (dashed) line is calculation by Boyle et al, reference 29. o, experimental results by Costello et al, reference 30.

multiplets $^5D, ^3P, ^3P, ^1S, ^1S$ all coupled to J=0. Calculations using this ground state are shown in Figure 7 and compared with experimental results[30] given by open circles. The large experimental peak near 37eV is not reproduced well by the calculations. The first type of calculation omitting spin-orbit effects in the initial state shows a deeper dip at 42 eV but is in poorer agreement with experiment beyond the peak at 47 eV.

Ideally, the resonances would have been calculated using the coupled equations method. However, in this case there are seventy-eight LSJ resonance states (including both 5p → 5d and 4f → 5d excitations) and forty-three possible decay channels. The excited spectrum of each channel was represented by forty-two continuum states. Therefore, the coupled equations matrix is of dimension 43 x 42 + 78 and must be inverted for approximately 1600 ω-values. The calculation was simplified using the "generalized resonance" method of Garvin[33] in which the energy denominator in the coupling between excited channels includes only the $-i\pi\delta(D)$ term of Eq. (3). This results in the inversion of a matrix of dimension forty-three (corresponding to the 43 final channels) for each ω value.

Because of the possibility that the tungsten atoms in the laser-ablation experiment of Kennedy et al[30] are not all in the lowest J level, calculations were carried out for all J-values of $5d^4(^5D_J)$ from J=0 to 4 and weighted by 2J+1. The resulting curve does not improve agreement with the experiment of Kennedy et al but does improve agreement with results of an experiment on tungsten metal. The main effect is a broadening and reduction of the resonance peak at 47eV.

R-MATRIX CALCULATIONS

The R-matrix method, developed in nuclear physics by Wigner[34] and Wigner and Eisenbud[35], was adapted to atomic physics by Burke and coworkers[7,8] and has been very successful in calculating atomic and molecular processes. As an example, in this

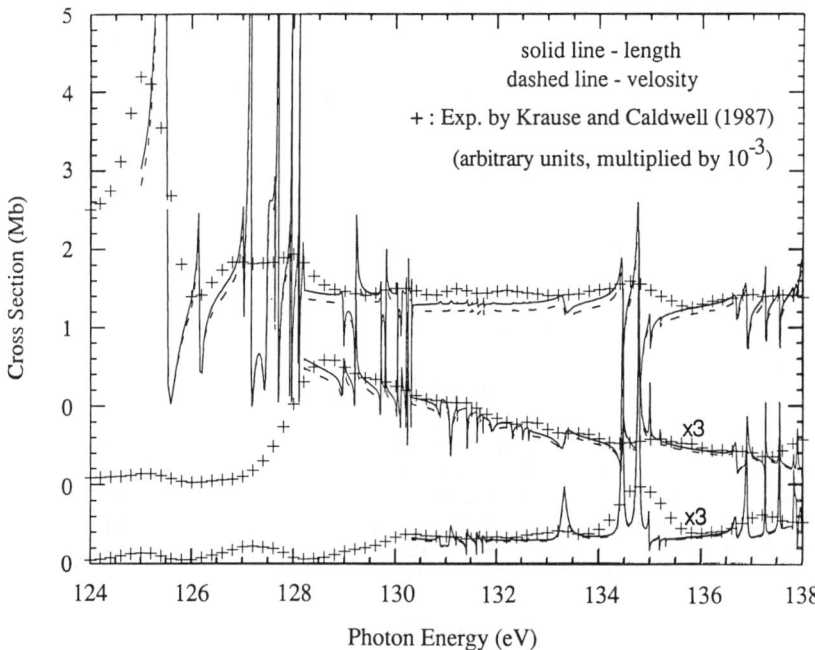

Fig.8. Photoionization of the neutral berylluim atom. Solid(dashed)lines are length (velocity) calculations using the R-matrix method by Vo Ky Lan et al, reference 36. +, experimental results by Krause and Caldwell, reference 38. Top curve represents ion left in the $1s2s^2$ level; middle curve ion in $1s(2s2p\ ^3P)$ level; bottom curve ion in $1s(2s2p\ ^1P)$ level.

section R-matrix results are presented of a recent calculation of inner-shell photoionization of the beryllium atom by Vo Ky Lan et al.[36]

In the region $r \leq a$ the wave function is expanded as follows[37]

$$\Psi_k = A \sum_{ij} \overline{\phi}_i\, u_j(r)\, a_{ijk} + \sum_i \chi_i\, b_{ik}\ , \qquad (9)$$

where $\overline{\phi}_i$ are channel functions constructed from ionic eigenstates Φ_i and the spin-angle function of the continuum electron, and A is the antisymmetrizer. The u(r) are radial basis functions which satisfy

$$u_j(0) = 0\ ,$$

and

$$\frac{a}{u_j}\frac{du_j}{dr}\bigg|_{r=a} = b\ , \qquad (10)$$

where b can be chosen arbitrarily (zero in this case). The radius a is chosen so that all bound functions are effectively zero (22.8 a.u. for the Be calculation). The χ_i are bound correlation functions, and the coefficients a_{ijk} and b_{ik} are determined by diagonalizing the Hamiltonian. The wave function for $r > a$ is determined by close-coupling methods. The bound functions Φ_i, χ_i, and the initial state Ψ_0 were obtained by configuration interaction calculations using bound orbitals 1s, 2s, 2p, 3s, 3p, 3d, $\overline{4s}$, $\overline{4p}$, $\overline{4d}$, and $\overline{4f}$, where the bar above the orbital indicates a pseudo-orbital. Thirty-two basis states $u_j(r)$

were calculated for each $\ell \leq 4$. For the Be^+ ion, 108 configurations were used, 183 configurations for Be states of ^1P symmetry and 100 configurations for Be states of ^1S symmetry.

Results of the R-matrix calculations are given in Fig. 8 and compared with experimental results by Krause and Caldwell.[38] The large resonance in the 1s $2s^2$ cross section near 125eV is due to the configuration 1s(2s2p^{3}P)3s. It is expected that the experimental resonance structure for higher resonances is limited by experimental resolution. Calculations of the Be 1s photoionization cross sections have been carried out by Saha and Caldwell[39] using the numerical multiconfiguration Hartree-Fock method.[9,10] The MCHF results are in reasonable agreement for $2s^2$ and 1s (2s2p^{3}P) cross sections but are too small for the 1s(2s2p^{1}P) cross section. This method does not include interchannel coupling between open channels which may account for the lack of agreement for the 1s (2s2p^{1}P) cross section.

CONCLUSIONS

In order to include the effects of electron correlations for atomic processes such as photoionization, it is necessary to carry out very extensive computations. It will be desirable in future calculations, particularly of photoionization with excitation, to carry out calculations which may couple as many as fifty channels or more and will require increased computational power. When these calculations include jj-coupling and relativistic orbitals, the number of channels will be effectively doubled.

ACKNOWLEDGEMENTS

This work was supported by the U.S. National Science Foundation. I wish to thank James J. Boyle, Dr. Vo Ky Lan, and Dr. Zuwei Liu for helpful discussions.

REFERENCES

1. X-Ray and Inner-Shell Processes, AIP Conf. Proc. 215, eds. T.A. Carlson, M.O. Krause, and S.T. Manson (NY, NY, 1990).
2. A.F. Starace, in Handbuch der Physik, Vol. 31 (ed., W. Mehlhorn, Springer, Berlin (1980)).
3. P.L. Altick and A.E. Glassgold, Phys. Rev. 133, A632 (1964).
4. M.Ya. Amusia and N.A. Cherepkov, Case Studies in Atomic Physics 5, 47 (1975).
5. W.R. Johnson and C.D. Lin, Phys. Rev. A14, 565 (1976).
6. A. Zangwill, Atomic Physics 8, 339 (1983).
7. P.G. Burke and W.D. Robb, Adv. At. Mol. Phys. 11, 143 (1975).
8. P.G. Burke and K.T. Taylor, J. Phys. B8, 2620 (1975).
9. J.R. Swanson and L. Armstrong, Jr., Phys. Rev. A15, 661 (1977).
10. H.P. Saha, Phys. Rev. A39, 628 (1989).
11. J. Tulkki, Phys. Rev. Lett. 62, 2817 (1989).
12. H.P. Kelly, Atomic Physics 8, 305 (1983); Physics Scripta T17, 109 (1987).
13. U. Fano and J.W. Cooper, Rev. Mod. Phys. 40, 441 (1968).
14. S. Huzinaga and C. Arman, Phys. Rev. A1, 1285 (1970).
15. E.R. Brown, S.L. Carter, and H.P. Kelly, Phys. Rev. A21, 1237 (1980).
16. P.R. Woodruff and J.A.R. Samson, Phys. Rev. A25, 848 (1982).
17. D.W. Lindle et al, Phys. Rev. A31, 714 (1985).
18. S. Salomonson, S.L. Carter, and H.P. Kelly, J. Phys. B 18, L149 (1985); Phys. Rev. 39, 5111 (1989).
19. A.-M. Mårtensson, J. Phys. B12, 3995 (1979).

20. W. Wijesundera and H.P. Kelly, Phys. Rev. A$\underline{39}$, 634 (1989).
21. M.Y. Adam, P. Morin, and G. Wendin, Phys. Rev. A$\underline{31}$, 1426 (1985).
22. J.A.R. Samson and J.L. Gardner, Phys. Rev. Lett. $\underline{33}$, 671 (1974).
23. R.G. Houlgate, J.B. West, K. Codling, and G.V. Marr, J. Phys. B$\underline{7}$, L470 (1974).
24. M.Ya. Amusia, N.A. Cherepkov, and L.V. Chernysheva, Phys. Lett. A$\underline{40}$, 15 (1972).
25. U. Becker et al, Phys. Rev. Lett. $\underline{60}$, 1490 (1988).
26. H. Kossmann, B. Krassig, V. Schmidt, and J.E. Hansen, Phys. Rev. Lett. $\underline{58}$, 1620 (1987).
27. J.A.R. Samson, Y. Chung, and E. Lee, Phys. Lett. A$\underline{127}$, 171 (1988).
28. M.Y. Adam et al, J. Phys. B$\underline{11}$, L413 (1978).
29. J. Boyle, Z. Altun, and H. P. Kelly, to be published.
30. J.T. Costello, E.T. Kennedy, B.F. Sonntag, and C.W. Clark, to be published.
31. Z.-D. Qian, S.L. Carter, and H.P. Kelly, Phys. Rev. A$\underline{33}$, 1751(1986).
32. R.D. Cowan, The Theory of Atomic Structure and Spectra (UC Press, Berkeley, 1981).
33. L.J. Garvin, Ph.D. Thesis, U. of Virginia, 1983, unpublished.
34. E.P. Wigner, Phys. Rev. $\underline{70}$, 15 (1946); $\underline{70}$, 606 (1946).
35. E.P. Wigner and L. Eisenbud, Phys. Rev. $\underline{72}$, 29 (1947).
36. Vo Ky Lan et al, to be published.
37. P.G. Burke, Atomic Physics $\underline{10}$, 243 (1987).
38. M.O. Krause and C.D. Caldwell, Phys. Rev. Lett. $\underline{59}$, 2736 (1987).
39. H.P. Saha and C.D. Caldwell, Phys. Rev. A$\underline{40}$, 7020 (1989).

COMPUTATION APPLIED TO PARTICLE ACCELERATOR SIMULATIONS*

W. B. Herrmannsfeldt
Stanford Linear Accelerator Center, Stanford University, Stanford, CA 94309
and
Yiton T. Yan
Superconducting Supercollider Laboratory, Dallas, TX 75237

ABSTRACT

The rapid growth in the power of large-scale computers has had a revolutionary effect on the study of charged-particle accelerators that is similar to the impact of smaller computers on everyday life. Before an accelerator is built, it is now the absolute rule to simulate every component and subsystem by computer to establish modes of operation and tolerances. We will bypass the important and fruitful areas of control and operation and consider only application to design and diagnostic interpretation. Applications of computers can be divided into separate categories including:
- component design,
- system design,
- stability studies,
- cost optimization, and
- operating condition simulation.

For the purposes of this report, we will choose a few examples taken from the above categories to illustrate the methods and we will discuss the significance of the work to the project, and also briefly discuss the accelerator project itself. The examples that will be discussed are:
(1) the tracking analysis done for the main ring of the Superconducting Supercollider, which contributed to the analysis which ultimately resulted in changing the dipole coil diameter to 5 cm from the earlier design for a 4-cm coil-diameter dipole magnet;
(2) the design of accelerator structures for electron-positron linear colliders and circular colliding beam systems (B-factories);
(3) simulation of the wake fields from multibunch electron beams for linear colliders; and
(4) particle-in-cell simulation of space-charge dominated beams for an experimental linear induction accelerator for Heavy Ion Fusion.

SSC APERTURE STUDY

One of the important issues for the SSC was the size of the superconducting dipole magnets to be used.[1] More protons can survive in the collider rings with dipoles of larger coil diameter because larger coil-diameter dipole magnets can provide more uniform bending magnetic field. However, larger dipole magnets are more expensive. Therefore, one must study the proton motion for each of the alternative magnet lattices under consideration. This was done by simulating the motion of the proton beam with numerical codes on supercomputers.

---

*Work supported by Department of Energy contract DE–AC03–76SF00515.

In these numerical studies, one starts with a well-designed linear lattice and then assigns systematic errors, random errors, and misalignment for the magnets, based on experience and measurement. Correction magnets may also be included. Ideally, protons are then tracked numerically for a limited number of turns to see if the motion is stable. At this stage, adjustment of the correction magnets is usually necessary (somewhat similar to the micro-tuning of a TV or a radio). After the accelerator is well tuned, one can start short-term tracking (say, 400 turns) to study some well-defined accelerator physics criteria to predict the behavior of the accelerator.

A typical short-term-tracking phase space plot is shown in Fig. 1. The variation in the amplitude traced out by given protons is greater for those protons of larger initial amplitude. Here, as shown in Fig. 1, the amplitude is defined as $\sqrt{x^2 + p_x^2}$, where x is a Floquet space coordinate and p_x is its corresponding Floquet space momentum; that is, they are normalized such that a proton with linear motion would trace out a circle in (x, p_x) phase space. This phenomenon serves as a diagnostic of accelerator nonlinearity. If the amplitude variation is considered too big for a certain desired amplitude, the corresponding accelerator design should be modified.

Generally, one would be more concerned with the long-term stability of the protons. One would like to track

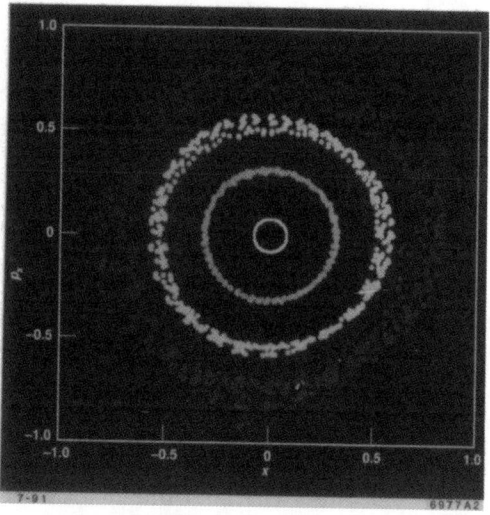

Fig. 1. Phase space plot p_x versus x for four protons with different initial amplitudes, where x is a Floquet space coordinate and p_x is its corresponding Floquet space momentum; that is, they are normalized such that a proton with linear motion would trace out a circle. The variation in the amplitude traced out by a given proton serves as a diagnostic of accelerator nonlinearity of that proton's motion. For example, the protons here with the smallest initial amplitude show so little nonlinearity that the data points merge into a solid line. However, the protons with the largest initial amplitude have correspondingly greater nonlinearity, so that the data points are more widely spaced in the circular band.

hundreds of protons (with appropriate initial amplitude distributions) around the ring element-by-element for 100,000 turns or more (0.5 minutes of SSC operation will be about 100,000 turns). Using a current scalar computer, this would require months of central processing unit (CPU) time, since there are more than 10,000 magnet elements in the SSC machine. Fortunately however, the protons in the beam may be considered to be independent from each other, so that a tracking code can be completely vectorized over the number of particles; thus, a supercomputer is ideal for this purpose. One can track many particles (say, 64 protons) simultaneously, saving enormous CPU time over what a scalar machine would require.

Figure 2 compares the tracking data up to 100,000 turns for a collider injection lattice (at 2 TeV energy) using 4-cm coil-diameter dipole magnets, with the corresponding data for the same lattice using 5-cm coil-diameter dipole magnets. None of the particles with initial displacement amplitude of less than 8.1 mm were lost in the 5-cm coil-diameter dipole magnet case; but in the 4-cm coil-diameter dipole magnet case, particles still get lost until their initial displacement amplitude is reduced to about 5.3 mm. The only difference between the two lattices was in the multipole content, due to the different size of the magnet aperture. With the increase in magnet aperture from 4-cm coil-diameter to 5-cm coil-diameter, the stable region of the proton motion for 100,000 turns enlarged from about 5.3 mm to about 8.1 mm in radius. Based on these numerical studies and many other investigations, the 5-cm coil-diameter superconducting dipole magnets have been chosen.

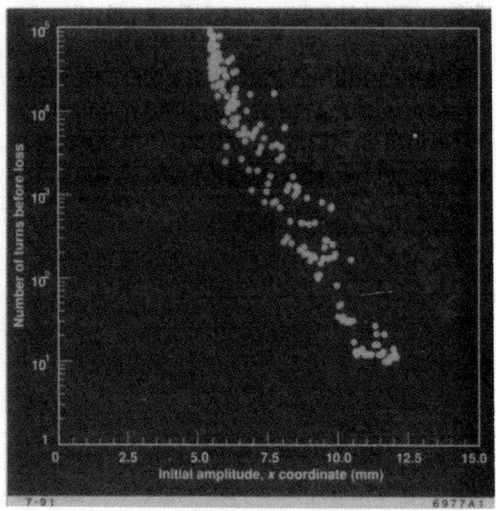

Fig. 2. A hundred-thousand-turn survival plot for a collider 2-TeV injection lattice, comparing the data for a 5-cm magnet aperture with the data for a 4-cm magnet aperture. With the 5-cm aperture, no particles with initial displacement amplitude of less than 8.1 mm were lost. By increasing the magnet aperture, the dynamic aperture for 100,000 turns enlarges from about 5.3 mm to about 8.1 mm in radius, which increases the machine's linearity. (This plot shows only the protons that were lost before 100,000 turns are reached.)

DESIGN OF ACCELERATOR STRUCTURES

The purpose that the accelerator structure serves is to provide a means for converting the electromagnetic power into fields that can efficiently and accurately couple power into the charged particle beams. Thus the structure must have an entrance, or port, through which electromagnetic energy can flow, and cooled walls to remove the heat generated by surface currents in the structure walls. Typically the structure must be designed to be resonant at the frequency that is chosen for the accelerating mode. It is often just as important that the structure not be resonant at frequencies that correspond to modes that can improperly steer the particles. From just the above conditions we have several requirements on the structure:

(1) dimensionally accurate to resonate at frequencies ranging from a few hundred megahertz to several million megahertz,

(2) dimensionally stable and properly cooled to maintain the dimensions needed to stay tuned to the drive frequency, and to keep away from damaging resonances, and

(3) non-cylindrical symmetry to allow for a port to permit the flow of electromagnetic power.

Structures can be either made of normal conductors, usually copper, or of superconducting materials, usually niobium alloys. Superconducting cavities are especially useful for continuous operation for installations such as the Continuous Electron Beam Accelerator Facility. For pulsed operation, with very high electric fields and high beam currents, normal conducting copper cavities are appropriate. In either case, it is important to consider fields left by the particles themselves, as they can affect particles coming through the structure later.

Figure 3 is a view of the parts of a test cavity[1] for a future electron-positron linear collider. To achieve very high electric fields, which are important if the accelerators are to be kept to a reasonable length, it is necessary to go to very high frequency RF power. The device in Fig. 1 is designed for 11.42 gigahertz, which corresponds to a wavelength of 2.62 cm, about one inch. This frequency corresponds to the range that is designated as X-band.

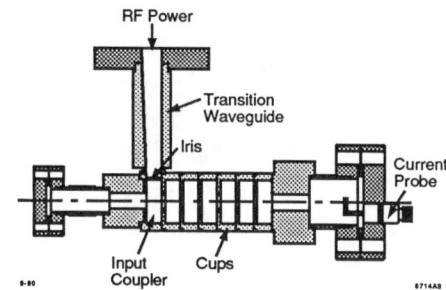

Fig. 3. Test structure for an X-band accelerator.

Figure 4 shows the three dimensional (3-D) mesh zoned for the structure shown in Fig. 1. In order to conserve on computation time and memory, only one half of the structure is modeled. With 500,000 zones, as shown in Fig. 3, 2 to 3 hours of Cray 2 time are needed to model the structure and find the necessary number of higher-order modes. A memory space of 10 million words is needed to make the simulation of the fields. Although higher resolution would be very useful, another factor of two increase in the number of mesh points is not currently possible with present facilities.

Figure 5 shows an accelerating structure[2] for the asymmetric storage rings operating at the production

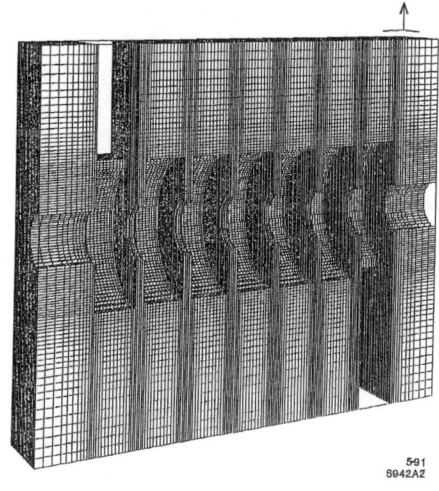

Fig. 4. Three-dimensional zoned model of the X-band structure shown in Fig. 1. About 500,000 zones are needed.

resonance for B particles, thus earning the designation B-factory. Although not at the highest center-of-mass energy, B-factory design is especially demanding because of the very high circulating electron currents that are required. Because the RF fields are provided continuously, the high average power needed to maintain the collision rate requires exceptional care in designing for heat dissipation and cavity cooling. The results of the design study with this simulation were used to provide input to a cavity heat-load study.

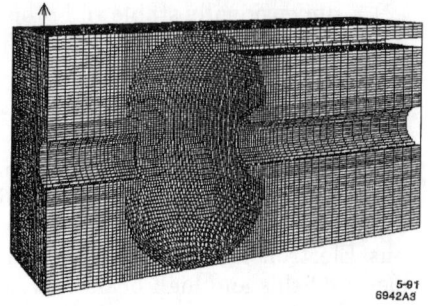

Fig. 5. The 3-D zone for the B-factory RF structure includes reentrant nose cones and an input coupler.

WAKE FIELD SIMULATION

The residual fields, or "wake" fields as they are named by analogy to the wake of a boat, can disrupt particles in bunches that follow in the wake of leading bunches. This problem can be controlled if the accelerating structures can be designed so that the damaging modes are sufficiently loaded, for instance by providing an escape path for the fields, or if the structures are designed so that a wide variety of resonant frequencies are present for higher modes.

The wake fields from a moving bunch of charged particles can be calculated using the programs BCI and TBCI (transverse beam-cavity interaction) by T. Weiland[3] These programs are similar to the large particle-in-cell programs such as ARGUS (in 3-D) Condor in (2-D) and are related to the more recent work by Weiland for the Mafia code group. Figure 6 shows the transient wake fields of a bunch of charged particles passing through a cavity, as calculated by Bane, Chao and Weiland.[4]

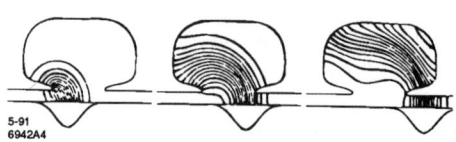

Fig. 6. The transient electromagnetic fields in a cavity as a bunch of electrons (shown in the position of the Gaussian curve below the axis) passes through the tube. The cavity in each view is the same cavity at a later instant.

A somewhat different problem is presented for the wakes within a single bunch of particles, by which particles off the axis cause fields which can displace particles within the same bunch. Single bunch effects were simulated by Bane.[5] The effect of these fields can be seen in the computer simulation shown in the left side of Fig. 7. Based on suggestions by Balakin et al.[6] Bane calculated the effects of introducing a small, controlled energy spread into the bunch. The effect known both as Landau Damping, and as BNS damping for the Soviet scientists who suggested this solution, is that the bunch remains well aligned as shown on the right side of Fig. 7. The energy differences remain correlated with position in the bunch so that manipulations of the phase of the RF power at the end of the accelerator can cancel the spread in energy.

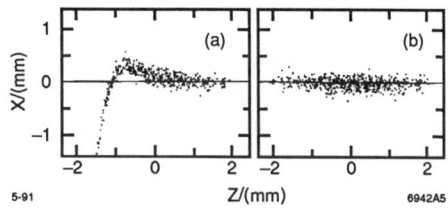

Fig. 7. Bunch shape without Landau Damping (on the left) and with the damping effect (on the right).

COMPRESSION OF HEAVY ION BEAMS FOR FUSION

Intense beams of heavy ions can be used to implode and ignite targets of deuterium and tritium to make Inertial Fusion Energy (IFE). Studies of focusing, bending, and especially the longitudinal compression necessary to increase the peak current in a bunch, are made with a new 3-D simulation program called WARP.[7]

In Fig. 8, a bunch of heavy ions is shown before beginning final compression and then again after undergoing some compression. Compression is accomplished by imposing a longitudinal velocity tilt on the ion beam, by accelerating the trailing (left) end of the bunch more than the leading end. After two-thirds of the compression process, the bunch profiles are as shown in the lower two figures. The shapes are controlled at any one point by the quadrupole focusing system which alternately focuses and defocuses the beam in the two orthogonal planes. Thus at any one point along the beam, the beam profile will be elliptical in shape. The primary concern in beam manipulations of this type is the degree to which the beam quality, or emittance, is disturbed by the compression.

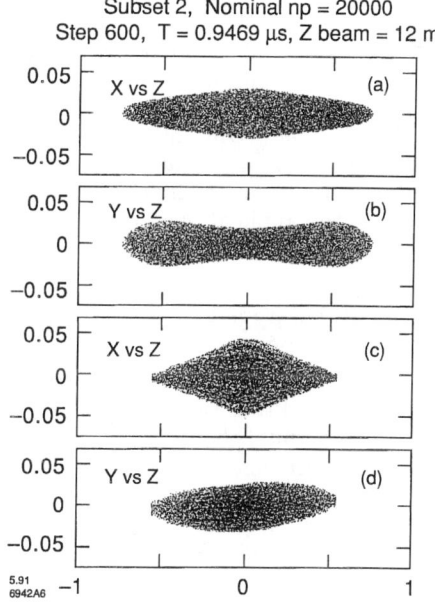

Fig. 8. Longitudinal compression is studied using the 3-D program WARP. The longitudinal spatial distribution is shown for both the X-Z and Y-Z projections in the upper pair of figures. The compressed bunch is shown in the two lower figures.

Thus it is necessary to use a great many particles (20,000 macro particles were used in this example) and very small time steps in order to maintain the accuracy of the calculation.

As the compression continues, the longitudinal line-charge density increases as shown in Fig. 9. Early experiments at LBL with the MBE-4 experiment have demonstrated longitudinal compression of this type. The simulations shown here are of the conditions as the beam is focused towards the target in the reactor vessel, and are thus far beyond what is available experimentally.

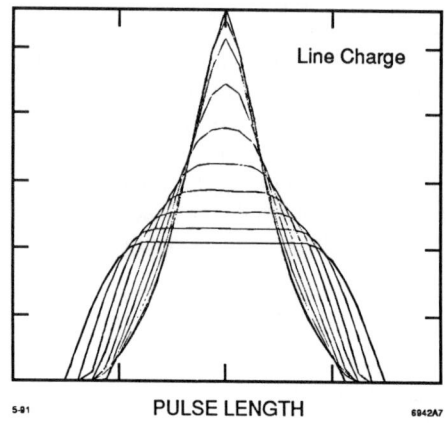

Fig. 9. Line-charge density of the compressed bunch shown at several different times during the compression process.

CONCLUSIONS

Computers play an ever-expanding role in the study of particle accelerators. Accelerators are growing in importance to research and industry, and may one day be an important part of the energy production industry. The economic arguments clearly favor the purchase of larger, faster computers which are far less expensive than the machines that they help design. In some areas, especially those involving 3-D modelling of RF cavities and intense beams, the applications are limited by the size and speed of the computers.

REFERENCES

1. Y. Yan, SSCL–239 (1990) and also L. Schachinger and Y. Yan, SSCL–N–664 (1989).
2. K. Ko, G. Loew and J. Wang, Transient Studies of High-Gradient Structures, Proc. IEEE 1991 Part. Accel. Conf., San Francisco, May 6–9, 1991, to be published.
3. R. Rimmer, F. Voelker, G. Lambertson (LBL), M. Allen, K. Ko, R. Pendleton, H. Schwartz (SLAC), and N. Kroll (UCSD), An RF Cavity for the B Factory, ibid., Ref. 1.
4. T. Weiland, Nuclear Instrum. and Methods **212**, 13 (1983).
5. K. Bane, A. Chao and T. Weiland, A Simple Model for the Energy Loss of a Bunched Beam Traversing a Cavity, Proc. 1981 Part. Accel. Conf., Washington, DC.
6. K. Bane, Landau Damping in the SLAC Linac, Proc. 1985 Part. Accel. Conf., Vancouver, B.C.
7. V. Balakin et al, Proc. 12th Int. Conf. on High Energy Accelerators, Fermilab, 1983, p. 119.
8. Alex Friedman, R. Bangerter, D. Callahan, D. Grote, A. B. Langdon (LLNL) and I. Haber (NRL), A 3-D Particle Simulation Code for Heavy Ion Fusion Accelerator Studies, ibid., Ref. 1.

A Discussion of Computational Methods and Applications in Molecular Physics

Kate P. Kirby
Harvard-Smithsonian Center for Astrophysics

I. Introduction

The processes and applications of molecular physics pervade many other fields of physics such as atmospheric physics and astrophysics. There is considerable overlap of interests and techniques with the field of quantum chemistry, however molecular physicists often focus on processes involving small, simple molecules which can be described more accurately mathematically than complex molecules with many nuclei and electrons. Theoretical studies of emisson or absorption of radiation by molecules, studies of molecular processes involving ionization or dissociation, and of scattering processes including collision excitation by atoms, ions and electrons, charge transfer and electron-ion recombination may all require as a first step a calculation of the molecular electronic structure of the relevant system.

Unlike many areas of physics, the equations of the molecular physicist are generally known and the Hamiltonian of the system can be specified to any required accuracy. Exact solutions, however, of these large many-body problems are very difficult to obtain. It is necessary to have a computational scheme that has physical simplicity for interpretive purposes and which can be modified, with added computational complexity, if need be, to obtain increased accuracy.

First I will review several examples of molecular processes and then review a standard computational scheme which has been used in a great many applications. Some recent work on metastability in small molecules will be described in the context of this method.

II. Some Molecular Processes

Although spectroscopy has provided much of the original insight into the quantum mechanical description of molecular structure, theoretical calculations of electronic energies and wavefunctions as a function of internuclear coordinates can provide a complementary picture which aids in the identification of transition frequencies and the prediction of line intensities or oscillator strenghts. A good *ab initio* calculation can provide a wealth of information which could be obtained in the laboratory only by a combination of several different experiments. Electronic wavefunctions, which can be used to calculate many physical properties, and potential energy surfaces, which describe atom-atom or atom-molecule interactions, are important in the description of many of the following examples of molecular scattering processes:

1) photodissociation — in which a photon is absorbed promoting a molecule to an excited state which is unbound with respect to nuclear motion:

$$OH \xrightarrow{h\nu} O + H$$

2) photoionization — in which the absorption of a photon produces a molecular positive ion and a free electron:

$$CH \xrightarrow{h\nu} CH^+ + e^-$$

3) Dissociative recombination — in which an electron collides with a positive ion forming a highly excited repulsive state of the neutral molecule which then dissociates:

$$CH^+ + e^- \rightarrow CH^{**} \rightarrow C + H$$

4) Charge transfer — in which an electron is transferred from one atomic center to another:

$$O^{++} + H \rightarrow O^+ + H^+$$

5) rotational excitation — in which a molecule is excited from one rotational level to another by impact of electrons, atoms or molecules:

$$H_2 + CO(J) \rightarrow H_2 + CO(J')$$

6) radiative association — in which two atoms collide and radiate a photon, forming a stable molecule:

$$C^+ + O \xrightarrow{-h\nu} CO^+$$

III. The Molecular Hamiltonian and the Born-Oppenheimer Approximation

The ordinary electrostatic Hamiltonian, H, for a system of electrons and nuclei can be written:

$$H = \frac{-\hbar^2}{2M_e} \sum_i \nabla_i^2 - \sum_A \frac{-\hbar^2}{2M_A} \nabla_A^2 - \sum_{A,i} \frac{Z_a e^2}{|\vec{r}_i - \vec{R}_A|}$$

$$+ \sum_{A>B} \frac{Z_A Z_B}{|\vec{R}_A - \vec{R}_B|} + \sum_{i>j} \frac{e^2}{|\vec{r}_i - \vec{r}_j|} \quad (1)$$

in which i and A refer to electrons and nuclei respectively, M_e and M_{Al} are the corresponding masses and Z_A the nuclear charge. A fundamental assumption usually made as a starting point in molecular calculations is the separability of the total wavefunction of the system into a product of two wavefunctions, one depending on electronic coordinates r, the other on nuclear coordinates R:

$$H \Psi_{mol} = E_{tot} \Psi_{mol} \quad (2)$$

and

$$\Psi_{mol}(\vec{r}, \vec{R}) = \Psi_{el}(\vec{r}; R) \chi_{nuc}(\vec{R}) \quad (3)$$

This separation, known generally as the Born-Oppenheimer Approximation, is based on the fact that the electrons move much faster than the nuclei — a reflection of their relative masses. By plugging the separated wavefunction above into the expression for H, collecting terms appropriately and excluding certain cross-terms, one can obtain two Schrödinger equations, one for the electron motion and one for the nuclear motion. The electronic Schrödinger equation assumes that the nuclei are fixed in space and the associated electronic energy and wavefunction depend parametrically on the nuclear coordinates R:

$$H_{el} \Psi_{el}(\vec{r};R) = E_{el}(R) \Psi_{el}(\vec{r};R) \quad (4)$$

$$H_{el} = \frac{-\hbar^2}{2M_e} \sum_i \nabla_i^2 - \sum_{A,i} \frac{Z_a e^2}{|\vec{r}_i - \vec{R}_A|} + \sum_{i>j} \frac{e^2}{|\vec{r}_i - \vec{r}_j|} \quad (5)$$

The nuclear Schrödinger equation can be written as

$$H_{nuc} \chi_{nuc}(\vec{R}) = E_{tot} \chi_{nuc}(\vec{R}), \quad (6)$$

in which

$$H_{nuc} = -\sum_A \frac{-\hbar^2}{2M_A} \nabla_A^2 + \sum_{A>B} \frac{Z_A Z_B e^2}{|R_A - R_B|} + E_{el}(R) \quad . \quad (7)$$

Note that the nuclei move in a potential provided by the electrons, $E_{el}(R)$.

In the separation, several cross-terms which are usually small have been neglected:

$$\left\{ -\sum_A \frac{-\hbar^2}{2M_A} 2(\nabla_A \Psi_{el})(\nabla_A \chi_{nuc}) + \chi_{nuc} \nabla_A^2 \Psi_{el} \right\} \quad . \quad (8)$$

These terms can become important when processes such as charge-transfer are considered and then special techniques must be employed to deal with them.

The recipe for calculating the electronic energy and wavefunction of a particular molecule is deceptively simple: 1) assume the nuclei fixed at a particular nuclear geometry R; 2) solve eq. (4) for $E_{el}(R)$ and $\psi_{el}(\vec{r};R)$; 3) change the nuclear geometry and solve eq. (4) again; and repeat 3) over the full range of nuclear geometries requred. In this way the energy as a function of nuclear coordinates is mapped out. For polyatomics one obtains a multi-dimensional potential energy <u>surface</u> and for diatomics one obtains a potential energy <u>curve</u>. Various properties, such as the dipole moment, vary with nuclear geometry and can be calculated as a matrix element of an operator once the electronic wavefunction has been obtained:

$$D(R) = <\Psi_{el}(\vec{r};R) | \sum_i e r_i | \Psi_{el}(\vec{r};R)> \quad . \quad (9)$$

As the internuclear separation between A and B becomes very large (R→∞) the energy of two separated noninteracting atoms is obtained. A bound electronic state is one in which the energy is negative with respect to the separated atom limit. A potential energy curve which is always positive with respect to separated atoms shows no binding and is called a repulsive state. Vibration and rotation of the nuclei take place within the bound state, and several vibrational levels are shown in Figure 1. The binding energy, D_e, is the difference between the energies at the minimum of the potential well and at the separated atom limit.

IV. <u>Several Computational Approaches: MO-SCF, MCSCF and CI</u>

Although atomic and molecular electronic structure calculations share a number of similarities and therefore similar computational approaches, the presence of more than one nucleus and the resulting multi-centered nature of the electron-electron and electron-nuclear interactions leads to significant differences in the way that the computational techniques are applied. In a description of bond formation and/or dissociation, the goal of a molecular structure calculation is more to provide a balanced description over a wide range of nuclear geometries than to obtain the greatest possible accuracy at a single geometry. As always in computational physics, there are the inevitable compromises between accuracy and computational effort, the latter determined largely by available computing resources.

Without further simplification, the solution of eq. (4) to obtain $\psi_{el}(\vec{r};R)$ which is a function of the coordinates of N electrons is very complicated. A fundamental concept, therefore, in molecular electronic structure, is molecular orbital theory in which each electron is described by a one-electron spin orbital

$$\psi_i = \phi_i(\vec{r}) \, \alpha \text{ or } \beta \ . \tag{10}$$

The simplest commonly used approximation is known as Hartree-Fock, in which the total wavefunction is approximated by a single configuration or "Slater determinant" consisting of an antisymmetrized product of the molecular orbitals:

$$\Phi = A \, |\phi_1 .... \phi_N| \ . \tag{11}$$

The spatial orbitals are usually expanded in a set of analytic functions called a basis set:

$$\phi_i(\vec{r}) = \sum_p \chi_p(\vec{r}) \, c_{pi} \ , \tag{12}$$

with $\chi_p(\vec{r})$ centered on the nuclei. Applying the vibrational principle to $E = \langle\Phi|H|\Phi\rangle$ leads to a set of equations

$$F \, \phi_i = \varepsilon_{ii} \phi_i \tag{13}$$

which can be solved iteratively from an initial guess to self-consistency. Solution consists of finding a set of c_{ip}'s which give an optimal description of the orbitals in the single Slater determinant. This method is known as the self-consistent field method (MO-SCF).

Independent particle models, such as Hartree-Fock MO-SCF, do not properly account for the strongly correlated motion of the electrons which is manifested in the $1/|r_i-r_j|$ terms in the Hamiltonian. The Hartree-Fock energy is a variational upperbound to the energy of the exact solution of the fixed-nucleus Schrödinger equation. The difference in energy between the Hartree-Fock solution and the exact solution is known as the "correlation energy," and many more sophisticated techniques have been developed to deal with the most important electron correlation effects. Only two methods will be briefly summarized here, but together they have been used in a wide range of applications of molecular electronic structure theory.

Another significant deficiency of the MO-SCF method is its inability to provide a good description of bond stretching or bond breaking. An example of this problem can be seen for H_2 in which a single determinant wavefunction for two electrons is written $1\sigma_g^2$. This description is reasonably good at the equilibrium geometry R_e, but at large internuclear separations, this wavefunction contains a mixture of ionic asymptotes $H^+ + H^-$ ($1s^2$), rather than the pure $H(1s) + H(1s)$ required for proper dissociation. Correct asymptotic behavior requires at least a two-configuration wavefunction, namely $1\sigma_g^2 - 1\sigma_u^2$.

The need for a better description of electron correlation and for accurate dissociation into products led to the development of the multi-configuration self-consistent field method (MCSCF) in which the wavefunction

$$\Psi_{mc} = \sum_I C_I \Phi_I \tag{14}$$

includes configurations to describe correctly bond distortion and dissociation. The configurations Φ_I are N-electron functions composed of products of the ϕ_i's which are part of a common set of orthonormal orbitals. In the MCSCF scheme ϕ_i's and C_I's are fully optimized variationally, thus limiting the size of the set I of configurations to a small number. In order to carry out MCSCF calculations on molecules with more than 6 or 8 electrons the orbital space is usually partitioned into a "core," "active," and "virtual" space. Core orbitals remain doubly occupied in all configurations, active orbitals are allowed to have different occupations in different configurations, and external orbitals may remain unoccupied in all configurations in an MCSCF calculation. The active space usually consists of the valence orbitals, and inclusion of configurations with all possible occupations of these orbitals leads to a wavefunction which describes proper dissociation. The MCSCF procedure with a limited number of configurations is still lacking in adequate treatment of electron correlation, however. The next step is usually large-scale configuration interaction.

The method of configuration interaction, CI, is simpler computationally than MCSCF. The wavefunction is written as in (14), but the Φ_I's are N-electron configuration state functions constructed from a common <u>fixed</u>

set of orbitals. The equation that needs to be solved,

$$(\underline{\underline{H}} - E\underline{\underline{1}}) \underline{C} = 0 \tag{15}$$

is a linear eigenvalue problem in which elements of the Hamiltonian matrix are $<\Phi_I|H|\Phi_J>$. Roots of this secular equation correspond to different electronic states. Typically one uses an optimized set of orbitals from an MCSCF calculation, and starting from a reference set of determinants which might describe proper dissociation, one might include all single and double excitations with respect to that reference set. Higher excitations such as triples and quadruples can be included but this usually make the problem too large to be computationally feasible. Today there are ways of handling up to 10^7 configurations, depending on the computer hardware and software used.

In a CI calculation the ultimate accuracy of $\Psi_{el}(\vec{r};R)$ and $E_{el}(R)$ is determined by several important elements. The first is the choice of basis set $\{\chi_{ip}\}$. While nodeless Slater-type functions, having the form

$$\chi = N_n r^{n-1} \exp(-\zeta r) Y_{lm}(\theta,\phi), \tag{16}$$

generally provide an excellent representation of both atomic and molecular orbitals, the multi-centered integrals which arise in the matrix elements become computationally difficult. In order to simplify these integrals, Gaussian type functions are commonly used, which take either a Cartesian form,

$$\chi = N x^a y^b z^c \exp(-\alpha r^2), \tag{17}$$

or a spherical form

$$\chi = N r^\ell \exp(-\alpha r^2) Y_{\ell m}(\theta,\phi). \tag{18}$$

Because the functional form of Gaussians is not a good description of molecular orbitals at both long and short range, much larger Gaussian basis sets are required to achieve a certain accuracy than could be achieved with a small Slater basis. Despite this, computational efficiency still dictates the use of Gaussians for molecules with more than two nuclei. A large and flexible basis set is essential for high accuracy, as no amount of configuration interaction or orbital optimization can make up for a deficient basis set.

A second important element is the determination of the set of molecular orbitals from which the wavefunction is constructed. If an

MO-SCF calculation is used to determine $\{\phi_i\}$, then the orbitals may not be well-determined at large r. Usually it is necessary to carry out some form of MCSCF calculation to obtain $\{\phi_i\}$.

Finally, in the construction of the CI wavefunction itself, it is important to consider which orbitals are active, what level of excitations can be included, and what choice of reference set of configurations makes good computational sense. These issues determine the size of the set $\{\Phi_I\}$ in the wavefunction, and must be resolved with the available computing resources in mind.

V. Applications to Metastable Molecular Systems

In order to give a feeling for the kind of problems which can be treated using molecular electronic structure methods, some recent work on metastable molecular systems will be briefly described. The prime focus of the work was to find ways of storing energy in a substance and of then extracting it by some perturbation. In the laboratory, metastable states are excited states with lifetimes, $\tau \gtrsim 1\mu s$. In these studies, the interest was in maximizing the excited state energy, minimizing the mass, and at the same time lengthening τ by orders of magnitude over laboratory timescales.

In order to ascertain the feasibility of certain molecular excited states acting as energy reservoirs, it was necessary to find the answers to four basic molecular physics questions:

1. Is there binding? i.e., is the proposed excited state bound with respect to nuclear motion?
2. Are any electric dipole transitions from this excited state to lower-lying states possible? Lifetimes for electric dipole tansitions are typically 1-10ns, and rapid decay by emission of radiation occurs whenever possible.
3) Are any predissociations possible? By interaction through small terms in the Hamiltonian which are often neglected, a repulsive state may couple to a bound excited state, causing dissociation.
4) Are any perturbations possible? Again, through small terms in the Hamiltonian, such as spin-orbit, spin-spin, or spin-rotation operators, coupling of the excited state of interest can cause population to "leak" into an adjacent state which can then radiate via a dipole transition to a lower-lying state.

If the answer to the first question is "yes," and the answer to each of the other three questions is "no" or "not likely," one may have discovered a long-lived metastable state. The problem is that a huge computational effort is required to answer these questions definitively for any particular system. Thus the recent work described here is "in progress," with question 4) being a very difficult possibility to rule out.

We have carried out calculations on three very different types of metastable systems -- a doubly charged molecular ion CH^{++}; high spin states of CN and NO; and two electronic states of CO which have long-lived vibrational levels.

In CH^{++}, states arising from asymptotic limits $C^+ + H^+$ are expected to be repulsive, totally dominated by the Coulomb repulsion of two positively charged nuclei. The only possibility for obtaining a small amount of binding is from asymptotes $C^{++} + H$ in which ion-atom polarization attraction is operative. Experimentally, CH^{++} is reported to have been observed[1-4] in charge-stripping experiments of CH_4, and a metastable excited state of this species was also seen.[5] The lowest $C^{++} + H$ asymptote lies 10.8eV above the lowest $C^+ + H^+$ asymptote and gives rise to the third $^2\Sigma^+$ state. The $3^2\Sigma^+$ and the $2^2\Sigma^+$ exhibit minima due to the $C^{++} + H$ polarization, but these states lie so far from the lowest $^2\Sigma^+$ state that no minimum is found in the $1^2\Sigma^+$ state.[6] The calculated electric dipole transition moments for $2^2\Sigma^+$-$1^2\Sigma^+$ and $3^2\Sigma^+$-$1^2\Sigma^+$ show that radiative decay occurs very rapidly from both of these states. Our calculations show the lowest-lying $^4\Sigma^+$ to be a strong contender for the metastable excited state reported.[5] This state correlates to an excited $C^{++} + H$ asymptote and its radiative lifetime is calculated to be 4μs. Possible perturbations and predissocations of this state have not yet been investigated.

High-spin states of several well-known molecules, CN and NO have been investigated to determine if binding exists in these states. Sextet states, with five unpaired electrons, arise from ground state asymptotes $C(^3P) + N(^4S)$ and $N(^4S) + O(^3P)$. If binding exists, no predissocation will be possible because no states dissociate to lower-lying asymptotes. Our calculations show significant binding for CN $^6\Sigma^+$ and marginal amounts of binding for NO $^6\Sigma^+$, but these calculations are not ready to be quantified because the degree of basis set superposition error has not been determined. If binding does exist, decay by electric dipole transitions will not occur because these are the lowest states of sextet symmetry. However spin-orbit mixing with adjacent quartet states may be a significant decay mechanism.

Finally, the v'=0 vibrational levels of the $D^1\Delta$ and $I^1\Sigma^-$ have been shown to have very long radiative lifetimes[7] because they are essentially energy-degenerate with the only lower-lying state to which they can radiate via an electric dipole transition. While these electronic states lie ~ 8 eV above the ground state and cannot predissociate because they correlate to ground-state asymptotes, significant perturbations such as spin-orbit interactions with the lower-lying $a^3\Pi$ may provide a "leak" for a metastable population. Such issues are under investigation.

Acknowledgment: The work on metastable molecular systems was supported by AFOSR #-88-0042.

References

1. T. Ast, C. J. Porter, C. J. Proctor and J. H. Beynon, Chem. Phys. Lett. 78, 439 (1981).

2. D. Mathur, C. Badrinathan, F. A. Rajgara, and U. T. Raheja, Chem. Physics 103, 447 (1986).

3. D. Mathur and C. Badrinathan, J. Phys. B. 20, 1517 (1987).

4. D. Mathur, Chem. Phys. Lett. 150, 547 (1988).

5. M. Hamdan, A. G. Brenton and D. Mathur, Chem. Phys. Lett. 144, 387 (1988).

6. W. Koch, B. Liu, T. Weiske, C. B. Lebrilla, T. Drewello, and H. Schwarz, Chem. Phys. Lett. 142, 147 (1987).

7. K. Kirby, M. E. Rosenkrantz and D. L. Cooper, submitted for publication.

A Dynamical Picture of Hadron-Hadron Collisions With the String-Parton Model

D. J. Dean[a,b], A. S. Umar[b], J.-S. Wu[a],
and M. R. Strayer[a]

*Center for Computationally Intensive Physics,
Physics Division, Oak Ridge National Laboratory, Oak Ridge, TN 37831*

[a] *Physics Division, Oak Ridge National Laboratory, Oak Ridge, TN 37831*
[b] *Vanderbilt University, Department of Physics & Astronomy, Nashville, TN 37235*

ABSTRACT

We introduce a dynamical model for the description of hadron-hadron collisions at relativistic energies. The model is based on classical Nambu-Gotō strings. The string motion is performed in unrestricted four-dimensional space-time. The string endpoints are interpreted as partons which carry energy and momentum. We study e^+e^-, $e-p$, and $p-p$ collisions at various center-of-mass energies. The three basic features of our model are as follows. An ensemble of strings with different endpoint dynamics is used to approximately reproduce the valence quark structure functions. We introduce an adiabatic hadronization mechanism for string breakup via $q\bar{q}$ pair production. The interaction between strings is formulated in terms of a quark-quark scattering amplitude and exchange. This model will be used to describe relativistic heavy-ion collisions in future work.

I. INTRODUCTION

It has been suggested that collisions of heavy ions at relativistic energies may achieve high enough temperatures and energy densities to induce a phase transition from ordinary hadronic matter to a new and novel form of matter called the quark-gluon plasma [1]. This form of matter may have been formed during the first few moments of the Universe and it is of significant importance for cosmological studies. It is believed that the conditions for the formation of the plasma may be attainable with heavy projectile beams at CERN and the AGS, and at the Relativistic Heavy-Ion Collider (RHIC) facility, under construction at the Brookhaven National Laboratory. Detection of the plasma formation also poses difficult experimental and theoretical questions. At the moment, there is no clear hadronic signal suggesting the formation of the plasma. The difficulty is mostly due to the confining nature of the strong interaction, which only allows hadronic final states, and thus a detailed understanding of all of the hadronic decay processes is necessary before the identification of the plasma can be achieved.

In this paper we introduce a real-time dynamical model for studying the inclusive properties of hadronic collisions. The model is based on the Nambu-Gotō string description of hadrons supplemented by extensions to incorporate the basic features of the parton model, together with a hadronization mechanism. The string action, being proportional to the invariant area swept by the string, results in a linear-confining potential in a way similar to the area dependence of the Wilson loop parameter of lattice QCD. In the string picture the quarks are attached at the ends of a string at all times, and the confinement is an integral part of the model. The Nambu-Gotō strings produce linearly rising Regge trajectories for the hadronic spectrum and reproduce the masses of hadronic resonances [2].

The real-time dynamics of boosted strings must be supplemented by a hadronization mechanism and an interaction mechanism. For the hadronization mechanism, we use the pair-creation followed by a string breakup method which is similar to flux-tube breaking of the strong-coupled QCD calculations [3]. Our procedure is consistent with the inside-outside cascade picture observed in relativistic collisions.

The primary focus of our approach to the string-string interactions is the string endpoints which are interpreted as dynamical quarks. The quarks on each string act as distinguishable particles [4] and scatter with a phenomenological amplitude. This interaction is followed by the exchange of the two quarks in analogy with quark-exchange mechanisms mentioned elsewhere [5,6]. The dynamics of boosted relativistic strings lead to a distribution of particles with a rapidity plateau, and with the intial quarks producing leading particle effects [7]. The result is a dynamical model with relatively few free parameters which could be applied to the study of relativistic heavy-ion collisions. The parameters of the model are fixed by comparing them to e^+e^-, $e-p$, and $p-p$ experiments. We also note that the real-time string-parton model is considerably different from other string based statistical models [8–12].

This paper is organized as follows. Section II gives a brief outline of the string-parton model. In Section III we discuss the decay mechanism. Section IV indicates how the quark distribution functions are introduced into the model, and in Section V we discuss interactions in the parton model. A conclusion follows in Section VI.

II. THE STRING-PARTON MODEL

In this section we outline the basic properties and equations for relativistic, open classical strings. Classical strings serve as a phenomenological tool to study the physics of extended confined objects [13]. The derivation of the string equations of motion is considerably involved. Here, we will only give an outline of the basic equations and concentrate on the physical picture of the string motion. More detailed derivations of the equations can be found in Refs. [14, 15]. We work in natural units where $\hbar = c = 1$.

The action of the relativistic strings is constructed in analogy with the action of a free particle. While the free particle action is proportional to the length of its world line, the string action is proportional to the invariant area swept by the string. The string is defined to be a finite curve in space which sweeps out a hypersurface in four-dimensional space-time. The two-dimensional surface can be parameterized in terms of the general coordinates τ and s as $x^\mu(\tau, s)$. For a two-dimensional surface embedded in four-dimensional space-time, the area element is [14]

$$dA = \left\{ \left(\frac{\partial x^\mu}{\partial s} \frac{\partial x_\mu}{\partial \tau} \right)^2 - \left(\frac{\partial x^\mu}{\partial s} \frac{\partial x_\mu}{\partial s} \right) \left(\frac{\partial x^\nu}{\partial \tau} \frac{\partial x_\nu}{\partial \tau} \right) \right\}^{\frac{1}{2}} ds d\tau , \qquad (1)$$

which is used to define the string action as

$$S = -\kappa \int dA \qquad (2)$$

where κ is the string tension and is equal to approximately $0.9\ GeV/fm$. The time over which the action is to be considered is defined by an initial time τ_i and a final time τ_f. The string ends are defined to be $s = 0$ and $s = S$, and κs represents the accumulative energy of the string from its zero endpoint to any other point along the string. Thus, the total energy of the string is κS. The initial and final configuration of the string will be that seen by a definite observer at a given instant of time in his Lorentz frame. This action is invariant under general string coordinate transformations and satisfies energy-momentum conservation.

General equations of motion and the boundary conditions are obtained by small variations of the surface that joins the initial and final configurations of the string. Due to the arbitrariness of the parameterization of the surface swept by the string, we can choose additional coordinate conditions (Virasoro gauge conditions) which simplify the equations of motion. We work with the orthonormal parameterization

$$\frac{\partial x^\mu}{\partial s}\frac{\partial x_\mu}{\partial \tau} = 0, \qquad (3)$$

which implies that the velocity of a point along the string is always perpendicular to the string. We also note that the motion perpendicular to the string is time-like, whereas motion along the string is space-like. With this choice, the equations of motion reduce to the wave equation

$$\frac{\partial^2 x^\mu}{\partial \tau^2} - \frac{\partial^2 x^\mu}{\partial s^2} = 0, \qquad (4)$$

with the additional coordinate conditions

$$\left(\frac{\partial x^\mu}{\partial \tau}\right)^2 + \left(\frac{\partial x^\mu}{\partial s}\right)^2 = 0, \qquad (5)$$

and Eq. (3). In the following, we choose the coordinates such that [14] $x^0(\tau, s) = \tau = ct$ and $\vec{x} = \vec{x}(\tau, s)$. Therefore, at the endpoints of the string, Eq. (5) gives

$$\frac{\partial \vec{x}(\tau, s)}{\partial t} = 1, \qquad (6)$$

which means that the endpoints move at the speed of light. We also obtain the spatial endpoint boundary conditions

$$\frac{\partial x^\mu(\tau, 0)}{\partial s} = \frac{\partial x^\mu(\tau, S)}{\partial s} = 0, \qquad (7)$$

which tell us that there is no energy-momentum transfer out from the string endpoints.

The solution of Eq. (4) subject to the above coordinate conditions and the first of the conditions in Eq. (7), give

$$x^\mu(\tau, s) = \frac{1}{2}[y^\mu(\tau + s) + y^\mu(\tau - s)], \qquad (8)$$

where we have defined $y^\mu(t) \equiv x^\mu(t, 0)$, which is the trajectory of a single endpoint. This indicates that the entire string can be constructed from the knowledge of the trajectory of a single endpoint. The application of the second boundary condition in Eq. (7) results in the endpoint periodicity equation

$$y^\mu(\tau + S) = y^\mu(\tau - S) + \frac{2P^\mu}{\kappa}, \qquad (9)$$

where P^μ is the total four-momentum of the string and is given by

$$P^\mu = \kappa \int_0^S ds \frac{\partial x^\mu}{\partial \tau}. \qquad (10)$$

Eq. (9) is a periodicity condition for the equations of motion. Thus, if at time $\tau = 0$ the trajectory values of $\vec{y}(s)$ are known from $s = -E/\kappa$ to $s = E/\kappa$, and the momentum $\vec{P}$ is also known, then one can compute $\vec{y}$ for all times. This general procedure is illustrated in Fig. 1. Here, the curve labeled $\vec{y}(\tau)$ describes the trajectory of one of the endpoints. To construct the entire string at time, say $\tau = 0$, we draw the vectors from $\vec{y}(0)$ to an arbitrary distance $\pm s$ along the trajectory. The location of the string point $\vec{x}(0, s)$ is then given by Eq. (8). Similarly, the vector from $\vec{y}(-s)$ to $\vec{y}(s)$ is proportional to the momentum of this string point via the generalized version of Eq. (9). At this point, we also note that linear segments of the endpoint trajectory will lead to multiply "hit" points. These points will have a larger energy-momentum content.

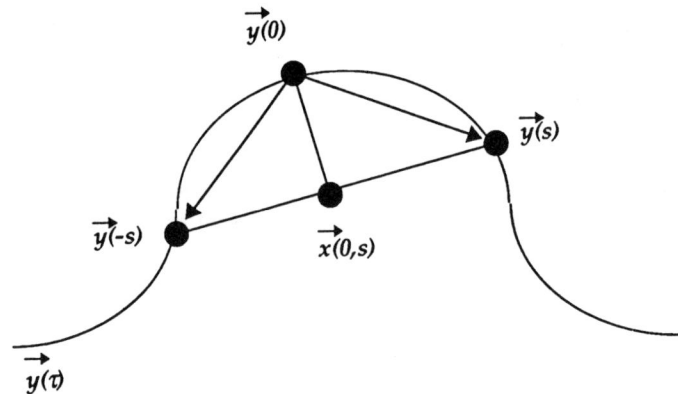

Figure 1. The construction of a string from the trajectory of one of its endpoints. The details are explained in the text.

Strings may be classified by their endpoint motion. The difference between various string structures manifests itself in the details of the string motion, and more importantly in the amount of the total string energy carried by the endpoints. In Section IV we utilize these properties to discuss the quark structure functions in the string picture.

III. HADRONIZATION MECHANISM

In order for the string picture to address the physics of strongly interacting high-energy particles, it must incorporate the hadronization process leading to color singlet final-state hadrons. The hadronization process is believed to take place via soft nonperturbative mechanisms which cannot be calculated by perturbation theory. Below, we will outline an adiabatic approach for string fragmentation. In addition we will introduce a method for assigning transverse momenta to the created $q\bar{q}$ pairs. The formalism will be used to study the hadronization process in e^+e^- collisions.

In our simulation of high-energy collisions, string-string interaction mechanisms lead to excited strings which stretch. These excited strings decay through string breaking. The decay process is outlined in three steps. First, only those strings which satisfy a minimal mass criterion may decay. The daughter strings of a decay must have a mass of at least M_{cq} for a meson, and M_{cqq} for a baryon. Second, at which point an excited string may decay is determined by a probabilistic decay law. The decay law depends on the invariant area that an infinitesimal piece of string δA_i sweeps out during a given time $d\tau$. A simple rate equation is given by

$$\frac{dP(\delta A_i)}{dA_i} = -\Lambda P(\delta A_i) , \qquad (11)$$

where Λ is the decay rate. Finally, $q\bar{q}$ pair production in a flux tube will produce transverse momentum. In the absence of any fundamental calculations, we choose to parameterize the transverse momentum assignment with a simple exponential distribution function

$$f(p_T)p_T dp_T \propto e^{-\alpha p_T} p_T dp_T , \qquad (12)$$

where α is related to the average transverse momentum produced during decay. This simple decay process, combined with the string dynamics, produces many of the gross features observed in high-energy fragmentation as follows: (i) the string breaking will commonly occur where the string stretches most, leading to the observed pionization, (ii) the pion distribution is somewhat uniform and produces a rapidity plateau, and (iii) since the primary quarks are at the ends of a string and fragmentation begins in the central region, the string pieces containing the primary quarks fragment latest thus producing the leading particle effect. Further details of the decay mechanism may be found in Ref. [15].

Parameters for decay were chosen as follows. The cutoff masses are chosen to obtain the experimental multiplicities in high-energy e^+e^- collisions. We have determined the cutoff mass M_{cq} to be $= 0.25\ GeV$. The cutoff mass M_{cqq} is set equal to $= 0.25\ GeV$ for the e^+e^- calculations, whereas in hadron-hadron collisions it is equal to the proton mass. The decay rate is chosen to be $1.0\ fm^{-2}$ which is commensurate with decay rates used in the LUND simulation programs. In principle, parameter Λ and the cutoff masses can be fine-tuned. This was not done for this work. Finally, we note that the average experimental transverse momentum is related to α in Eq. (12) by $<p_T> = 2/\alpha$. We have found that $\alpha = 3.88 \, (GeV/c)^{-1}$ produces the experimental low transverse momentum distributions fairly accurately with $<p_T> \approx 0.350 \, GeV/c$. This difference is due to the restrictions arising from the cutoff masses. The inclusion of the string-string interactions will be discussed in Section V. First we must consider the initial string states for the nucleons.

IV. STRUCTURE FUNCTIONS

In building the phenomenology of the dynamical string-parton model description of relativistic heavy-ion collisions, it is desirable to start from a description which entails many of the features observed for more elementary high-energy processes. One of the most important properties of hadrons is their substructure observed mainly via deep-inelastic, charged lepton-hadron collisions [16, 17]. The quark distribution functions obtained from deep-inelastic $e - p$ collisions [18] can be parameterized for convenience. Here, we use the Eichten, Hinchliffe, Lane, and Quigg, ($EHLQ$) parameterization [19]

$$xu_v(x) = 1.78 x^{0.5} \left(1 - x^{1.51}\right)^{3.5}$$
$$xd_v(x) = 0.67 x^{0.4} \left(1 - x^{1.51}\right)^{4.5}, \qquad (13)$$

where u_v and d_v are the valence quark distribution functions, and x is the Bjorken scaling variable. The u_v distribution integrates to 2, and d_v distribution to 1, giving

$$\frac{1}{3} \int_0^1 [u_v(x) + d_v(x)] \, dx = 1, \qquad (14)$$

and thus yielding a flavor-averaged *proton* distribution function

$$q_p(x) = \frac{1}{3} [u_v(x) + d_v(x)]. \qquad (15)$$

It can be shown in the infinite momentum frame that the scaling variable x is the fraction of the momentum of the nucleon carried by the struck parton. This relation is true only in this frame; however, it is approximately valid in other frames if the partons are assumed to be massless. Corrections arising from the finite parton mass are usually neglected, as well as the small differences between neutron and proton distribution functions [20].

In the relativistic string picture, the string endpoints are interpreted as massless quarks moving at the speed of light. For the description of baryons, one end represents a single quark, whereas the other end a diquark. Each quark carries a baryon number of $1/3$ thus giving $B = 1$ for baryons. The description of mesons involves a quark at one end and an antiquark at the other. In addition, as we have discussed in a previous section, the endpoints contain a varying amount of the total string energy. Thus, for relativistic strings, it is natural to define a fractional momentum variable associated with the string endpoints. Assuming collinear motion along the z-direction, which will be the boost axis, we define the string longitudinal momentum fraction in terms of the ratio of the light-cone variables,

$$x_s = \frac{k_0 + k_3}{P_0 + P_3}, \quad 0 \le x_s \le 1, \tag{16}$$

where k is the endpoint four-momentum and P is the total string four-momentum. The variable x_s is Lorentz invariant for boosts in the longitudinal direction.

In order to construct a distribution function, we consider the probability for finding a string endpoint with a momentum fraction x_s. As we have seen in a previous section, the string endpoints move in time and change their energy-momentum content. Furthermore, Eq. (16) also indicates a dependence on the orientation of the endpoint trajectory with respect to the boost axis. Thus, the probability depends on two independent variables, time t and the orientation angle θ, and can be described by the function $\mathcal{P}_i(x_s; t, \theta)$, where the index i identifies strings with different endpoint dynamics. θ is defined as the angle between the normal to the plane in which the endpoint trajectory lies and the boost axis in the rest frame of the string. Averaging this distribution over one time period and over all possible orientations defines the distribution function for a particular string:

$$q_i(x_s) = \frac{1}{2T} \int_0^T dt \int_0^\pi sin\theta d\theta \mathcal{P}_i(x_s; t, \theta). \tag{17}$$

The functions $q_i(x_s)$ approach frame independence for large γ, in accordance with the parton model. Actually $\gamma \ge 5$ yields nearly frame independent distribution functions. Since strings with different endpoint dynamics correspond to different values for $(x_s)_{max}$, an ensemble of such strings can be used to reproduce the proton structure function $q_p(x)$ of Eq. (15). We denote this ensemble by

$$q_{sp}(x_s) = \sum_i \rho_{i/p} = \sum_i C_i q_i(x_s), \tag{18}$$

where q_{sp} is the calculated valence quark distribution function for the proton p, and C_i denotes the weight of the type i string in our ensemble. We choose for our ensemble endpoint dynamics of the yo-yo ($i = 1$), isosceles triangles with sides whose lengths are in the ratio 0.4 : 0.4 : 0.2 ($i = 2$) and 0.45 : 0.45 : 0.1 ($i = 3$), and equalateral triangles ($i = 4$). Since the range of x_s values for strings executing different endpoint motion do not fully overlap, the coefficients C_i can be determined by first requiring the equivalence of $q_p(x_s)$ and $q_{sp}(x_s)$ at various values of x_s. The details are given in Ref. [15]. The coefficients are $C_1 = 0.091$, $C_2 = 0.44$, $C_3 = 0.226$ and $C_4 = 0.243$. The corresponding agreement between $q_p(x_s)$ and $q_{sp}(x_s)$ is very reasonable. This ensemble will be used when performing string-parton model calculations of collisions involving protons.

V. HADRON-HADRON COLLISIONS

For the simulation of relativistic hadron-hadron collisions, an interaction mechanism between two evolving strings must be introduced. In the large Q^2 limit (where Q denotes the four-momentum transfer), there is considerable evidence that constituent quarks behave as point-like particles and undergo hard scatterings [16, 21, 22]. Such hard scatterings are believed to be the main source of high p_T production in hadron-hadron collisions [23]. Similarly, in the low Q^2 regime, the hadrons can interact via quark or other exchange mechanisms and still preserve their color singlet nature. Such mechanisms have been applied to construct potential models for the hadron-hadron interactions [5]. Experimental studies of jets in hadron-hadron and e^+e^- collisions also indicate that the hadronization mechanism occurs from color neutral objects [9] suggesting that the color singlet nature of the strings must be preserved during the interaction. In this section we incorporate these features into the string model via the introduction of phenomenological interactions between the string endpoints.

In previous work that involved parton-parton collisions, a classical radius of interaction was chosen [24]. In the following we present a brief outline of parton-parton scattering theory which has been used in this work. We consider the scattering of two nucleons for which the interaction takes place between constituent partons. Factorization [25] may be used to write the nucleon-nucleon cross section in terms of the parton-parton cross sections. The nucleon-nucleon cross section is then

$$\omega_{AB}d\sigma = \sum_{ij} dx_A dx_B \rho_{i/A}(x_A)\rho_{j/B}(x_B)\omega_{ij}d\sigma_{ij} , \qquad (19)$$

where

$$d\sigma_{ij} = \sum_{\alpha\beta} |\langle f'_\alpha f'_\beta | S | f_i f_j \rangle|^2 d^2 b . \qquad (20)$$

The expression $\rho_{i/A}(x_A)dx_A$ is the probability of finding a parton of type i in nucleon A with a momentum fraction between x_A and $x_A + dx_A$. The normalization is given by $\omega_{AB} = 2\omega_A 2\omega_B$, and $\omega_{ij} = 2\omega_i 2\omega_j$. The cross section $d\sigma_{ij}$ denotes an inclusive observable summed over all final states, $|f'_\alpha f'_\beta\rangle$. In principle, we could evaluate the scattering into definite final states (f'_a, f'_b); however, in our treatment it is advantageous to use completeness of these states and replace them with a complete set of free particle states such that

$$d\sigma_{ij} = \sum_{\alpha\beta}\langle f_i f_j | S^\dagger | f'_\alpha f'_\beta\rangle\langle f'_\alpha f'_\beta | S | f_i f_j\rangle d^2 b$$

$$= \sum_{p'_1 p'_2}\langle f_i f_j | S^\dagger | p'_1 p'_2\rangle\langle p'_1 p'_2 | S | f_i f_j\rangle d^2 b$$

$$= \sum_{p'_1 p'_2} |\langle p'_1 p'_2 | S | f_i f_j\rangle|^2 . \qquad (21)$$

In order to calculate the cross sections, we must evaluate the scattering matrix, which is carried out in Ref. [26]. The final expression for the impact parameter dependent cross section for nucleon-nucleon scattering is given by

$$\omega_{AB}d\sigma = \sum_{ij} dx_A dx_B \rho_{i/A}(x_A)\rho_{j/B}(x_B)\omega_{ij} \sum_{p'_1 p'_2} \frac{1}{|\vec{\beta}_i - \vec{\beta}_j|}$$

$$\times F_{ij}(\vec{b})(2\pi)^4 \delta^4(\bar{p}_i + \bar{p}_j - p'_1 - p'_2)|\mathcal{M}(\bar{p}_i, \bar{p}_j; p'_1, p'_2)|^2 d^2 b , \qquad (22)$$

where $F_{ij}(\vec{b})$ describes the impact parameter dependence of the overlaps of the interaction wavefunctions, and $\mathcal{M}$ describes the point like scattering of the interacting partons. $\bar{p}$ are entrance channel momenta and $\bar{p}'$ are exit channel momenta of the scattering partons.

The invariant matrix element $\mathcal{M}$ may be calculated in terms of the Mandelstam variables for the quarks s, t, u, which are defined as

$$s = (\bar{p}_i + \bar{p}_j)^2, \quad t = (\bar{p}_i - p'_2)^2, \quad u = (\bar{p}_i - p'_1)^2 . \tag{23}$$

In the string model, since all interactions at this stage are mediated by the quarks, we use a phenomenological scattering, which includes the first Born order terms, for the quark scattering. The invariant matrix element is then given by:

$$|\mathcal{M}(\bar{p}_i \bar{p}_j; p'_1 p'_2)|^2 = \alpha_t^* \frac{s^2 + u^2}{(t - m_t^2)^2} + \alpha_u^* \frac{s^2 + t^2}{(u - m_s^2)^2} , \tag{24}$$

where α_t^*, α_u^* are strength parameters for the two scattering channels and m_t, m_u are effective range parameters. At this time $\alpha_t^* = \alpha_u^* = 0.7$ and $m_t = m_u = 1/r_\perp$, where $r_\perp$ is an effective range of the interaction between two quarks. In future work, as we move the model up in collision energy, we will be able to use these parameters to obtain the correct jet cross sections in proton-proton collisions. In the present work, $r_\perp = 0.7 \; fm$ which gives a proton-proton cross section of 30 mb.

We have performed simulations for $p - p$ collisions at two different center-of-mass energies, $\sqrt{s} = 19.44$, and 53 GeV. For these energies, strings were discretized using 1,300 and 1,990 points, respectively. Various distributions were obtained including the rapidity and transverse momentum. Rapidity is defined as

$$y = \frac{1}{2} \ln \left(\frac{E + p_l}{E - p_l} \right) ,$$

where the longitudinal direction is along the jet axis. The value of the quark cutoff mass was increased to $M_{cq} = 0.36 \; GeV$ in order to obtain the correct total multiplicity of final particles. The diquark cutoff mass was set equal to the proton mass, $M_{cqq} = 0.94 \; GeV$. This readjustment is expected since the charged particle multiplicities in $p-p$ collisions are approximately 20% lower than the e^+e^- collisions at the same center-of-mass energy [27]. In Fig. 2 we plot the rapidity distribution for $\sqrt{s} = 53 \; GeV$. The data show the experimental positive- and negative- charged particle distributions [28] in comparison to our calculations. Similar to the e^+e^- case, the agreement in the central rapidity region is good whereas the large rapidity tails are not reproduced due to the large value of the cutoff mass, which inhibits the production of low-mass large rapidity particles. Fig.3 shows the calculated transverse momentum distribution at $\sqrt{s} = 19.4 \; GeV$ and at $\sqrt{s} = 53 \; GeV$ in comparison with the experimental results [28] at 53 GeV. The results are in good agreement up to the transverse momentum range considered in these calculations ($p_T < 1.1 \; GeV/c$). The scattering model discussed in the previous section would allow comparisons at much higher values for transverse momentum; however, the required statistical accuracy for the reproduction of the exponentially decaying distribution would require 50,000 − 100,000 collisions. These calculations will be performed in the future.

VI. CONCLUSION

We have presented a model based on the relativistic Nambu-Gotō string formalism for studying inclusive hadronic processes in high-energy collisions. The

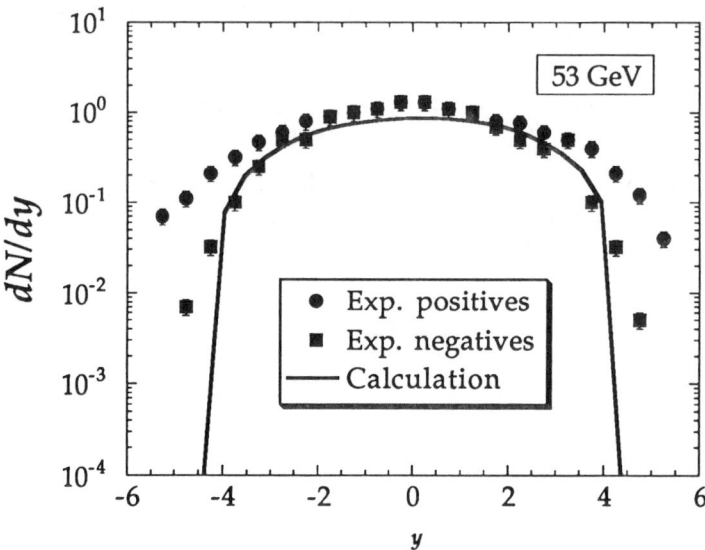

Figure 2. Rapidity distributions of charged particles of positive and negative particles for pp collisions $\sqrt{s} = 53\ GeV$. Data are from Ref. [28].

calculations are performed in four-dimensional space-time with no restrictions. We have shown that an ensemble of different string structures can be used to approximately reproduce the experimentally observed valence quark structure functions. This was achieved by the interpretation of string endpoints as quarks or diquarks. The interaction mechanism employs a $u-$ and $t-$ channel parton exchange interaction with phenomenological ranges and strengths. Excited strings decay according to an adiabatic decay scheme, and the decay is interpreted as being due to $q\bar{q}$ production along the string. The produced quarks are assigned transverse momenta based on a phenomenological distribution. In this form the formalism successfully reproduces the gross features of the data from relativistic e^+e^-, $e-p$, and $p-p$ collisions. These include the rapidity distributions, transverse momentum distributions, longitudinal momentum fraction distributions, and total charged particle multiplicities. Due to the time-dependent nature of the model, we also obtain meson and baryon formation times and the time development of various particle and energy densities. In Table I we tabulate the set of parameters used in the model and the criteria for their choice.

The string phenomenology has been a successful tool for performing simulations of strong interaction physics. Confinement is conceptually incorporated into the model, and other connections with QCD model calculations, as mentioned in the introduction, make the string formalism an attractive approach for studying strongly interacting extended objects. An important caveat to stress at this point is that the connection of the model to the true nature of the strong interaction dynamics is rather tenuous. This is particularly true for the formulation of quark-quark interactions and the $q\bar{q}$ production which are among the unresolved problems of quantum chromodynamics. Our goal is to develop an effective dynamical formalism which could be used for studying the hadronic interactions taking place during the collisions of relativistic heavy ions. The hope is that a better understanding of the hadronic debris would improve our chances of isolating the signals coming from the decay of the quark-gluon plasma. As it stands, the model is readily extendable to the studies of $e - A$, $p - A$, and $A - A$ collisions. These computer- intensive calculations are currently underway.

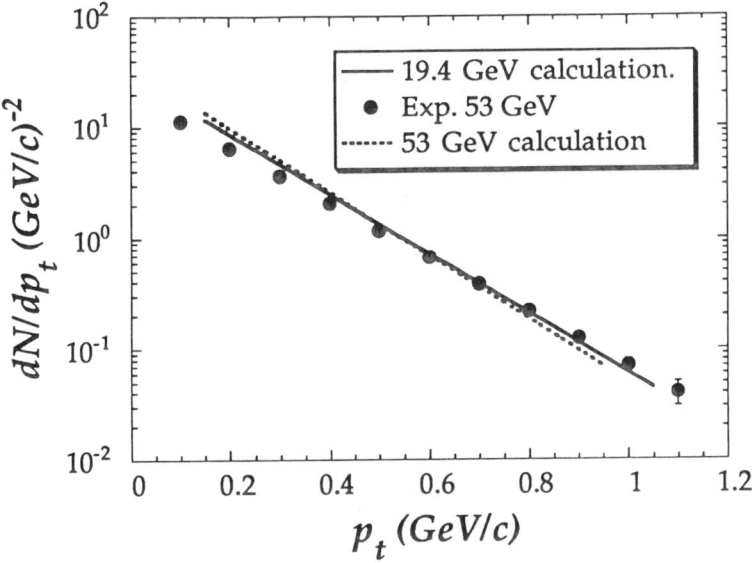

Figure 3. The calculated transverse momentum distributions are shown for pp collisions at $\sqrt{s} = 19.4$ and $\sqrt{s} = 53$ GeV. Data for the two energies in this momentum range are similar, and only the 53 GeV data are plotted [28].

ACKNOWLEDGMENTS

This research was sponsored in part by the U.S. Department of Energy under contract No. DE-AC05-84OR21400 with Martin Marietta Energy Systems, Inc. and under contract No. DE-FG05-87ER40376 with Vanderbilt University. The numerical calculations were carried out on CRAY-2 supercomputers at the National Center for Supercomputing Applications, Illinois, at the National Energy Research Supercomputer Center, Livermore, and on the Intel iPSC/860 hypercube at the Oak Ridge National Laboratory. The authors also wish to acknowledge the Institute for Nuclear Theory at the University of Washington, Program III: Hard QCD Probes of Dense Nuclear and Hadronic Matter, 1990, where some of the development of the present work took place.

TABLE I. Parameters of the Model

Symbol	Name	Method of Determination	Value
κ	String tension	Regge slope	$0.88\ GeV/fm$
$M_{cq}(e^+e^-)$	Meson mass cutoff	Total multiplicity	$0.25\ GeV$
$M_{cq}(p-p)$	Meson mass cutoff	Total multiplicity	$0.36\ GeV$
M_{cqq}	Baryon mass cutoff	Proton mass	$0.94\ GeV$
Λ	Decay rate	LUND	$1.0\ fm^{-2}$
$f(p_T),\ \alpha$	p_T distribution	fit to e^+e^- data	$3.88\ (GeV/c)^{-1}$
$r_\perp$	Interaction range	$p-p$ cross section	$0.7\ fm$
α_s^*	Effective coupling constant	$p-p$ cross section	0.71

REFERENCES

[1] *Quark Matter '88*, Proceedings of the Seventh International Conference, Lenox, Massachusetts, 1988, edited by G. A. Baym, P. Braun-Munzinger and S. Nagamiya, Nucl. Phys. **A498**, (1989), and references therein; J. D. Bjorken, Phys. Rev. D **27**, 140 (1983).

[2] S. Mandelstam, Phys. Rep. **13C**, 261 (1974); J. Scherk, Rev. Mod. Phys. **47**, 123 (1975).

[3] R. Kokoski and N. Isgur, Phys. Rev. D **35**, 907 (1987); G. A. Miller, Phys. Rev. D **37**, 2431 (1988); P. Geiger and N. Isgur, Phys. Rev. D **41**, 1595 (1990).

[4] G. A. Miller, Nucl. Phys. **A497**, 277c (1989).

[5] N. Isgur, Nucl. Phys. **A497**, 91c (1989); K. Maltman and N. Isgur, Phys. Rev. D **29**, 952 (1984).

[6] P. J. Mulders and A. E. L. Dieperink, Nucl. Phys. **A483**, 461 (1988); A. H. Mueller, Nucl. Phys. **A498**, 41c (1989).

[7] R. Anishetty, P. Koehler, and L. McLerran, Phys. Rev. D **22**, 2793 (1980).

[8] B. Andersson and G. Gustafson, Z. Phys. C **3**, 223 (1980); B. Andersson, G. Gustafson, G. Ingelman, and T. Sjostrand, Phys. Rep. **97**, 33 (1983).

[9] P. Mättig, Phys. Rep. **177**, 141 (1989).

[10] B. Andersson, G. Gustafson, and B. Nilsson-Almqvist, Nucl. Phys. **B281**, 289 (1987).

[11] K. Werner, Z. Phys. C **42**, 85 (1989).

[12] J. Ranft, Phys. Rev. D **37**, 1842 (1988).

[13] X. Artru and G. Mennessier, Nucl. Phys. **B70**, 93 (1974); X. Artru, Nucl. Phys. **B85**, 442 (1975); X. Artru, Phys. Rep. **97**, 147 (1983).

[14] T. Gotō, Prog. Theo. Phys. **46**, 1560 (1971); P. Goddard, J. Goldstone, C. Rebbi, and C. B. Thorn, Nucl. Phys. **B56**, 109 (1973).

[15] D. J. Dean, Ph.D. Thesis, Vanderbilt University, 1991, (unpublished); D. J. Dean, A. S. Umar, J.-S. Wu, and M. R. Strayer, submitted to Phys. Rev. C, (1991).

[16] V. D. Barger and R. J. N. Phillips, *Collider Physics*, (Addison-Wesley, New York, 1987) and references therein; P. D. B. Collins and A. D. Martin, *Hadron*

Interactions, (Adam Hilger, Bristol, 1984).
[17] T. Sloan, G. Smadja, and R. Voss, Phys. Rep. **162**, 45 (1988).
[18] D. H. Perkins, *Introduction to High Energy Physics*, (Addison-Wesley, MA, 1987), 3rd edition.
[19] E. Eichten, I. Hinchliffe, K. Lane, and C. Quigg, Rev. Mod. Phys. **56**, 579 (1984); A. D. Martin, R. G. Roberts, and W. J. Stirling, Phys. Rev. D **37**, 1161 (1988).
[20] A. Bodek et al., Phys. Rev. D **20**, 1471 (1979).
[21] G. Altarelli, Phys. Rep. **81**, 1 (1982).
[22] F. E. Close, *An Introduction to Quarks and Partons*, (Academic Press, London, 1979).
[23] S. M. Berman, J. D. Bjorken, and J. B. Kogut, Phys. Rev. D **4**, 3388 (1971); R. F. Cahalan, K. A. Geer, J. Kogut, and L. Suskind, Phys. Rev. D **11**, 1199 (1975); B. L. Combridge, J. Kripfganz, and J. Ranft, Phys. Lett. **70B**, 234 (1977); R. Cutler and D. Sivers, Phys. Rev. D **17**, 196 (1978).
[24] K. Werner, Phys. Rev. Lett. **62**, 2460 (1989).
[25] L. Magnea and G. Sterman, Proc. of *Workshop on Hadron Structure Functions and Parton Distributions*, ed. D. F. Geesaman, *et.al.*, (World Scientific, Singapore, 1990) p.423.
[26] J.-S. Wu, D. J. Dean, A. S. Umar, and M. R. Strayer, to be published.
[27] W. Thome, et al, Nucl. Phys. **B129**, 365 (1978).
[28] A. Breakstone et al, Phys. Lett. **132B**, 458 (1983); Phys. Rev. D **30**, 528 (1984).

HIGH Tc SUPERCONDUCTORS: NUMERICAL STUDIES OF THE HUBBARD MODEL

Elbio Dagotto

Institute for Theoretical Physics
University of California at Santa Barbara
Santa Barbara, CA 93106, USA

ABSTRACT

Recent numerical work on strongly correlated electronic models using the Lanczos approach is reviewed. In particular, static and dynamical properties of the Hubbard and $t-J$ models are presented attempting to make a connection with experiments. We conclude that numerical methods like the Lanczos technique are providing useful information in the study of these models.

I. INTRODUCTION

High Tc superconductivity[1] is a very active field of research. Hundreds of papers have addressed this subject from many different points of view and using different approximations and formulations. The potential technological importance of room temperature superconductors contributes to this excitement. The "physics" behind these materials seems to be highly nontrivial and to explain their behavior represents a major challenge to theoretical physics. In this article I will review numerical work done in the context of very particular two dimensional models which are believed to have some relation with the new oxide superconductors. To be more concrete, I will try to present a status report of the $t-J$ and Hubbard models discussing static and dynamical properties. This review is basically a summary of references and previously published results in different papers and journals presented in a unified way. A basic knowledge of the phenomenology of the high Tc superconductors is assumed. An early version of this review[2] has already appeared.

Many electronic models are being currently studied in the context of high Tc superconductivity. In constructing model Hamiltonians, the kinetic energy term corresponds to electrons "jumping" from atom to atom with a probability, in principle, proportional to the overlap of their localized wave functions. Without a first principles calculation it is not clear if the probability of hopping at distances larger than one lattice spacing is relevant or not. The interaction between electrons is electromagnetic and in principle slowly decaying with distance r as $1/r$ but screening due to the other electrons may drastically reduce the range of interaction. It is widely assumed that this interaction is basically local i.e., two electrons "see" each other only when they are at the same atom (with different spins, of course).

This discussion should be applied, in principle, to a model where both Cu's and O's are included on a square lattice similar to the observed two dimensional (2D) structure of the Cu-O planes in the new materials (which are assumed to be responsible for their superconducting properties). Cu's are on the sites and O's on the links. The hopping term is mainly between Cu and O but other terms may be present. For example, a large overlap between O atoms would lead to a direct hopping among them. Although major simplifications have been done to arrive at this description of the Cu-O planes, it would be interesting, and quite useful, to find an even simpler Hamiltonian containing the basic physics of the new superconductor. Anderson and others[3] have claimed that a one band Hubbard model contains the basic physics of the Cu-O planes. This model is defined by the Hamiltonian,

$$H = -t \sum_{i,\hat{\delta},\sigma} \left(c^\dagger_{i,\sigma} c_{i+\hat{\delta},\sigma} + h.c. \right) + U \sum_i \left(n_{i,\uparrow} - \frac{1}{2} \right) \left(n_{i,\downarrow} - \frac{1}{2} \right), \quad (1)$$

where $c^\dagger_{i,\sigma}$ creates an electron of spin σ at site i of a two dimensional square lattice, $n_{i,\sigma}$ is the number operator and $\hat{\delta}$ is a unit vector connecting nearest-neighbor sites. t is the hopping parameter and U the on site Coulomb repulsion. Note that in this model there are no variables on the links of the lattice i.e., the O atoms have basically disappeared. Some experimental results suggest that the realistic regime of parameter space is U/t large (but that issue is not completely clear yet). In this strong coupling limit and using second order perturbation theory it is possible to further simplify the Hubbard model to what is called the $t - J$ model which is defined by the Hamiltonian

$$H = J \sum_{i,\hat{\delta}} \mathbf{S}_i \cdot \mathbf{S}_{i+\hat{\delta}} - t \sum_{i,\hat{\delta},\sigma} \left(\bar{c}^\dagger_{i,\sigma} \bar{c}_{i+\hat{\delta},\sigma} + h.c. \right), \quad (2)$$

where $\bar{c}^\dagger_{i,\sigma}$ is a *hole* creation operator acting in the space where there is no double occupancy, the spin variable is $\mathbf{S}_i = \frac{1}{2} c^\dagger_{i,\alpha} \vec{\sigma}_{\alpha,\beta} c_{i,\beta}$ and the rest of the notation is as in the Hubbard model. The coupling constants are related through $J = \frac{4t^2}{U}$. Recently, some work has been devoted to a possible derivation of the $t - J$ model based on the more general Cu-O Hamiltonian.[4] Below I present my understanding of the present status of numerical studies of the $t - J$ and Hubbard models. It is certainly not claimed that the references of this short review exhaust the literature on this very active topic of research.

II. HOLES IN THE $t - J$ MODEL: STATIC AND DYNAMICAL PROPERTIES

At half-filling the $t-J$ model is identical to the Heisenberg model. This is a welcome result since we know the Heisenberg model has a Néel ground state[5] with infinite range antiferromagnetic correlations at zero temperature and experimentally the existence of antiferromagnetism has been clearly established in the new materials. Then, the most important case to study corresponds to the analysis of the $t - J$ model with doping. If this model is realistic then at low temperatures and small but nonzero doping a superconducting phase should exist.

What numerical technique is suitable for the $t - J$ and Hubbard models with holes? In principle Quantum Monte Carlo techniques have sign problems at low temperatures and finite doping. However, there is another powerful method that allows us to get *ab initio* information of electronic and spin models. This method is the Lanczos technique in which the lowest lying eigenvalues and eigenvectors of a finite cluster of atoms can be obtained with high accuracy. There are many variations and generalizations of this technique. In particular we use a modification of the Lanczos method[6] which basically consists of the following steps: 1) choose an arbitrary initial state (for example the classical Néel state. The method works as long as this initial guess has a nonzero overlap with the actual ground state.); 2) apply H to this state. In such a way a vector orthogonal to the initial state is generated; 3) a combination of the old vector plus the new vector is proposed to minimize the energy; 4) iterate this procedure many times. This technique is easy to program and provides the ground state energy and wave function with high accuracy. Other techniques can also be used like the standard Lanczos or the conjugate gradient method.

The problem of all the Lanczos-like methods is that in their implementation it is necessary to deal with vectors of length the size of the Hilbert space of the cluster under consideration. Since these sizes usually increase exponentially with the number of sites of the lattice, the technique is restricted to small clusters. For example currently the limit for the Heisenberg model is about 32 sites (although 36 sites can be studied with some effort). For more complicated systems with more states per site, the largest lattices that can be analyzed are smaller. Typically a 4×4 lattice is used for the $t - J$ model as shown below (although slightly larger lattices of 18 and 20 sites are also reachable nowadays). For the Hubbard model the current record is a 4×4 lattice.[7] Are these lattices too small for our purposes? For the special case of the unfrustrated Heisenberg model a Lanczos study on a 4×4 lattice produced semi-quantitative correct results[8] compared with Monte Carlo simulations (no sign problems in this particular case) and there are no obvious reasons why including holes the situation will be worse at least at low doping. Then, Lanczos studies on 4×4 lattices should be regarded as a reasonable first principles calculation which provides unbiased results for electronic models.

Summarizing, the main differences between Lanczos and Monte Carlo methods are the following: the Lanczos approach is *exact*, it provides information about the ground state (*zero temperature*) and there are *no sign* problems but it is constrained to work on small lattices (up to 16 or 32 sites depending on the model). On the other hand, the Quantum Monte Carlo (QMC) method can attack *larger* lattices (for the Hubbard model at half-filling it is possible to analyze a 16×16 lattice). Its main restrictions are that it has sign problems, works at small but nonzero temperature and there are statistical errors. Depending on the problem at hand one should decide whether to use one technique or the other. Below and in Sections III and IV we will describe various results obtained using the Lanczos method.

a) **Quantum numbers of one hole:** Some of the properties of one hole are by now well established while others are controversial. It has been shown numerically[9] (on small lattices) that one hole in the $t - J$ model has a ground state with momentum either $\mathbf{k} = (\pm\frac{\pi}{2}, \pm\frac{\pi}{2})$ or $\mathbf{k} = (0, \pi), (\pi, 0)$. These two sets of momenta are part of the Fermi "surface" of the Hubbard model in the noninteracting case ($U = 0$). This result shows that turning on the Coulombic interaction and moving to the strong coupling regime does not alter much the quantum numbers of a hole (a smooth connection between weak and strong coupling seems to exist in this model).

b) **Binding of holes and phase separation:** Since the Hubbard model with $U > 0$ contains repulsive forces at the level of the Hamiltonian interaction, the claim that the model has a superconducting phase somewhat implies that effective attractive forces have to be generated leading to the condensate of pairs we are all looking for. Then, it would be interesting to find an indication of these types of forces in strongly correlated systems. In the $t - J$ model interesting work has been done on 4×4 lattices with many holes showing that there is actually a region where two holes prefer to be at short distance forming a bound state.[10,11] This conclusion was obtained by analyzing the difference in energy between the state of two holes and a state of two infinitely separated holes (both measured with respect to the "vacuum" which is the half-filled case). In other words, if $\Delta = (E_2 - E_0) - 2(E_1 - E_0)$ (E_n = energy of the ground state in the n-hole subspace) is negative then two holes minimize the energy by forming a bound state. The fact that an effective attraction is obtained in this model is not too surprising at least at large J/t where it is clear that in order to minimize the number of broken bonds in the antiferromagnet from 8 to 7, two holes should be at a distance of one lattice spacing.

However, this seemingly interesting situation where an attraction appears out of a repulsive Hamiltonian leads to another potential problem which is *phase separation*.[12] Like in a system of neutrons and protons in nuclear physics, the short range interactions are so strong that the ground state of many holes is simply a cluster (like a nucleus). This is the way in which the number of broken bonds is minimized. This state is not desirable if we want to find superconductivity out of the $t - J$ model. Of course, the Coulombic interaction between holes which has been neglected so far may influence this result but this is still under discussion.

What do we know about binding of two holes in the one band Hubbard model? It is difficult to perform a Lanczos study of this model on a 4×4 lattice. The main difficulty is the size of the Hilbert space for this model (4 states per site rather than 3 as in the $t - J$ model). A study of binding of holes for the Hubbard model has been presented using a new MC method that relies on an *imaginary* chemical potential for a 4×4 lattice.[13] It was found that two holes in a half-filled system bind with a small energy $\Delta \approx -0.10t$ at $U/t = 4$ although with large error bars coming from the fact that a binding energy arises from the combination of three bulk energies. Is there phase separation in the Hubbard model? This is a point that only recently has been addressed numerically by Moreo et al.[14] There are *no* numerical indications of phase separation for this model. The study is based on an analysis of the chemical potential versus density of particles looking for discontinuities.

III. DYNAMICAL PROPERTIES OF HOLES

The analysis of dynamical properties in the context of the $t - J$ model allows a direct comparison between theory and experiments. In the case of one hole it is first important to know if there is a quasiparticle in the spectrum. Diagrammatic calculations have been presented[15] for 1D and 2D showing the presence of such a quasiparticle. The rest of the spectrum is conjectured to be incoherent. In the limit of $J/t = 0$ the quasiparticle disappears. For finite J the hole acquires an effective mass $\sim 1/J$ due to the interactions with the spins. This effect is similar to photons renormalizing the mass of electrons in QED.

These very interesting results can be obtained by numerical methods like the Lanczos approach and others. Let us calculate the spectral function $A(\mathbf{k}, \omega)$ of one hole with momentum $\mathbf{k}$ and energy ω (with respect to the ground state of zero holes) in the $t - J$ model.[16] To calculate $A(\mathbf{k}, \omega)$ numerically we used the definition:

$$A(\mathbf{k}, \omega) = -\frac{1}{\pi} Im\{G(\mathbf{k}, \omega + E_0 + i\epsilon)\}, \quad (3)$$

where $G(\mathbf{k}, x) = \langle \psi_0 | \bar{c}_{\mathbf{k},\sigma} (x - H)^{-1} \bar{c}^\dagger_{\mathbf{k},\sigma} | \psi_0 \rangle$ and $\bar{c}^\dagger_{\mathbf{k},\sigma} = \sum_i e^{i\mathbf{k}\cdot\mathbf{i}} \bar{c}^\dagger_{i,\sigma}$ i.e., the spectral function can be related to the imaginary part of the retarded Green's function of a hole. $|\psi_0\rangle$ is the ground state of the $t - J$ model for zero hole (having energy E_0) that we obtained using a Lanczos method. ϵ is a small parameter that gives a finite width to the delta-functions appearing at each pole of G. Any state of the one hole subspace which has a nonzero projection over the state $|1\rangle = \bar{c}^\dagger_{\mathbf{k}} |\psi_0\rangle$ will contribute to the spectral function. Note that using Eq.(3) we only need to evaluate matrix elements of appropriately chosen operators in the state $|1\rangle$. Then, it is not necessary to explicitly obtain all the excited states of the one hole sector as naively required to calculate dynamical properties. The next step is to realize that G admits a continued fraction expansion as

$$G(\mathbf{k}, x) = \cfrac{1}{x - a_0 - \cfrac{b_1^2}{x - a_1 - \cfrac{b_2^2}{x - a_2 - \ldots}}}. \quad (4)$$

We know that G has poles only on the real axis and Eq.(4) has this property. The coefficients a_m, b_m appearing in Eq.(4) are functions of the matrix elements $\langle 1 | H^n | 1 \rangle$ $(n = 1, ..., m)$. These matrix elements can be evaluated numerically for increasing m by using a Lanczos program (for details see Ref.16).

In Fig.1 we show a typical result for the spectral function of one hole in the $t - J$ model using a 4×4 lattice[16] (results for the Hubbard model on a 10 site lattice are also discussed in Ref.16). At the bottom of the spectrum there is a large peak showing that the overlap between the state $|1\rangle$ and the actual ground state is large. This peak is usually a signal of the existence of a quasiparticle in the model. Note also that at least for the 4×4 lattice the rest of the spectrum is not completely incoherent but it presents structure with some peaks clearly more important than others (there are a couple of hundred spikes in Fig.1 although just a few are clearly visible). In the region $0.2 \leq J/t \leq 1.0$ the quasiparticle energy (measured with respect to the zero hole ground state energy) is well approximated by $E_h = -3.17 + 2.83 J^{0.73}$ on the 4×4 lattice. Similar power law results have been obtained for higher excited states.[16] The physical origin of the additional structure after the quasiparticle peak has been extensively discussed in Ref.16.

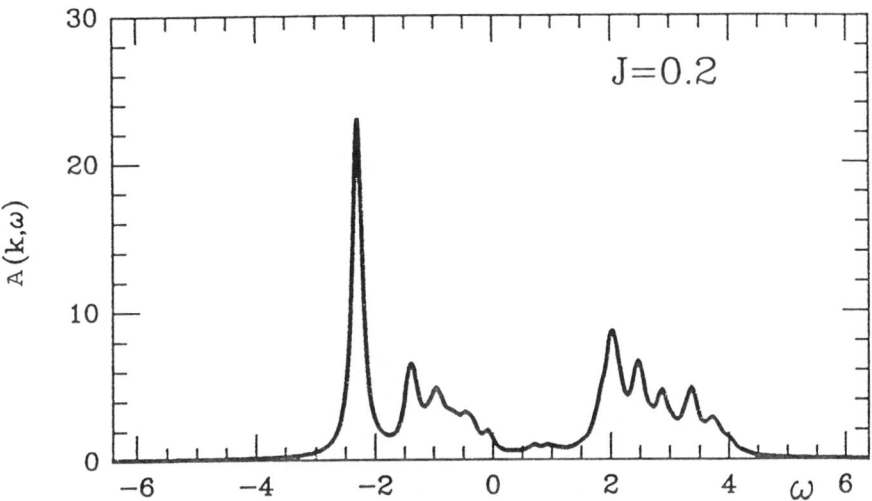

Figure 1) Spectral function of one hole in the $t - J$ model at $\mathbf{k} = (\frac{\pi}{2}, \frac{\pi}{2})$ and $J = 0.2$, $t = 1$ on a 4×4 lattice.

IV. COMPARING THEORY WITH EXPERIMENTS

The behavior of the new superconducting materials in the normal phase is not standard. Many features would be difficult to explain with a simple Fermi liquid theory which was so successful in the description of BCS superconductors. Recently, these anomalous behaviors have been listed and it was claimed that, phenomenologically, they can be explained by a "marginal" Fermi liquid theory.[17] Perhaps the $t - J$ model already presents this new behavior so it is certainly worthwhile to calculate "measurable" quantities in this model and compare with experiments.

Consider for example the *optical conductivity*. There are many experimental results[18] with measurements of the optical conductivity of the new materials in the normal and superconducting phases. An unusual result was found in the normal state where in addition to the Drude peak a broad "mid-infrared" band was observed. Can these results be understood in the context of strongly correlated electronic models? Based on the numerical results for the spectral function of one hole described before, it is very tempting to associate that mid-infrared band with the structure found after the quasiparticle peak in Fig.1. To make this idea quantitative it is necessary to evaluate the conductivity using the Lanczos technique.[19] For that purpose we define the current operator in the x-direction at zero momentum as

$$j_x = it \sum_{\mathbf{i},\sigma} \left(c^\dagger_{\mathbf{i},\sigma} c_{\mathbf{i}+\hat{x},\sigma} - c^\dagger_{\mathbf{i}+\hat{x},\sigma} c_{\mathbf{i},\sigma} \right), \qquad (5)$$

and then, through the usual linear response theory, the optical absorption can be written as,

$$\sigma(\omega) = -\frac{1}{\omega\pi} Im\left\{\langle\psi_{gs}|j_x\frac{1}{\omega + E_{gs} + i\epsilon - H}j_x|\psi_{gs}\rangle\right\}, \qquad (6)$$

where $|\psi_{gs}\rangle$ is the ground state of the Hubbard or $t - J$ models with a given number of holes and energy E_{gs}, which we can obtain with a Lanczos method. ϵ is a small parameter that moves away from the real axis the poles that otherwise would appear in Eq.(6). Using the Lanczos method adapted to calculate dynamical quantities, a 4×4 lattice was studied for the $t - J$ model with one hole and for the Hubbard model on a 10 site lattice. At low energy a large peak was found that is very tempting to associate with a Drude peak. Finite size effects move the peak to a finite frequency but it will converge to zero in the thermodynamic limit. Another important feature was found in this calculation: after the Drude peak we observed a large accumulation of spectral weight which resembles the experimentally found "mid-infrared" band (we are assuming that in the normal state holes behave like free (renormalized) particles and thus we can compare numerical results for one hole with experiments). Similar good qualitative agreement between theory and experiments has been found for other normal state properties like Raman scattering and regarding the shift in the antiferromagnetic peak due to doping.[2]

Of course, the most important property of the new cuprates that we want to reproduce theoretically is the presence of a *superconducting* phase at low temperatures and small doping. Regretfully, this is a very elusive subject. Preliminary numerical results by Moreo et al.[20] suggest that there is enhancement of some pairing correlations (s and d waves) but there are no indications of divergent susceptibilities with the size of the system. At this point there are many alternatives: i) it may occur that the Hubbard model, while reproducing some aspects of the normal state properties of materials, still lacks terms that will lead to a superconducting phase; ii) the lattices and temperatures studied so far may be not suitable for the superconducting condensate; iii) the operators used in Monte Carlo simulations may have a small overlap with the actual quasiparticle operators and thus the signal of superconductivity may be hidden in the statistical errors of the Monte Carlo simulation.[21]

V. CONCLUSIONS

The last year has been very important in the study of spin and electronic models of high Tc. As emphasized in this review the existence of some powerful computational methods allowed us to clarify in part what is fact and what is just fancy in the study of ground state properties of these models. Much progress has been done in the analysis of dynamical properties and at least some rough comparisons between first principles theoretical calculations and experiments are now possible. From the results of the previous subsection we can conclude that some experimental features of the new superconductors may have a good qualitative description within the Hubbard and $t - J$ models. Of course, these results were obtained using finite lattices and they certainly should be verified by other numerical or analytical methods, but at least the rough predictions are not far away from experiments. There are problems however, related with i) phase separation which is an undesirable effect which may be alleviated by introducing explicitly a hole-hole repulsion and, obviously, ii) the lack, so far, of a convincing proof of superconductivity in this model. Perhaps "questioning" the small clusters in a better way we may find some indications of this phenomenon. New algorithms and realistic mean-field techniques should be developed to improve our understanding of strongly correlated electrons. Future work along these lines should be strongly encouraged.

I thank the organizers of the Computational Quantum Physics conference for the invitation to this meeting. I also thank my many collaborators and colleagues for discussions and comments about the physics of spin and electronic models. This project was supported in part by the NSF grant PHY89-04035. Most of the numerical results have been obtained using the CRAY-2 supercomputer at the National Center for Supercomputing Applications (NCSA), Urbana, Illinois. I thank NCSA for its support.

REFERENCES

1. J. G. Bednorz and K. A. Müller, Z. Phys. **B64**, 188 (1986).
2. E. Dagotto, Int. J. Mod. Phys. **B5**, 77 (1991).
3. P. W. Anderson, Science **235**, 1196 (1987); J. R. Schrieffer, X. G. Wen and S. C. Zhang, Phys. Rev. **B 39**, 11663 (1989).
4. F. C. Zhang and T. M. Rice, Phys. Rev. **37**, 3759 (1988).
5. J. Reger and A. P. Young, Phys. Rev. **B 37**, 5978 (1988); T. Barnes and E. Swanson, ibid **37**, 9405 (1988).
6. E. Dagotto and A. Moreo, Phys. Rev. **D 31**, 865 (1985); E. Gagliano et al., Phys. Rev. **B 34**, 1677 (1986).
7. G. Fano, F. Ortolani and A. Parola, Bologna preprint.
8. J. Oitmaa and D.D. Betts, Can. J. Phys. **56**, 897 (1978); E. Dagotto and A. Moreo, Phys. Rev. **B 38**, 5087 (1988).
9. E. Dagotto, A. Moreo and T. Barnes, Phys. Rev. **B 40**, 6721 (1989). See also J. Bonca, P. Prelovsek and I. Sega, Phys. Rev. **B 39**, 7074(1989).
10. Y. Hasegawa and D. Poilblanc, Phys. Rev. **B 40**, 9035 (1989).
11. J. Riera and A. Young, Phys. Rev. **B 39**, 9697 (1989). See also E. Kaxiras and E. Manousakis, Phys. Rev. **B 38**, 866 (1988).
12. See for example V. Emery, S. Kivelson and H. Q. Lin, Phys. Rev. Lett. **64**, 475 (1989); F. Nori, E. Abrahams and G. Zimanyi, Phys. Rev. **B**, to appear.
13. E. Dagotto, A. Moreo, R. Sugar and D. Toussaint, Phys. Rev. **B 41**, 811 (1990).
14. A. Moreo et al., Phys. Rev. **B 43**, 11442 (1991); E. Dagotto et al., NSF-ITP-91-54 preprint.
15. C. Kane, P. Lee and N. Read, Phys. Rev. **B 39**, 6880 (1989).
16. E. Dagotto, A. Moreo, R. Joynt, S. Bacci and E. Gagliano, Phys. Rev. **B 41**, 2585 (1990); *ibid* **B 41**, 9049 (1990) and references therein.
17. C. Varma et al., Phys. Rev. Lett. **63**, 1996 (1989).
18. S. L. Cooper et al., Phys. Rev. **B 40**, 11358 (1989) and references therein.
19. A. Moreo and E. Dagotto, Phys. Rev. **B 42**, 4786 (1990). See Ref.2 for a complete list of references.
20. A. Moreo and D. Scalapino, UCSB preprint.
21. E. Dagotto and J. R. Schrieffer, Phys. Rev. **B 43**, 8705 (1991).

The Classical Scattering of Waves: Some Analogies with Quantum Scattering

Michael F. Werby, Theoretical Acoustics and Simulation,
NOARL, Stennis Space Center, MS 39529

Abstract

The scattering of waves in classical physics and quantum scattering theory have many dissimilarities, but also many things in common. Many of the modern developments in classical wave theory have their origin in quantum scattering despite the later development of quantum physics. Although each field has diverged from the other over time, there are many analogies between the two disciplines and much in one area may enhance the other. In this work an outline is given of some aspects of the classical scattering theory of waves which have some relation with quantum theory. In addition, some numerical techniques are presented that may be of use in both areas.

Introduction

As in the field of quantum scattering, one employs the scattering of waves on objects and interfaces to gain knowledge of the scatterer. From targets in the sky to those under the sea, to oil domes under the ground, we use electromagnetic, acoustic, and elastic waves to gain knowledge of what would otherwise elude us. Remarkably, much of the mathematical methodology and some of the physical events, such as resonances in classical scattering occur in quantum scattering and it is likely that the base of knowledge in one area will promote "hybrid vigor" in the other. The purpose of this article is to describe some aspects of the classical scattering of waves from the viewpoint of one familiar with quantum scattering theory and quantum phenomena. The mathematical development to describe scattering from targets that will be emphasized here is based on an exact numerical technique by Peter Waterman, who in a series of beautiful papers,[1-3] outlined the course of treatment that constitutes a unified theory of the classical scattering of waves. The T-matrix or more properly, the extended boundary condition (EBC) method due to Waterman[1-3] is in part an algorithmic method which, in my view, is as powerful in an algorithmic sense as that of Hamilton's principle or the Euler-Lagrange equations. This method (largely overlooked by the classical scattering community) along with numerical or structural improvements has enabled researchers to perform enormously complicated calculations and understand physical phenomena previously not possible.

Along with the formal and numerical procedures outlined here I will discuss some physical phenomena in wave scattering that has some similarities with quantum scattering. In the area of acoustical phenomena such as resonance scattering, Herbert Überall[4-8] has, in my view, been the major contributor, both in introducing techniques and in describing phenomena in the context found in quantum physics. Of particular note is his development of a resonance scattering theory[4] (RST) along lines parallel to that of Briet and Wigner[9] and Kapur and Peierls[10] as well as his realization that resonances excited on elastic targets were mainly due to circumferential waves excited on elastic surfaces that due to phase matching conditions form standing waves at distinct frequencies resulting in the resonance phenomena.[6] Analogies between quantum and classical scattering and propagation are numerous.[11] The treatment of some scattering from submerged particles such as sediment, due to Foldy and Lox, is similar to the adiabatic theory of scattering developed by Foldy. The Born approximation and the related Kirchhoff approximation play an important role in wave scattering, particularly from surfaces. Both a T-matrix and an S-matrix can be defined in wave scattering where the important concept of symmetry, unitarity, and the generalized optical theory play a role.[12,13] Überall has shown that even Reggie poles[8] find their place in the classical scattering of waves and that one can divide scattering into a superposition of a form of direct scattering (called the background) and resonance scattering just as in quantum scattering. This was done for elastic spheres by Überall in which he employed a rigid scatterer as the background for elastic solids[4] and by Werby using the concept of entrained mass for a shell.[14,15] The analogy with quantum scattering occurs at low energy where shape elastic scattering adds coherently with resonance scattering due usually to collective motion of the nucleus. Due to space limitations, I will limit my discussions to topics that I have had direct experience with, with it being understood that many more analogies exist between the two areas than presented here. A comprehensive development of the vast area of resonance scattering from elastic targets may be found in a new book by Überall.[7]

Several techniques are available for describing waves that scatter from objects of known constitution and geometry. Many of them are rather specific or have intractable numerical pitfalls. It is therefore desirable to obtain a formulation that allows for general objects, ranges of frequency, and boundary conditions, and that overcomes numerical difficulties commonly encountered in several broad classes of numerical methods. In this paper we briefly describe the Waterman method which proves to be a consistent, unified and manageable numerical approach useful for researchers interested in solving any of a wide class of scattering problems. It is based on the coupling of the exterior and interior solutions of the surface boundary representation of the Helmholtz or elastodynamic equations and yields the EBC method of Waterman.[1-3] The EBC method avoids numerical problems often encountered by other techniques. We present numerous physical examples which are chosen not only for their intrinsic interest but also because they represent comparatively difficult problems to solve by other means.

We then focus on some of the physical events in wave scattering that have similarities with quantum mechanical events. We first start with a review of the Extended Boundary Condition method for determining the field scattered from a rigid spheroidal object. We next extend this development to include elastic objects. Finally, we describe the time domain solution, partial wave analysis, and additional physical phenomena.

The Extended Boundary Condition Method in Review

The EBC method forms the basis of the equations used here to develop the eigen-expansion and transformation methods that follow. Since emphasis is on the mathematical basis and properties of these two procedures (with the exception of scattering from impenetrable targets, details of EBC equations can be both intricate and extended), we will indicate only the EBC equations for the simplest case, and list the generalizations for more complicated scenarios. The Helmholtz-Poincare' integral representation for a field exterior to a bounded object can be expanded as follows.

$$U(r) = U_i(r) + \int_S \left[U_+(r') \frac{\partial G(r,r')}{\partial n} - G(r,r') \frac{\partial U_+(r')}{\partial n} \right] ds \qquad (1)$$

where r is chosen to be at an exterior point to the object; i.e., r is a member of D, where D is the set of all points exterior to the object, G is an outgoing Green's function, D' is the set of points in the interior of the bounded object, and S is the set of all points on the object surface. The surface of the object is assumed to be piecewise continuous. It is redolent of Gauss' law that when r is in the interior the total wavefield $U(r)$ is zero (extinguished or nulled), hence the terms "extinction theorem" or, "null-field condition." This fact was generally thought inconsequential, but Waterman[1] took specific advantage of this condition to employ a constraint on the exterior solution to eliminate the surface terms U_+ or $\partial U_+(r')/\partial n$, which arise in the exterior solution. Here n is a unit outward normal to the surface. Although Waterman[1] employed the condition algorithmically to eliminate the surface term, he developed a method that also produced a unique solution for all positive frequencies of the exterior problem. This is of considerable importance, because exterior solutions have often suffered from "spurious resonances" at the so-called "irregular values" of incident frequency. These irregular values correspond to the eigenfrequency of a problem related to the interior problem considered here. It has been established that by coupling the interior points in the solution the irregular values are eliminated. Thus, Waterman's equations actually serve two computational purposes. For completeness we list the interior problem as follows

$$0 = U_i(r'') + \int_s \left[U_+(r') \frac{\partial G(r'',r')}{\partial n} - G(r'',r') \frac{\partial U_+(r')}{\partial n} \right] ds \qquad (2)$$

where r" is an interior point. The above equations constitute the extended boundary condition equations. Since in their present form they are not directly useful, we now reduce them to a form amenable to numerical computation. For convenience of presentation we simplify the problem, to that of an impenetrable object (although solution of the fluid target case is quite similar). Elastic targets submerged in a fluid require far greater mathematical detail, as indicated by Waterman.[3] Let us assume that $\partial U_+(r')/\partial n = 0$ so that we obtain the expressions

$$U(r) = U_i(r) + \int_s \left[U_+(r') \frac{\partial G(r,r')}{\partial n} \right] ds \qquad (3)$$

$$0 = U_i(r'') + \int_s \left[U_+(r') \frac{\partial G(r'',r')}{\partial n} \right] ds . \qquad (4)$$

To solve these expressions, it is convenient to represent $U_i(r)$, $U_+(r')$ and $G(r,r')$ in some convenient series expansion, which upon truncation would lead to matrix equations that can then be solved using digital computers. The Green's function G is a normal operator, and thus can be represented by the biorthogonal series

$$G(r,r') = i\kappa \Sigma \text{Re}\varphi_i(r_<)\varphi_i(r_>) \qquad (5)$$

where $r_<$ and $r_>$ are the greater and lesser of the two points r and r relative to the origin of the object, respectively. The quantity U_i, the incident wavefield, is known. In a manner similar to that of the Hilbert-Schmidt theorem for symmetric kernels, it can be shown that

$$U_i(r) = \Sigma a_n \text{Re}\varphi_n(r) . \qquad (6)$$

For incident plane waves, the a's are known. We now have the relation

$$\sum_n a_n \operatorname{Re}\varphi_n(r) = i\kappa \sum_n \int \operatorname{Re}\varphi_n(r) U_+(r) \frac{\partial \varphi_i(r')}{\partial n} dS \tag{7}$$

where it follows that

$$a_n = i\kappa \int U_+(r) \frac{\partial \varphi_i(r')}{\partial n} dS . \tag{8}$$

We now wish to represent the above matrix form. This can be achieved by writing $U_+(r')$ in some complete set of known functions so that

$$U_+(r) = \sum_n b_n \operatorname{Re}\varphi_n(r) \tag{9}$$

where b_n is the only unknown. This reduces to an expression in which the expansion coefficients b_n are the only quantities to be determined. Note that we are not really concerned with the b_n's as such, but rather are eliminating the unknown surface quantities U_+. We get:

$$a_n = \sum_m b_m \int \operatorname{Re}\varphi_m(r') \frac{\partial \operatorname{Re}\varphi_n(r')}{\partial n} dS \tag{10}$$

where Q_{nm} is an element of some known matrix. The $\operatorname{Re}\varphi_m$'s are not the most efficient functions to employ. Our intention is to determine the most efficient expansion functions, i.e., those that would form an orthonormal (ON) basis set on the surface of the bounded object. We then obtain the most effective expansion of U_+ on the object surface.[16, 17] To do so, we premultiply the above equation by the adjoint of Q, namely $Q^\dagger$, where the latter quantity is the complex transpose of Q.

$$Q^\dagger a = iQ^\dagger Qb = i\,Hb \tag{11}$$

where the matrix H can easily be shown to be self-adjoint or Hermitian (where $H^\dagger = H$). The advantage pursuing this course is that we can easily find the eigenvalues and eigenfunctions of H that have known and computationally desirable properties. In particular, the eigenvalues here are real, positive, and increase monotonically, and form an orthonormal set of functions on the surface. The eigenfunctions can be obtained as follows:

$$H\beta_i = \lambda_i \beta_i. \qquad (12)$$

Here, the adjoint of β is $\beta^\dagger$, so that

$$\beta_i \beta^\dagger_j = \delta_{ij} \qquad (13)$$

where δ_{ij} is the Kronecker delta function.

We also have the ordering $\lambda_1 < \lambda_2 < \lambda_3 \ldots$ where the dimension of H is any desired order and relates to the number of surface quantities required in expanding U_+. One can show that the β_i's are an alternate representation of the $\partial \varphi_i(r')/\partial n$'s, with the desired property that they are an orthogonal representation (with another computationally desirable property to be discussed). Thus, we have $b = \Sigma \alpha_i \beta_i$ so that

$$Q^\dagger a = i \Sigma H \alpha_i \beta_i = i \Sigma \alpha_i \lambda_i \beta_i. \qquad (14)$$

Thus
$$\alpha_i = -i\beta_i Q'A/\lambda_i \text{ and } b = -i \Sigma \beta_i Q'A/\lambda_i. \qquad (15)$$

We also expand the exterior problem as follows

$$f_i = ik \Sigma_i b_j \int \text{Re } \varphi_j(r') \text{Re} \partial \varphi_i(r)/\partial n \ dS = ik \ \Sigma_i b_j Q_{ij}. \qquad (16)$$

The final expression for the scattered field in terms of the incident wave field is thus

$$f = -\Sigma \text{Re} Q \beta_i Q'A/\lambda_i \beta_i. \qquad (17)$$

Although the above expression has proven computationally efficient, it is also possible and sometimes necessary to obtain an alternate T-matrix representation. This may be done by means of the following derived relation:

$$T = -\text{Re} Q \beta (1/\lambda) \ \beta^\dagger Q^\dagger \qquad (18)$$

where $\beta \beta^\dagger = \beta^\dagger \beta = I$. The set of eigenvectors β form a unitary matrix and therefore the above expression may be viewed as having been obtained by transforming Q via a unitary matrix obtained from Q times its adjoint. This offers a generalization of this method to one more complicated and that cannot be posed in such a simple form. For the general problem a T-matrix may be written in the form

$$T = -PQ^{-1} \qquad (19)$$

with solutions of the form

$$T = -P\beta \ (1/\lambda) \ \beta^\dagger Q^\dagger \qquad (20)$$

where $QQ^\dagger \beta_i = \lambda_i \beta_i$.

The Elastic T-Matrix

We now consider the case of an elastic object. The equation of motion in a fluid is

$$\nabla^2 \vec{U} + k^2 \vec{U} = 0 \tag{21}$$

where k is the acoustic wavenumber and U is related to the particle displacement. The equation for motion in an elastic body is

$$\left(k_0^2\right)^{-1} \nabla\nabla \cdot \vec{U} - \left(\kappa_0^2\right)^{-1} \nabla \times \nabla \times \vec{U} + \vec{U} = 0$$

$$k_0^2 = \frac{\rho_0 \omega^2}{\lambda_0 + 2\mu_0} \quad \text{and} \quad \kappa_0^2 = \frac{\rho_0 \omega^2}{\mu_0}, \tag{22}$$

where k_0 and κ_0 are the longitudinal and transverse wavenumbers respectively. The boundary conditions at an interface are:

$$\tilde{n} \cdot \vec{U}_+ = \tilde{n} \cdot \vec{U}_- \quad \tilde{n} \cdot \vec{t}_+ = \tilde{n} \cdot \vec{t}_- \quad \nabla \times \vec{t}_- = 0. \tag{23}$$

Where the traction t for the outer (+) and inner (−) surfaces are:

$$\vec{t}_+ = \lambda \tilde{n} \nabla \cdot \vec{U}_+ \quad \vec{t}_- = \lambda \tilde{n} \nabla \cdot \vec{U}_- \, 2\mu_0 \tilde{n} \frac{\partial \vec{U}_-}{\partial n} + \mu_0 \tilde{n} \times (\nabla \times \vec{U}_-). \tag{24}$$

The Green's Functions G(r, r') and Green's stress triadic Σ(r, r') are:

$$\left(k_0^2\right)^{-1} \nabla\nabla \cdot \overline{G} - \left(\kappa_0^2\right)^{-1} \nabla \times \nabla \times \overline{G} + \overline{G} = -\left(\kappa_0^3\right)^{-1} \overline{I}\delta(r - r') \tag{25}$$

$$\Sigma_+ = \lambda \overline{I} \nabla \cdot \vec{U}_+ \quad \Sigma_- = \lambda \overline{I} \nabla \cdot \overline{G}_- + 2\mu_0 \tilde{n} \frac{\partial \overline{G}_-}{\partial n} + \mu_0 \overline{I} \times (\nabla \times \overline{G}_-).$$

The boundary integrals are in the fluid and in the elastic body[18-20]

$$\vec{U}^i + \frac{k^3}{\rho\omega^2}\int\int_s \left[\vec{u}_+ \cdot (\tilde{n} \cdot \tilde{\Sigma}) - t_+ \cdot \overline{G}\right] ds = \begin{cases} U & \text{Outside the object} \\ 0 & \text{Inside the object} \end{cases}$$

$$\frac{k_0^3}{\rho_0\omega^2}\int\int_s \left[\vec{u}_- \cdot (\tilde{n} \cdot \tilde{\Sigma}_0) - t_- \cdot \overline{G}_0\right] ds = \begin{cases} U & \text{Outside the object} \\ 0 & \text{Inside the object} \end{cases} \quad (26)$$

We need to solve these equations subject to the boundary conditions at the interfaces and with the appropriate asymptotic boundary conditions. The partial wave expansion functions in the fluid and the elastic body are:[21]

$$\phi_n = \frac{1}{k}\nabla h_n(kr)Y_l^m(\theta,\varphi) \quad \psi_n^1 = [n(n+1)]^{-1}\nabla \times \left[rh_n(k_0 r)Y_l^m(\theta,\varphi)\right]$$

$$\psi_n^2 = \frac{1}{\kappa_0}\nabla \times \psi_n^1 \quad \psi_n^3 = \left(\frac{k_0}{\kappa_0}\right)^{3/2}\frac{1}{k_0}\nabla h_n(k_0 r)Y_l^m(\theta,\varphi). \quad (27)$$

Here the Y's are spherical harmonics and the h's are outgoing spherical Hankel functions. Expand everything into partial wave,

$$\vec{U}^i = \sum_{n=0}^{N} a_n \text{Re}\phi_n \quad \vec{U}^f = \sum_{n=0}^{N} f_n \phi_n \quad \vec{U}^b = \sum_{n=0}^{N} (\alpha_n \text{Re}\psi_n + \beta_n \psi_n). \quad (28)$$

The Green's functions are also expanded using biorthogonal expansions for normal operators:

$$\overline{G}(r,r') = i\sum_n \text{Re}\phi_n(r)\phi_n(r') \quad \overline{G}_0(r,r') = i\sum_n \text{Re}\psi_n(r)\psi_n(r'). \quad (29)$$

Expansion Method to Reduce the BIE to Matrix Form

The T-matrix for the most general case is:[20]

$$T = -\{Q_{rr} - Q_{ro}T^2\}\{R_{ro}T^2 + R_{rr} + iT^2\}^{-1}P_{rr}\{Q_{or} - Q_{oo}T^2\}$$

$$\{R_{ro}T^2 + R_{rr} + iT^2\}^{-1}P_{rr}\}^{-1}. \quad (30)$$

T^2 corresponds to a reflection from the inner face of a shell. For a solid $T^2 = 0$ so that[19]

$$T = -Q_{rr}[R_{rr}]^{-1}P_{rr} \{Q_{or}[R_{ro}]^{-1}P_{rr}\}^{-1}. \tag{31}$$

For fluid or sound soft or sound hard objects one has:

$$T = -Q_{rr}\{Q_{or}\}^{-1} \tag{32}$$

where the matrices Q, R, P are as follows:

$$Q_{nn'} = \frac{k^3}{\rho\omega^2} \oint \left[\tilde{n} \cdot \text{Re}\psi_{n'} \lambda \nabla \cdot \phi_n - \tilde{n} \cdot t(\text{Re}\psi_{n'})\tilde{n} \cdot \phi_n \right] ds$$

$$R_{nn'} = \frac{k^3}{\rho\omega^2} \oint \left[\text{Re}\psi_{n'} \cdot t(\text{Re}\psi_n) - \tilde{n} \cdot t(\text{Re}\psi_{n'})\tilde{n} \cdot \psi_n \right] ds$$

$$P_{nn'} = \frac{k^3}{\rho\omega^2} \oint \left[\tilde{n} \cdot \text{Re}\phi_{n'} \tilde{n} \cdot t(\text{Re}\psi_n) \right] ds . \tag{33}$$

Unitarity, Symmetry and the S-Matrix

The above, equations are difficult to solve because the matrices are often poorly conditioned. We obviate this problem by a method we refer to as the unitary method which we briefly outline. We know from reciprocity that T is symmetric. Also S is unitary if the target and fluid are not energy absorbing which we assume. We can then write $T = -RP^{-1}$ which is its most general form. Then

$$S = 1 + 2T = 1 - 2RP^{-1} = UP^{-1} \tag{34}$$

where $U = P - 2R$. S now becomes

$$S = S' = P'^{-1}u' . \tag{35}$$

Now write $U = M\mathbb{U}$ $P = N\mathbb{P}$ where $\mathbb{P}$ and $\mathbb{U}$ are unitary and N and M are upper triangular. Then $S = \mathbb{P}'^{-1} L\mathbb{U}'$ where $L = N'^{-1}M$. But $SS'^* = \mathbb{P}*LL'*\mathbb{P}*^{-1} = 1$ which implies $LL'^* = 1$. Which implied L is unitary. That means that L which is a product of two upper triangular matrices is upper triangular. But it has to be lower triangular too and therefore it has to be diagonal. The diagonal elements have to be real and therefore L is the unit matrix. Thus, $S = UP'^*$ and that implies $T = (UP'^* \pm 1)/2$. This expression is much easier to calculate than expressions dependent upon matrix inversion.[13]

Time Domain Resonance Scattering Theory

The partial wave series that emerges from normal mode theory for separable geometries can be represented in distinct partial waves or modes. It has been shown[4,5] that a representation due to a distinct mode {n} can be written in the form:

$$f_n(\theta) = \frac{2}{ka} e^{2i\xi_n^{(r)}} \left\{ \frac{\left(\frac{1}{2}\right) s\Gamma_n^{(r)}}{\chi - \chi_n^{(r)} + \left(\frac{i}{2}\right)\Gamma_n^{(r)}} + e^{-i\xi_n^{(r)}} \sin\xi_n^{(r)} \right\} \tag{36}$$

where $\chi = ka$, $\chi_n^{(r)}$ is the nth resonance and $\left(\frac{1}{2}\right)\Gamma_n^{(r)}$ the half-width.

Where $e^{2i\xi_n^{(r)}} = -\dfrac{h_n^{(2)'}(x)}{h_n^{(1)'}(x)}$.

Here, the factor $2n + 1$ is absorbed in the expansion coefficient. For the pulse form, a continuous wave (cw) ping is used which corresponds to a very broad frequency range. For each time domain modal component, one has that

$$\text{Re} \int_{-\infty}^{\infty} \frac{\left(\frac{1}{2}\right)\Gamma_n^{(r)}}{\chi - \chi_n^{(r)} + \left(\frac{i}{2}\right)\Gamma_n^{(r)}} e^{-ixs} dx = 2\pi \left(\frac{1}{2}\right)\Gamma_n^{(r)} \sin\left(\chi_n^{(r)} s\right) e^{-\left(\frac{1}{2}\right) s\Gamma_n^{(r)}}. \tag{37}$$

That is, at a resonance, the time-domain solution is simply the product of the half-width times a sinusoidal function times an exponential damping factor. From the time-domain solution for a nest of resonances (N-m) for a cw ping, one obtains the form

$$p(s) \approx 2\pi \sum_{n=m}^{N} \left(\frac{1}{2}\right) \Gamma_n^{(r)} \sin\left(\chi_n^{(r)} s\right) e^{-\left(\frac{1}{2}\right) s \Gamma_n^{(r)}}. \tag{38}$$

The remaining contributions from backscatter are small due to phase averaging.

It is assumed that calculations are performed in a resonance region for which the resonance widths are fairly constant and the resonance spacing is fairly uniform.[22,23] This assumption leads to the important expression

$$P(s) \approx 2\pi \, 2^M \left(\sin(\chi_{ave}^{(r)} s)\right) \left\{\cos(\Delta\chi_{ave}^{(r)} s/2)\right\}^M e^{-s\Gamma/2}, \tag{39}$$

where $\chi_{ave}^{(r)} = \dfrac{1}{2M} \sum_{i=n}^{n+2M} \chi_i^{(r)}$.

Here one sets $n - m = 2M$. It is seen from the above expression:

— The half-width is associated with the decay of the response in the time domain solution: the response decreases exponentially with increasing value of the half-width.

— When the number of adjacent resonances (2M) sensed increases, the return signal becomes more sharply defined and the envelope function (the beats) are more enhanced and clearly defined.

— For larger carrier frequencies, the signal is more oscillatory within the envelope.

Applications of the EBC Method to Various Problems

We first treat scattering from rigid impenetrable objects in a free space. The two simplest cases are for spheroids and for cylinders with hemispherical caps. We focus on spheroids.

Application to Rigid Target

There are two classes of targets for impenetrable problems, i.e., soft and hard scatterers. They do not support body resonances; therefore; we examine acoustic quantities appropriate for nonresonant targets, such as circumferentially diffracted or creeping waves. These arise when scattering end-on from a spheroid in which one observes the return signal at the origin of the signal. The values of the incident wavefield frequency are expressed using the dimensionless quantity $kL/2$, where L is the object length and k the total wavenumber ($k = 2\pi/L$).

Bistatic angular distributions correspond to measurement of a scattered field at any point in space for some incident wave fixed relative to some source-object orientation. In Figure 1 we examine a rigid spheroid of aspect (length-to-width) ratio of 30:1. Figure 1a and 1b represent scattering from the object along the axis of symmetry (end-on) (a) and 90-degrees relative to the symmetry axis (broadside). The value of kL/2 in Figure 1a and 1b is 200, which implies that the object is about 70 wavelengths long and thus in the intermediate- to high-frequency region where neither low nor high frequency approximations apply. In all figures, frequency is sufficiently high that wave diffraction effects are significant in the forward scattering direction.

There are two competing mechanisms in the backscatter case. One arises from specular scattering (geometric) and the other arises from the creeping waves. The result is a coherent effect in which the two waves add constructively at some point leading to a maximum value when they are in phase and destructively leading to a minimum when they are out of phase. This can be seen in Figure 1 for a spheroid of aspect ratio (c) 4 to 1 (d) 8 to 1 and (e) 16 to 1. The more pronounced dips with increasing aspect ratio is due to the greater grazing angular region for higher aspect ratio targets.

Applications to Elastic Targets

We now examine a phenomenon observed frequently when scattering from elastic objects with smooth boundary conditions surrounded by an acoustic fluid, namely, body resonances. The resonances examined for the elastic solid case originate from the curved-surface equivalents of seismic interface waves of pseudo-Rayleigh or Scholte type, propagating circumferentially to form standing waves on a bounded object or from bending modes when scattering at oblique angles. These types of resonances occur at discrete values of kL/2 and manifest themselves in a characteristic manner. For elongated elastic solids, three distinct resonance types occur. The first kind (at lower frequencies) are due to leaky Rayleigh waves and have been shown to be related to both target geometry and material parameters (notably shear modulus and density). Resonances can, in this case, be best observed by examining the backscattered echo amplitude and phase response plotted as a function of kL/2, often referred to in acoustic scattering literature as a form function. We illustrate this for WC spheroids of aspect ratios of 6, 8, and 10 to 1 end-on incidence in Figure 2a, 2b, and 2c respectively. Here we see two resonances superimposed on the semi-periodic pattern due to Franz waves associated with rigid scattering. If we were to subtract rigid scattering (in partial wave space) from the elastic response then we would be left with the resonance response alone. Note the slight upward shift in kL/2 value with increasing aspect ratio (L/D) which can be explained in terms of standing waves on the surface. In addition to the above wave phenomena, it is also possible to excite "whispering gallery" resonances, which for these examples occur at higher kL/2 values.

In Figure 3 we examine broadside resonances for 2, 3, 4, and 5 to 1 steel spheroids. Here we can excite three phenomena. At the lowest value we can see a spike representing a bending resonance (Werby and Gaunaurd[24]) discussed below. The second lowest spike corresponds to the lower order Rayleigh resonance seen end-on, corresponding to a standing wave, circumnavigating the largest meridian of the spheroid. We also see weak Franz waves similar to those excited on a cylinder, and then we see the lowest order Rayleigh and Whispering Gallery resonances corresponding to circumferential waves around the smallest meridian. The third kind we wish to illustrate has to do with bending modes or flexural resonances. For unsupported spheroids, a plane incident wave at 45 degrees relative to the axis of symmetry can excite these modes illustrated in Figure 4a through 4d for aspect ratios of 2:1 through 5:1. It can be shown that the lowest mode corresponds to 2, and thereafter 3, 4, etc. The interesting thing about these resonances is they can be predicted by exact bar theories and coincide nicely with results here. Of particular interest is the effect that with increasing aspect ratio, the onset of resonances occur at lower kL/2 values, the opposite observed in Rayleigh resonances.

Finally, we examine scattering from a thin elastic aluminum spheroidal shell. Figure 5a, 5b, and 5c illustrates scattering end-on, at 45 degrees relative to the axis of symmetry and broadside. As noted earlier by Werby and Gaunaurd,[25] one can only excite resonances due to modal vibrations corresponding to standing waves about the largest meridian (end-on) and modal vibrations corresponding to standing, to modal vibrations about the shortest meridian (broadside). Further, it is possible to excite bending modes at oblique angles. In fact, the lowest nulls in Figure 5b and 5c correspond to the lowest (n = 2) bending mode. Evidence of bending modes can be seen at higher frequencies as slight nulls in the two figures and are the thin shell analogues of the elastic solid case. Here they appear as nulls instead of spikes due to a change of phase of 180 degrees in acoustic background (from rigid to soft).

Resonance Phenomena, Time Domain, and Partial Waves

Flexural waves do not yield resonances from fluid-loaded shells until the phase velocity of the flexural wave is about equal to the speed of sound in the ambient fluid.[22,23,26] The value in frequency for which this happens is referred to as the coincidence frequency; however, some subsonic fluid-borne waves produce sharp[19,20] resonances below coincidence frequency. These waves are referred to as pseudo-Stoneley waves and the related resonances as pseudo-Stoneley resonances.[22,23,27] The pseudo-Stoneley resonances are well defined in partial wave space; they usually correspond to only one partial wave mode number and a very narrow half-width with a dispersive phase velocity, which approaches the speed of sound in the fluid with increasing frequency. The pseudo-Stoneley resonances diminish in significance at the point where the flexural resonances begin to dominate. It can be determined that a phase change occurs in the pressure field in the transition region from subsonic to supersonic. This change accounts for the envelope of the resonance curve at coincidence frequency where the waves are in phase until coincidence, and are out of phase afterwards. Our interest here is in examining the time-domain response, since one expects the conditions previously described to be partially met over a broad frequency range, and thus to yield a strong coherent response with a carrier frequency in the neighborhood of the frequency at coincidence. Accordingly, the case of cw pings for two examples—for which coincidence resonances are expected to arise—is examined. This is certainly suggested by the strong responses in Figure 6b at the ka value 45, for WC. Further, in this analysis, the Mindlin-Timoshenko[28] thick plate theory is used to determine the value for which the flexural phase velocity will equal the ambient speed of sound in water. The phase and group velocities are determined from flat plate theory, which proves to be quite reliable in predicting the phase velocity for the curved surfaces of the spheres at the coincidence frequency.

The time-domain calculations are now examined. The example is a WC shell of 1% thickness. In this case, a well-defined envelope (illustrated in Fig. 6a) with pronounced oscillations within the envelope, is consistent with Eq. 39. The enhancement due to the factor 2^M is obvious here for the WC case. The group velocity can be obtained from the peak-to-peak distance of the adjacent envelopes. The result leads to a value of 2.33 km/sec. Both flexural and pseudo-Stoneley resonances compete in this region. A mixture of pseudo-Stoneley waves, as well as flexural waves, must be leaking into the fluid. For flexural waves, the group velocity is 2.65 km/sec at coincidence frequency with a range between 2.49 and 2.78 km/sec over the ka range of 30–60, where the strong flexurals are significant. In that range the phase velocity varies from 1.37 to 1.58 km/sec. The value of the extracted group velocity does not agree well with the flexural group velocity; the discrepancy is 12%. This variation suggests that the flexural resonances are of little importance for the time sequence presented here. The group velocity of the pseudo-Stoneley waves for this case has been determined[27] to be 2.65 km/sec based on plate theory. The phase velocity is in the range from 88% to 98% of the speed of sound in the fluid. This value of group velocity is within 3% of the extracted value from the time-domain solution. Moreover, the pseudo-Stoneley resonances have very narrow widths, while the flexural resonances are quite large. The conditions in a previous section would indicate that the flexural resonances would rapidly dampen due to the large half-widths, while the pseudo-Stoneley resonances would attenuate slowly in time. Thus, based on the similarity of the extracted group velocity and that of the pseudo-Stoneley wave and the conditions in the previous section on level widths, one may conclude that the time-domain calculations in Figure 6a represent pseudo-Stoneley resonances.

It is of some interest to discuss resonance scattering from elastic targets because of the close analogy to low energy nuclear resonance scattering. It was mentioned earlier that one can describe the resonance return signal as a function of the nondimensionalized frequency ka (excitation function for the nuclear case) as a resonance term and a background term (shape elastic scattering in low energy nuclear scattering). To show this, Figure 7a illustrates the total backscattered response from a steel spheroid (length to width of 3 to 1) end-on incidence. Figure 7b illustrates the resonance return signal obtained by subtracting the rigid background. The resonances are labeled according to the fundamental group ([n, 1] with n = 2, 3, . . .) and the higher-order group ([n, l] with l = 2, 3, . . . and n = 0, 1, 2, . . .) in analogy to the ground state and excited states of a nucleus. The series [n, 1] is referred to as the Rayleigh series R_l, while the higher order series has been labeled as a Whispering Gallery resonance because of the presumed analogy with the phenomenon at St. Paul's Cathedral in London. To illustrate that these resonances form standing waves on the surface of the object and to suggest the origin of the labeling, we plot the residual bistatic angular distribution (illustrated in Figure 8) as a function of the angle in a plane of the object. It is clear that we observe dipole, quadruple, etc. terms according to the "N" designation for both classes of waves consistent with the Überall notion of this class or resonances. Although is has been commonly assumed that Überall's notions for elastic solid spheres are accepted, these calculations form the basis for establishing that the notions are also valid for a spheroid.[29]

The theory of partial wave analysis often used in nuclear physics has found a useful place in resonance analysis of elastic scattering. Large resonance returns have been noted in scattering from elastic shells. In Figure 9a the analysis used[30] to resolve the matter of the origin of these resonances is illustrated. It has been determined that the sharp spikes correspond to waterborne waves, referred to as pseudo-Stoneley waves, superimposed on broad overlapping flexure resonances. Figure 9b illustrates this effect by examining the contributing partial waves for a fixed frequency. N = 32 corresponds to the sharp waterborne wave and n = 28 corresponds to the broad flexural resonance. N = 6 relates to a fast symmetric mode.

Finally we illustrate a "level diagram" in Figure 10 for WC spheroids for aspect ratios ranging from 1 to 4 in steps of 0.25. We see that the Rayleigh resonances gradually shift upward with increasing ratio while the Whispering Gallery resonances shift up more rapidly for fixed index N. Eventually, the Whispering Gallery resonances shift upward to the extent that they cross over[31] the Rayleigh resonances. We have referred to this as "level crossing" in analogy with a similar event for prolate nuclei.[32]

Acknowledgments

I wish to thank NOARL management and the Office of Naval Research for support of this work. I am indebted to M. Strayer, C. Bottcher and S. Umar for the invitation to their stimulating conference on Quantum Scattering. Shortly after the conference and during the time I was preparing this paper, my father, Michael Moises Werby, passed away. I wish to dedicate this paper in honor of him and the dear memories that will linger on. NOARL contribution number JA 221:010:92.

References

1. P. C. Waterman, J. Acoust. Soc. Am. **45**, 1417 (1969).
2. P. C. Waterman, *Proc. IEEE*, 53(3), 802 (1965).
3. P. C. Waterman, J. Acoust. Soc. Am. **63**(6), 1320 (1977).
4. H. Überall, in *Proceedings of the IUTAM Symposium: Modern Problems in Elastic Wave Propagation*, edited by J. Miklowitz and J. D. Auchenbach, (Wiley Interscience, New York, 1978), pp. 239–263.
5. L. Flax, G. C. Gaunaurd, and H. Überall, in *Physical Acoustics*, edited by W. P. Mason and R. N. Thurston, (Academic Press, 1981), vol. 15, ch. 3, pp. 191–294.
6. H. Überall, L. R. Dragonette, and L. Flax, J. Acoust. Soc. Am. **61**, pp. 711–715 (1977).
7. H. Überall, Editor, *Acoustical Resonance Scattering*, (Gordon and Breach, New York, in press).
8. H. Überall, et al., Appl. Mech. Rev. **43**(10) (1990).
9. G. B. Briet and E. P. Wigner, Phys. Rev. **49**, 519 (1936).
10. P. L. Kapur and Peierls, Proc. Roy. Soc., London, A166, 277 (1938).
11. J. B. Keller, "Progress and Prospects in the Theory of Linear Wave Propagation", SIAM SEREV **21**(2), 229–245 (1979).
12. P. C. Waterman, Phys. Rev. **D 3**, 825–839 (1971).
13. M. F. Werby and L. R. Green, J. Acoust. Soc. Am. **74**(2), 625 (1983).
14. M. F. Werby, Acoustic Lett. **15**, (4), 65–69 (1991).
15. M. F. Werby, J. Acoust. Soc. Am., Dec. (1991).
16. M. F. Werby and S. Chin-Bing, Int. J. Comp. Math. Appls. **11**(7/8), 717 (1985).
17. M. F. Werby, G. Tango, and L. H. Green, in *Computational Acoustics: Algorithms and Applications*, edited by D. Lee, R. L. Sternberg, M. H. Schultz (Elsevier Science Publishers B. V., North Holland, 1988), vol. 2, pp. 257–278.
18. Y.-H. Pao and V. Varatharajulu, J. Acoust. Soc. Am. **60**(7), 1361 (1976).
19. A. Bostrom, J. Acoust. Soc. Am. **67**(2), 390 (1980).
20. B. A. Peterson, V. V. Varadan, and V. K. Varadan, J. Acoust. Soc. Am. **74**(5), 1051 (1983).
21. M. F. Werby and G. J. Tango, Eng. Analysis, **5**(1), 12–20 (1988).
22. M. F. Werby and H. B. Ali, in *Computational Acoustics* edited by D. Lee, Cakmak, R. Vichnevetsky (Elsevier Science Publishers B. V., North Holland, 1990), vol. 2, pp. 133–158.
23. M. F. Werby, Acoustic Lett. **15**(3) 39–42 (1991).
24. M. F. Werby and G. C. Gaunaurd, J. Acoust. Soc. Am. **85**, 2365–2371 (1989).
25. M. F. Werby and G. C. Gaunaurd, J. Acoust. Soc. Am. **82**, 1369 (1987).
26. M. F. Werby and G. C. Gaunaurd, SPIE, Automatic Object Recognition, vol. 1471, 2–17 (1991).

27. M. Talmant, H. Überall, R. D. Miller, M. F. Werby, and J. W. Dickey, J. Acoust. Soc. Am. **86**, 278–289 (1989).
28. D. Ross, *Mechanics of Underwater Noise*. New York: Pergamon Press (1976); S. P. Timoshenko, Philos. Mag. **43**:125–131 (1922).
29. M. F. Werby, et al., J. Acoust. Soc. Am. **84**, 1425 (1988).
30. G. C. Gaunaurd and M. F. Werby, J. Acoust. Soc. Am. **90**, 2539–2550 (1991).
31. M. F. Werby, et al., J. Acoust. Soc. Am. **88**, 2822–2929 (1990).
32. S. G. Nilsson, K. Dan. Vidensk. Selsk. Mat. Fys. Medd. **29**, No. 16 (1955).

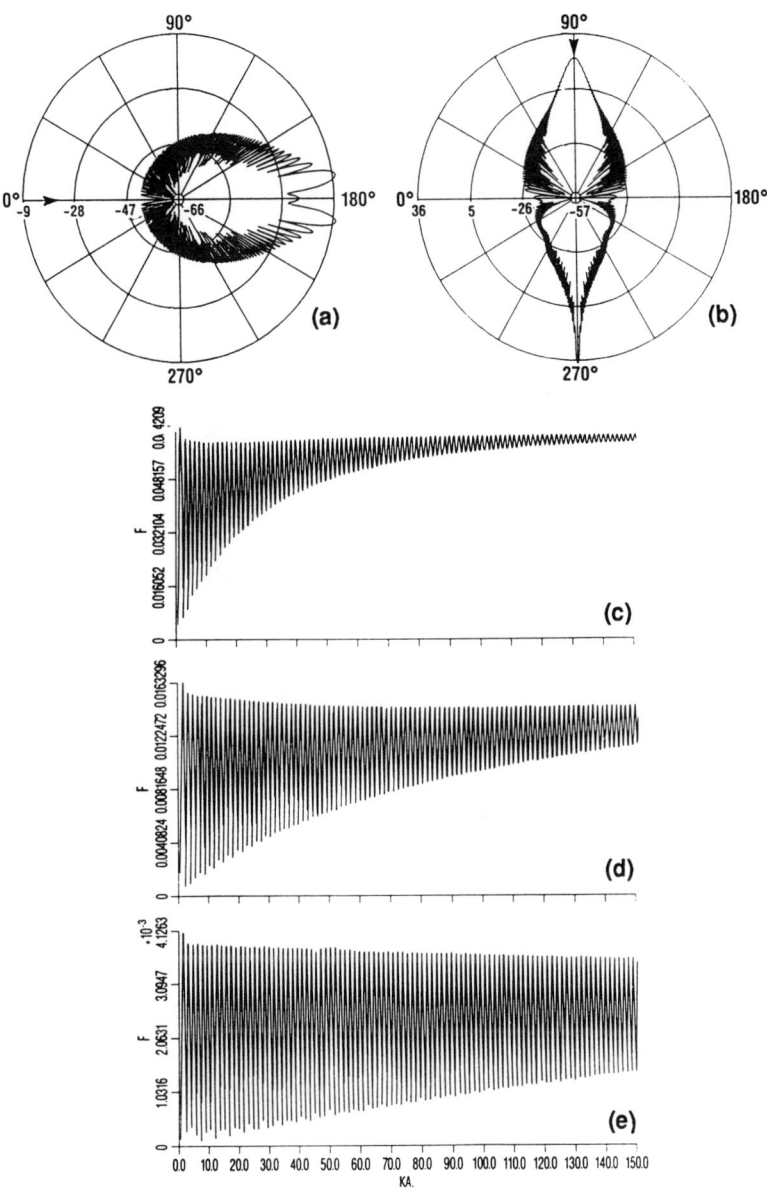

Figure 1. (a) Bistatic scattering from a 30-to-1 aspect ratio rigid spheroid end on; (b) broadside incidence for kL/2 = 200; backscatter from spheroid of aspect ratio of; (c) 4 to 1; (d) 8 to 1; and (e) 16 to 1 for kL/2 = 0 to 150.

198 The Classical Scattering of Waves

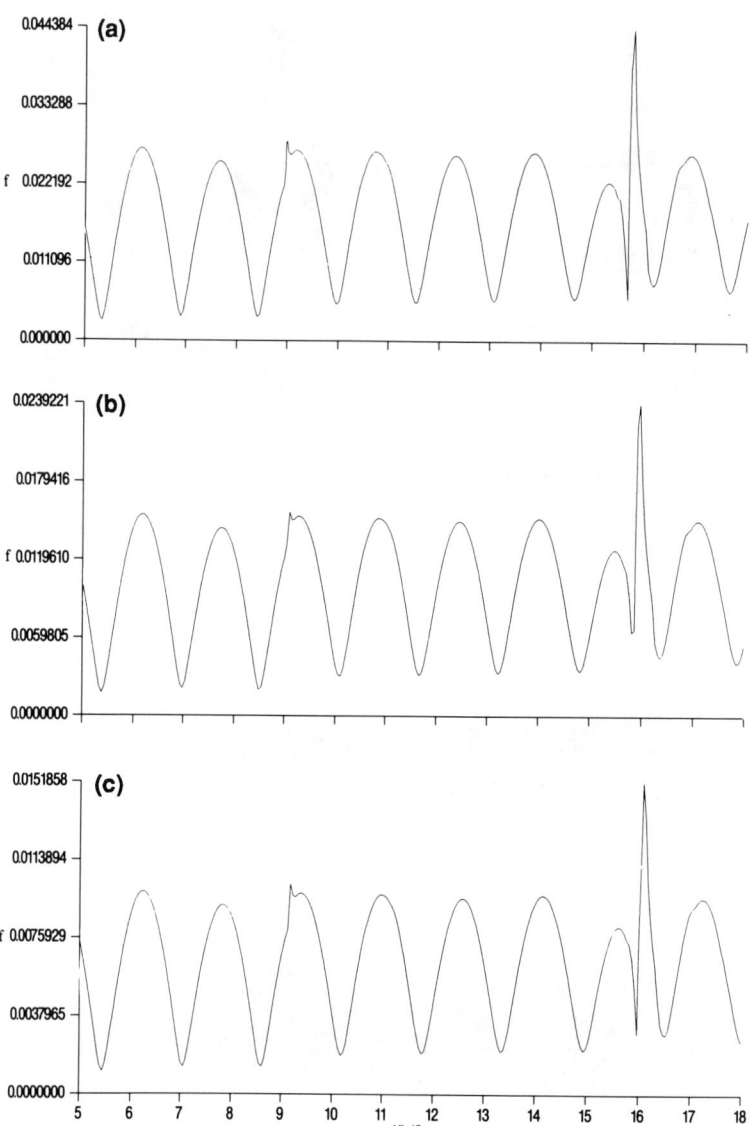

Figure 2. Backscatter from solid WC spheroid end on incidence for aspect ratio of (a) 6 to 1; (b) 8 to 1; and (c) 10 to 1 for kL/2 = 5 to 18.

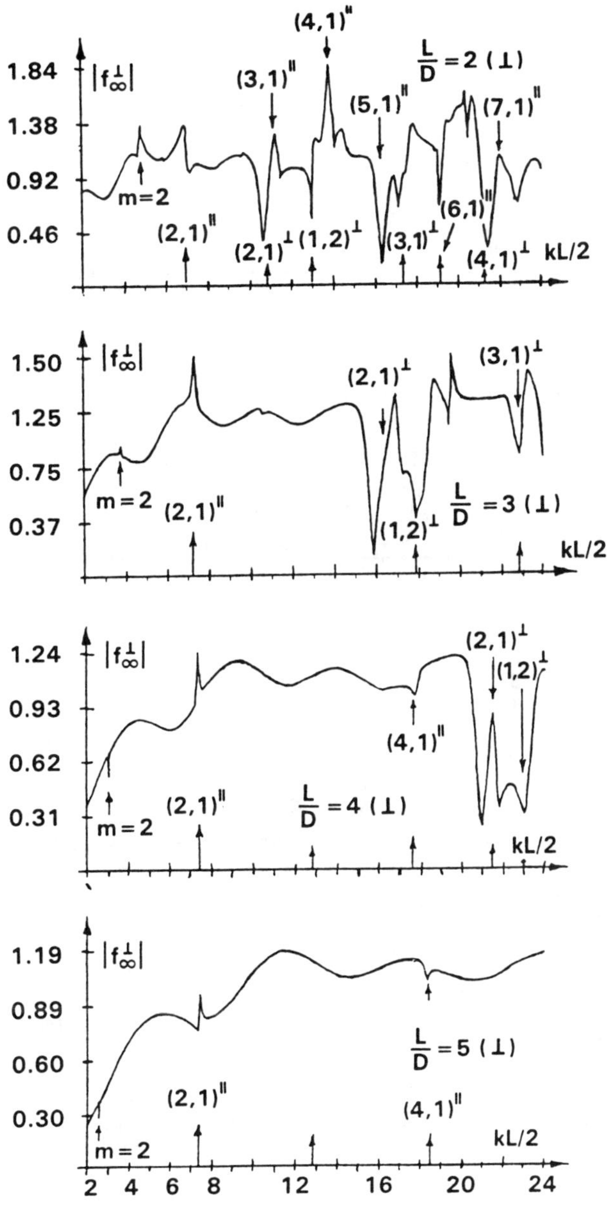

Figure 3. Backscatter from solid steel spheroid broadside incidence for aspect ratio of (a) 2 to 1; (b) 3 to 1; (c) 4 to 1; and (d) 5 to 1 for $kL/2 = 2$ to 24.

200 The Classical Scattering of Waves

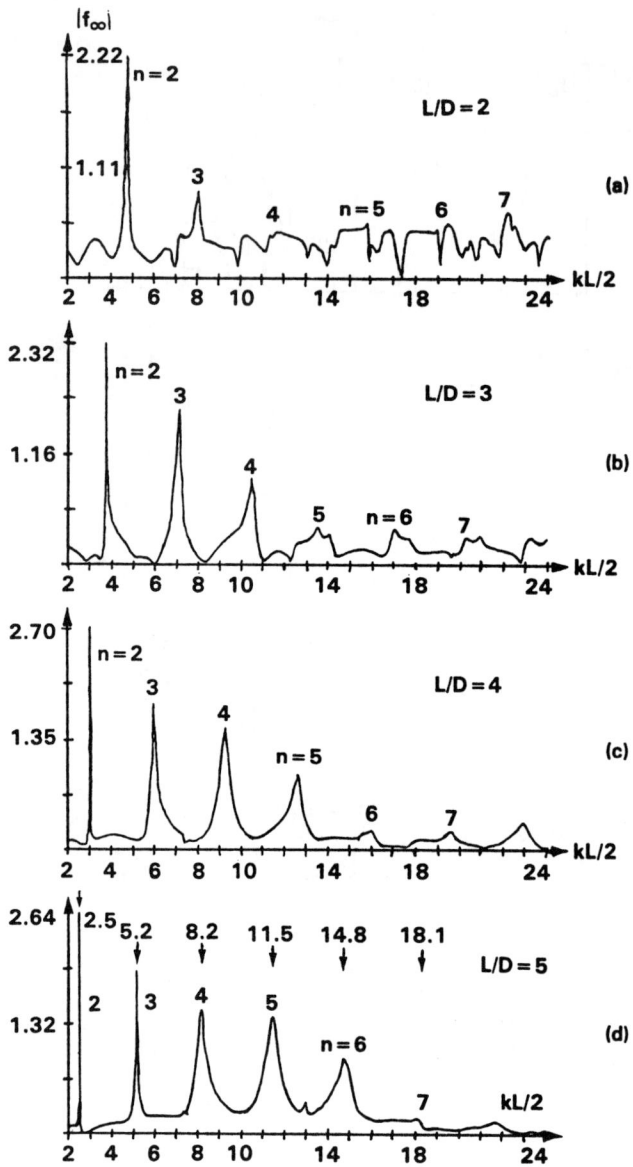

Figure 4. Backscatter from solid steel spheroid at incidence of 45° relative to axis of symmetry for aspect ratio of (a) 2 to 1; (b) 3 to 1; (c) 4 to 1; and (d) 5 to 1 for kL/2 = 2 to 24.

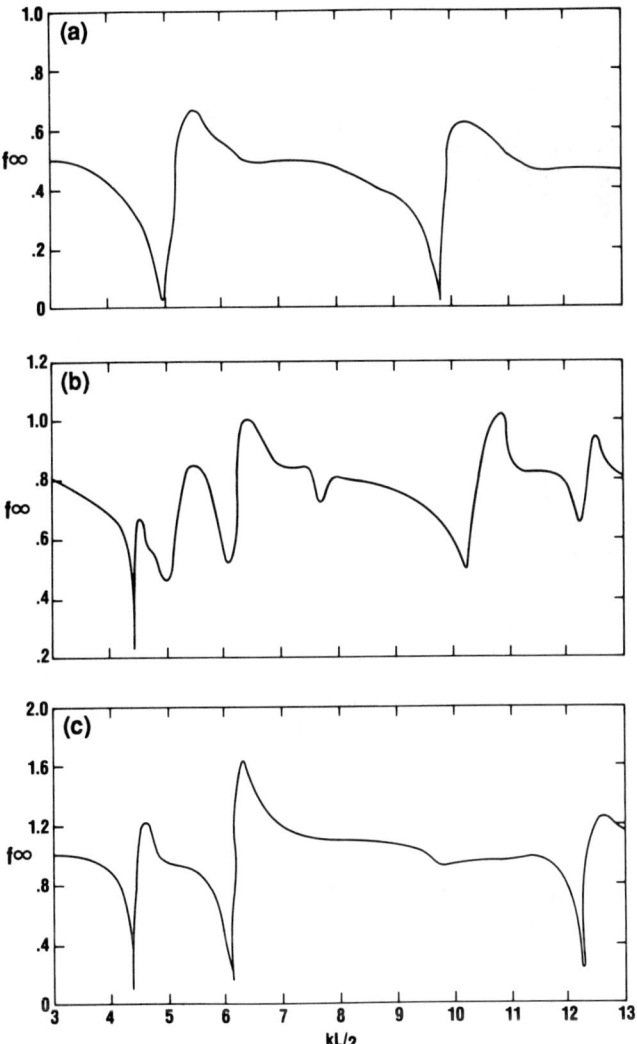

Figure 5. Backscatter from steel spheroidal shell of aspect ratio of 1.5 to 1 (a) backscatter; (b) 45° relative to the axis of symmetry; and (c) broadside.

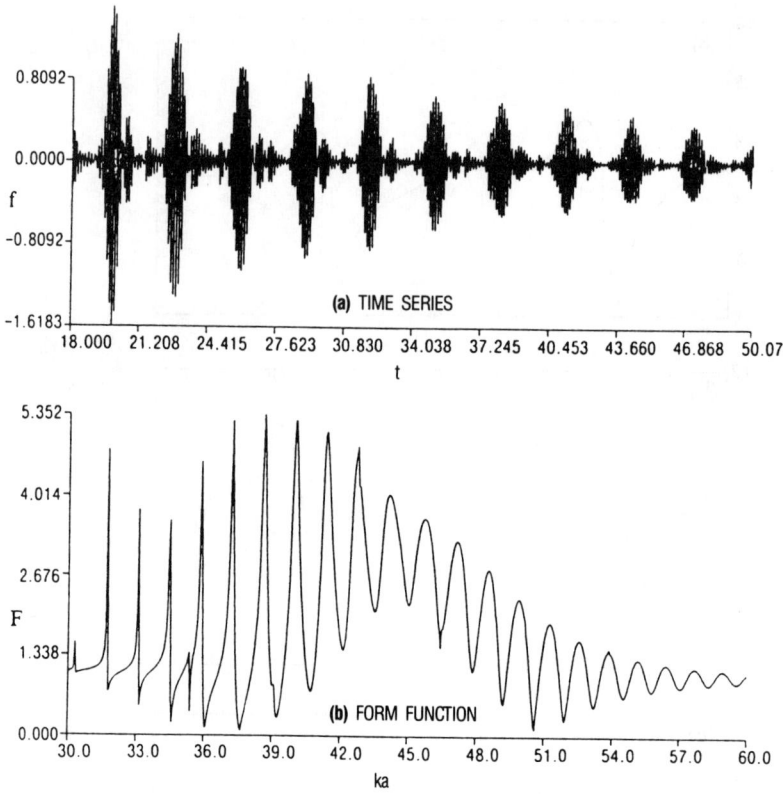

Figure 6. (a) Time domain scattering from a 2.5% thick WC shell from 18 microseconds to 50.07 and (b) backscattered echoes from a 2.5% WC shell from ka = 30–60.

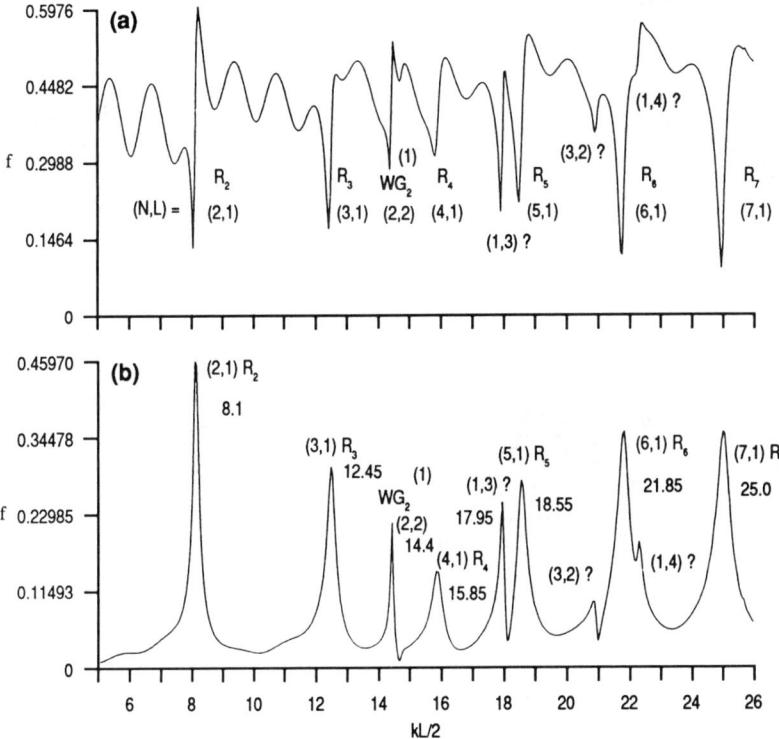

Figure 7. (a) Resonance scattering from a steel spheroid from $kL/2 = 0$–26 and (b) resonance scattering from a steel spheroid from $kL/2 = 0$–26 minus a rigid background.

204   The Classical Scattering of Waves

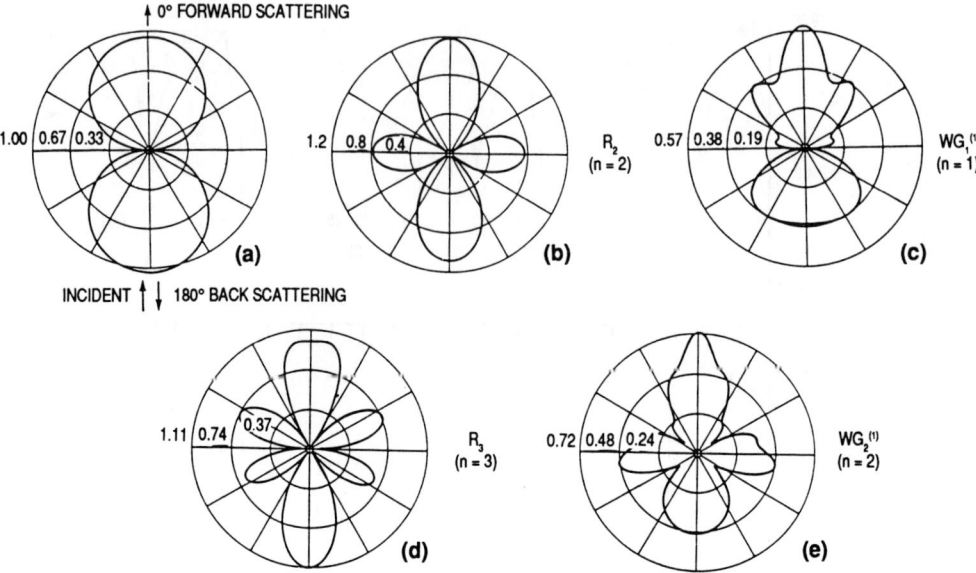

Figure 8. Bistatic scattering of the residual echoes from resonances listed in Figure 7b.

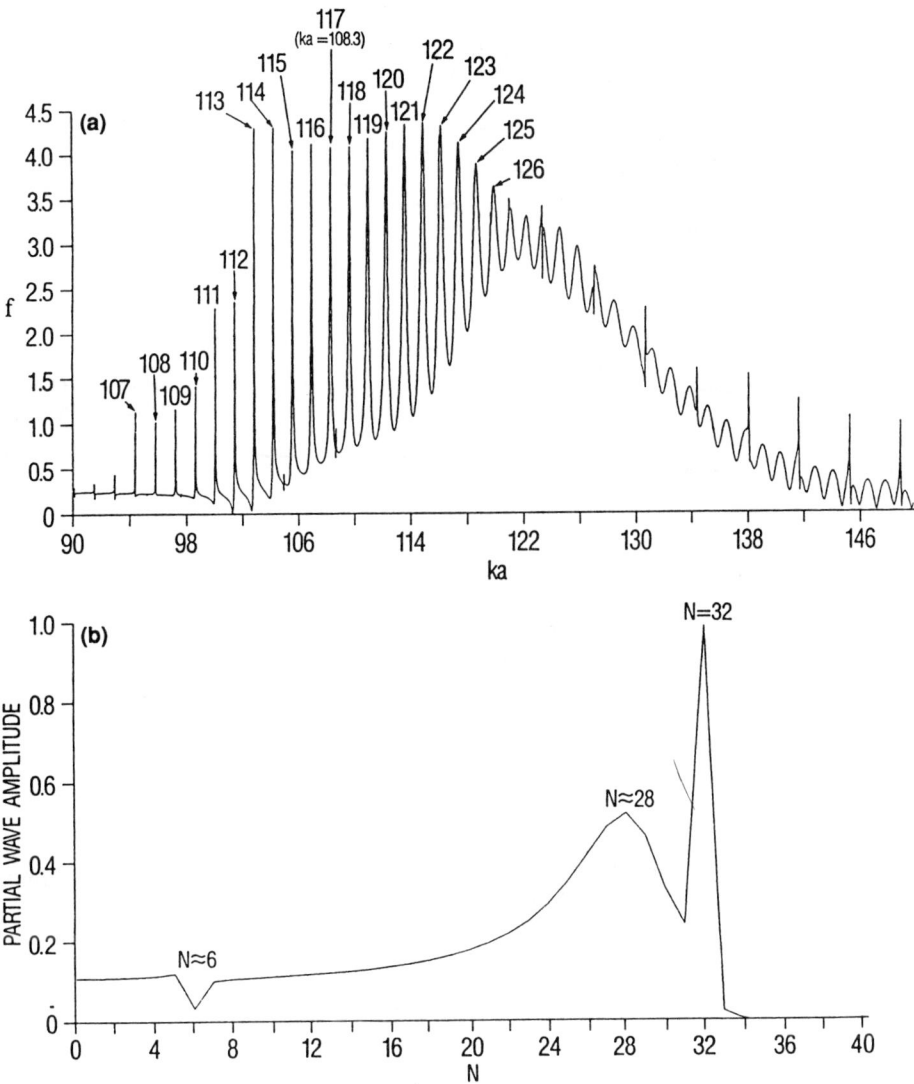

Figure 9. (a) The residual backscattered echo from a 1% thick steel shell and (b) example of a partial wave analysis for the coincidence frequency case for a 2.5% thick WC shell for ka = 33.

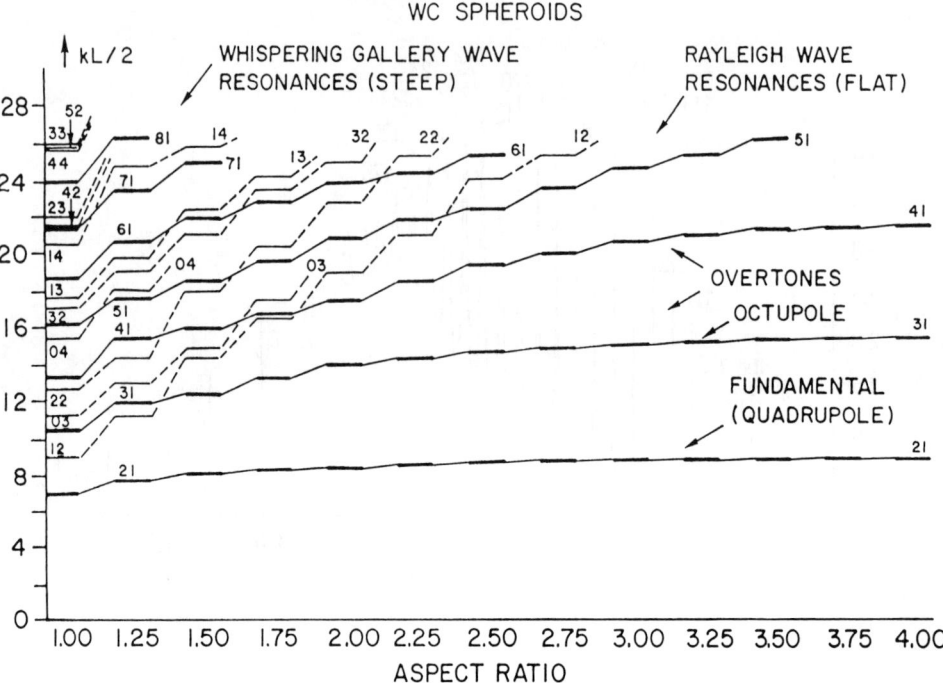

Figure 10. Spectroscopic diagram as a function of aspect ratio of WC spheroidal shells for Rayleigh and Whispering Gallery resonances.

THE R-MATRIX APPROACH TO ATOMIC COLLISIONS

N.R. Badnell and M.S. Pindzola
Department of Physics, Auburn University, Auburn, Alabama 36849

D.C. Griffin
Department of Physics, Rollins College, Winter Park, Florida 32789

ABSTRACT

We review briefly the R-matrix approach to atomic collisions. We then discuss its application to the electron impact excitation of Si^{3+} and Zn^+, and to the excitation-autoionization of Ca^+, and discuss the results in comparison with experiment.

INTRODUCTION

Low-energy electron scattering by atoms and ions is of intrinsic interest due to its many-body nature as well as being of fundamental importance to our understanding of non-equilibrium plasmas viz. magnetic fusion plasmas, laser-produced plasmas, the solar corona, planetary atmospheres and gaseous nebulae. The close-coupling method[1] is the main approach that is used to study the low-energy electron-impact excitation of atoms and ions when the coupling between reaction channels is too strong to be treated perturbatively. The total wavefunction for the system is expanded in terms of an infinite set of eigenfunctions for the target. The close-coupling approximation arises from retaining only a finite number of terms in the expansion. In the next section, we discuss the use of the R-matrix method[2] for the solution of said equations. We then apply this approach to the electron impact excitation[3,4] of Si^{3+} and Zn^+, which have been studied experimentally[5,6] recently using an electron energy-loss spectroscopy technique; and to the calculation of the excitation-autoionization contribution to the ionization[7] of Ca^+, which has long been a source of disagreement between theory[8] and experiment[9] - we reveal the solution.

THE R-MATRIX METHOD

The R-matrix method was introduced originally in the context of nuclear physics[10,11] and was subsequently[2] applied to atomic physics where the long-range nature of the coulomb potential drastically changes the picture. In the R-matrix approach[2], configuration space is divided-up into two regions, an inner-region containing all of the short-range interactions and an outer-region containing just the coulomb interaction and, maybe, some long-range channel couplings. In the inner-region, the continuum wavefunction is expanded in terms of a finite basis set of continuum orbitals whose expansion coefficients are determined by a single diagonalization of the Hamiltonian for the complete system and the imposition of suitable boundary conditions. Thus, the main overhead in generating a solution in the inner-region is essentially independent of the scattering energy. This is the powerful feature of the R-matrix method, not found in earlier methods, which enables the resonance features, which so dominate low-energy scattering, to be mapped-out economically. A solution is still required at each energy in the outer-region but, for non-neutrals, the (long-range) coupling terms can be treated perturbatively[12] and so the zero-order solutions are just coulomb functions (in the outer region). The most time consuming part of a large calculation is still the inner-region solution, in particular, the diagonalization of the total Hamiltonian.

The standard R-matrix approach described above is just a particularly efficient solution of the close-coupling equations and its accuracy is limited accordingly by the accuracy of the close-coupling approximation. In particular, and unlike in nuclear physics, the target eigenfunction expansion can be slowly convergent. An essentially exact solution to the scattering problem in the inner-region is obtained with the intermediate energy R-matrix method[13] (IERM) whereby a second R-matrix expansion is now applied to the active target electron. The incomplete expansion in the outer-region (i.e. a standard close-coupling expansion) still models the loss of flux into omitted channels provided that the resulting pseudo-resonances are suitably averaged-over[14,15]. So far, the IERM has only been applied to a few partial waves for electron-hydrogen collisions[16], the rest of this article deals with the standard R-matrix method.

The R-matrix method has been implemented in a number of large-scale computer codes which utilize non-relativistic[17], Breit-Pauli[18] or Dirac[19] Hamiltonians. We will be concerned here only with the non-relativistic approach. In particular, we use the version[20] of the code developed for the Opacity Project. The Opacity Project[21] is a large scale international effort to generate radiative data for all of the important bound-bound, bound-free and free-free transitions for all of the astrophysically important elements for all stages of ionization. The requirement for such a large amount of atomic data stimulated the development of a non-relativistic R-matrix code with faster angular momentum codes, faster asymptotics (the perturbative solution), the ability to take (suitable) radial wavefunctions from any atomic structure code, and routines to "top-up" the high angular momentum contribution to dipole transitions for electron-impact excitation based-on angular momentum sum rules (see Ref. 20 and references therein). The main emphasis of the Opacity Project being on radiative data, there has been less application to electron-impact excitation (although, see Ref. 22). Our interest lies in seeing how well the results of the standard R-matrix method compare against experiment, in particular, the first experimental results obtained using the new high-resolution electron energy-loss spectroscopy technique[5,6]; and, indeed, interpreting and assessing the experimental results in the first place. To this end, we now consider three examples, Si^{3+}, Zn^+ and Ca^+.

EXCITATION OF Si^{3+}

In figures 1 and 2, we present our results[3] for the 3s-3p excitation of Si^{3+} obtained from a seven-state (3s,3p,3d,4s,4p,4d,4f) close-coupling calculation using the R- matrix method. The bound-state orbitals were obtained from a single-configuration Hartree-Fock frozen-core ($1s^2 2s^2 2p^6$) calculation. There is a wealth of resonance structure above 10 eV (see figure 1) but near threshold there is only a single very strong broad feature due to the 3d4p ^1F and ^1P terms which is only apparent from a detailed study of the source of the various contributions to the theoretical cross section. The cross sections in figure 2 were convoluted with a 0.2 eV FWHM Gaussian to facilitate comparison with experiment. Also shown in figure 2 are the results obtained from an electron energy-loss spectroscopy measurement[5] made at ORNL, the two sets of results are in good agreement. In the experiment[5], electron and ion beams were merged in a solenoid field, allowed to interact, and then were demerged. The inelastically scattered electrons were detected with a position sensitive detector. The experiment is limited to near threshold energies, at higher energies it becomes impossible to separate-out the inelastically scattered electrons from the elastically scattered electons. In fact, the experimental results[5] could have been extended in energy another 2 eV, or so, by measuring only a partial inelastic cross section (forward scattering).

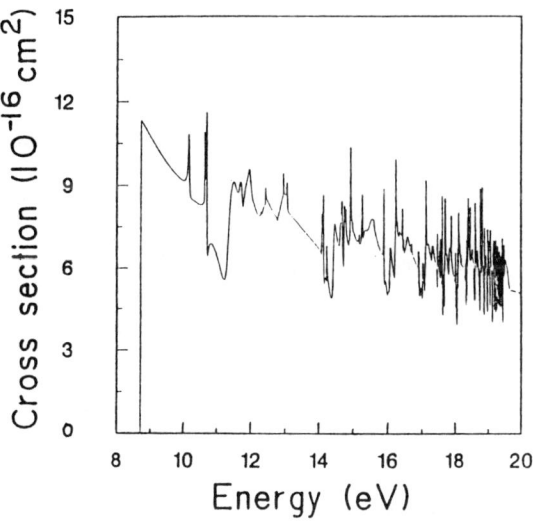

FIG. 1. Electron-impact excitation cross section for the 3s-3p transition in Si^{3+} evaluated from a seven-state close-coupling approximation (Ref. 3).

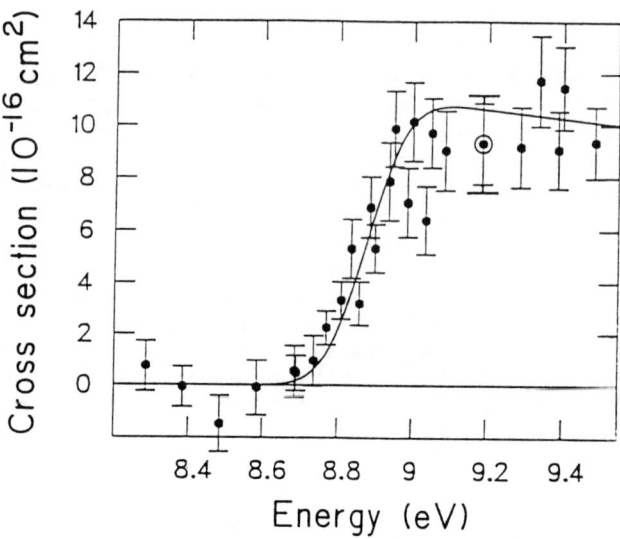

FIG. 2. A comparison of theory (Ref. 3) and experiment (Ref. 5) for the 3s-3p transition in Si^{3+} in the near-threshold region.

EXCITATION OF Zn^+

A second group[6], this time at JPL, has also made electron energy-loss spectroscopy (EELS) measurements[6], this time for the 4s-4p transition in Zn^+. Although similar to the ORNL apparatus[5], this group has a different experimental arrangement which limits them to measuring only forward scattering but, which they claim, does not restrict them to near-threshold energies. We have carried-out[4] a 15-state close-coupling calculation for Zn^+ using the R-matrix method, using bound-state wavefunctions obtained from the SUPERSTRUCTURE code. In figure 3, we present both total cross sections ($\theta = 0\text{-}180°$) and partial cross sections ($\theta = 0\text{-}90°$) for the 4s-4p transition and compare them with the experimental results[6]. Near threshold, we see that there is good agreement between experiment and theory (for $\theta = 0\text{-}90°$). By 12 eV there is little difference between the partial and total theoretical cross sections and so the experimental results can probably be regarded as a total cross section above 12 eV. However, theory and experiment start to diverge above 12 eV. In figure 4, we present our total 4s-4p cross section over a wider energy range, along with the JPL results[6]. We also show the results of an earlier experiment[23] which detected the photon emitted by the subsequent decay of the collisionally excited $4p\,^2P$ state. Above about 12 eV, these experimental results are affected by cascades from excitations to higher-energy states. The dashed line shows our theoretical results[4] which allow for the cascade enhancement.

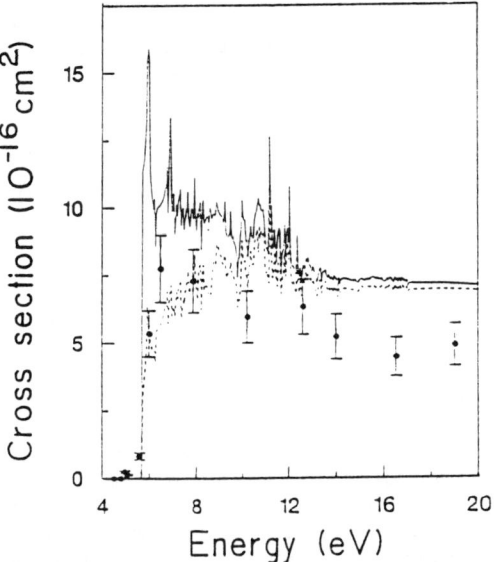

FIG. 3. Total and partial cross sections for the 4s-4p transition in Zn$^+$. Solid curve, $\theta = 0\text{-}180°$ total cross section; dashed curve, $\theta = 0\text{-}90°$ partial cross section; both evaluated from a fifteen-state close-coupling approximation (Ref. 4). Experimental points from Ref. 6.

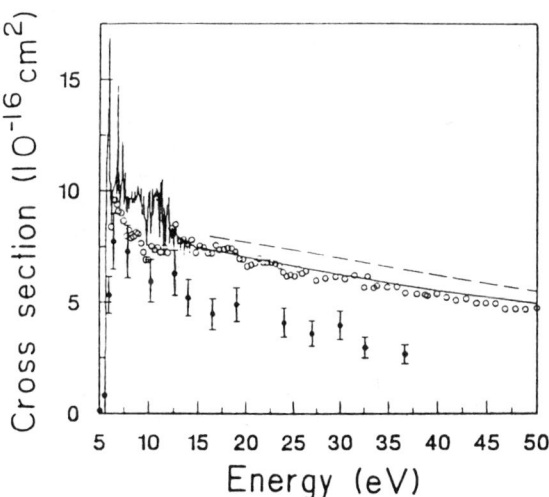

FIG. 4. Excitation cross sections for the 4s-4p transition in Zn$^+$. Solid curve, fifteen-state close-coupling total cross section; dashed curve, plus cascade enhancement. Open circles, cascade affected crossed-beams experiment (Ref. 23); solid circles, cascade-free merged-beams experiment (Ref. 6).

However, from a detailed study[4] of the 4s-4p oscillator strength, we believe that our theoretical results overestimate the 4s-4p cross section and that both theoretical curves should be lowered by about 15%. The resulting cascade-corrected results are then in very good agreement with the optical measurements[23], but the cascade-free results are still somewhat larger than those of the EELS experiment[6], above 12 eV.

IONIZATION OF Ca^+

The low-energy electron impact ionization of Ca^+ is dominated by the indirect contribution from excitation-autoionization, namely

$$3p^64s + e^- \rightarrow 3p^53d4s + e^- \rightarrow 3p^6 + e^- + e^- .$$

The first calculation[8], using the R-matrix method, included only three of the nine autoionizing terms of the $3p^53d4s$ configuration and found a very large feature in the cross section due to dielectronic capture into autoionizing states of Ca followed by sequential double autoionization to Ca^{2+}. This feature was not observed by experiment[9]. A subsequent close-coupling calculation[24] and high-resolution experiment[25] did not change the picture. Distorted-wave calculations[26] showed that it was important to include all nine autoionizing terms of the $3p^53d4s$ configuration. However, these distorted-wave calculations[26] did not allow for the dielectronic capture process discussed above and interchannel coupling is too strong to be treated as a perturbation.

We have carried-out[7] a 13-state close-coupling calculation for Ca^+, using the R-matrix method, that included all nine autoionizing terms of the $3p^53d4s$ configuration as well as the $3p^64s$, $3p^63d$, $3p^64p$ and $3p^54s^2$ configurations. The bound-state wavefunctions were generated using the SUPERSTRUCTURE code and the 3d term-dependence was modelled by configuration mixing with the $3p^54s4d$. We multiplied each excitation cross section by the appropriate Auger yield[26], summed them, and added the result to a direct ionization cross section determined from a single parameter Lotz formula, scaled to agree with experiment at the lowest energies. We compare the results of this calculation[7], convoluted with a 0.2 eV FWHM Gaussian, with experiment[25] in figure 5. We see that there is no evidence of a large dielectronic capture resonance and our results are in much better agreement with experiment than the results we obtained[7] by including only three of the autoionizing terms of $3p^53d4s$, although, they are still somewhat high. The main effect of retaining 13 terms in the close-coupling expansion was to redistribute the collision strength of the resonant feature over all nine autoionizing terms.

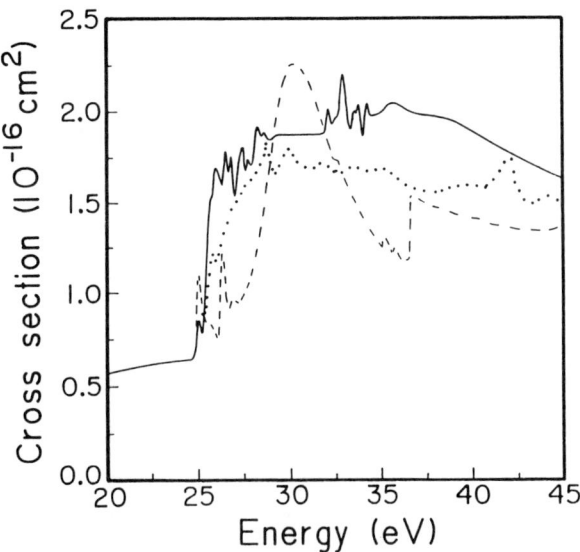

FIG. 5. Electron-impact ionization cross sections for Ca$^+$. Solid curve, thirteen-state close-coupling calculation; dashed curve, four-state close-coupling; both Ref. 7. Dotted line, experiment from Ref. 25.

CONCLUSION

We have briefly reviewed the R-matrix approach to atomic collisions and have looked in detail at its application to several problems which have also been studied experimentally. In the case of the 3s-3p transition in Si^{3+}, there is very good agreement between theory and the near-threshold EELS measurements made at ORNL. In the case of the 4s-4p transition in Zn$^+$, there is good agreement near threshold with the JPL EELS experiment; but, at higher energies theory is somewhat larger than the EELS experiment, but in good agreement with measurements based on an optical technique. For Ca$^+$, we discussed the resolution of the long-standing disagreement between theory and experiment for the low-energy excitation-autoionization contribution to the ionization cross section.

ACKNOWLEDGEMENTS

We would like to thank the members of the Opacity Project for the use of the R-matrix code developed for said project. This work was supported by the Office of Fusion Energy, U.S. Department of Energy under Contract No. DE-FG05-86ER53217 with Auburn University and Contract No. DE-AC05-84OR21400 with Martin Marietta Energy Systems, Inc.

REFERENCES

1. I.C. Percival and M.J. Seaton, Proc. Camb. Phil. Soc. 53, 654 (1957).
2. P.G. Burke and W.D. Robb, Adv. At. Mol. Phys. 11, 143 (1975).
3. N.R. Badnell, M.S. Pindzola, and D.C. Griffin, Phys. Rev. A43, 2250 (1991).
4. M.S. Pindzola, N.R. Badnell, R.J.W. Henry, D.C. Griffin, and W.L. van Wyngaarden, Phys. Rev. A (1991).
5. E. Wahlin, J.S. Thompson, G.H. Dunn, R.A. Phaneuf, D.C. Gregory, and A.C.H. Smith, Phys. Rev. Lett. 66, 157 (1991).
6. S.J. Smith, K.F. Man, R.J. Mawhorter, I.D. Williams, and A. Chutjian, Phys. Rev. Lett. (1991).
7. N.R. Badnell, D.C. Griffin, and M.S. Pindzola, J. Phys. B24, L (1991).
8. P.G. Burke, A.E. Kingston, and A. Thompson, J. Phys. B16, L385 (1983).
9. B. Peart and K. Dolder, J. Phys. B8, 56 (1975).
10. E.P. Wigner, Phys. Rev. 70, 15 and 606 (1946).
11. A.M. Lane and R.G. Thomas, Rev. Mod. Phys. 30, 257 (1958).
12. M.J. Seaton, unpublished (see also Ref. 20).
13. P.G. Burke, C.J. Noble, and P. Scott, Proc. R. Soc. Lond. A410, 289 (1987).
14. T.T. Scholz, J. Phys. B24, 2127 (1991).
15. P.G. Burke, K.A. Berrington, and C.V. Sukumar, J. Phys. B14, 289 (1981).
16. M.P. Scott, T.T. Scholz, H.R.J. Walters, and P.G. Burke, J. Phys. B22, 3055 (1989).
17. K.A. Berrington, P.G. Burke, M. Le Dourneuf, W.D. Robb, K.T. Taylor, and Vo Ky Lan, Comput. Phys. Commun. 14, 367 (1978).
18. N.S. Scott and K.T. Taylor, Comput. Phys. Commun. 25, 347 (1982).
19. P.H. Norrington and I.P. Grant, J. Phys. B20, 4869 (1987).
20. K.A. Berrington, P.G. Burke, K. Butler, M.J. Seaton, P.J. Storey, K.T. Taylor, and Y. Yan, J. Phys. B20, 6379 (1987).
21. M.J. Seaton, J. Phys. B20, 6363 (1987).
22. D. Luo and A.K. Pradhan, Phys. Rev. A41, 165 (1990).
23. W.T. Rogers, G.H. Dunn, J. Ostgaard Olsen, M. Reading, and G. Stefani, Phys. Rev. A25, 681 (1982).
24. M.S. Pindzola, C. Bottcher, and D.C. Griffin, J. Phys. B20, 3535 (1987).
25. B. Peart, J.R.A. Underwood, and K. Dolder, J. Phys. B22, 2789 (1989).
26. D.C. Griffin, M.S. Pindzola, and C. Bottcher, J. Phys. B17, 3183 (1984).

Nonperturbative Electromagnetic Muon-Pair Production with Capture in Peripheral Relativistic Heavy-Ion Collisions

J. C. Wells[a,b], V. E. Oberacker[b], A. S. Umar[b],
C. Bottcher[a], M. R. Strayer[a] & J.-S. Wu[a]

*Center for Computationally Intensive Physics, Oak Ridge National Laboratory
Oak Ridge, TN 37831*

[a] *Physics Division, Oak Ridge National Laboratory, Oak Ridge, TN 37831*
[b] *Department of Physics & Astronomy, Vanderbilt University, Nashville, TN 37235*

ABSTRACT

We discuss preliminary calculations of impact-parameter-dependent probabilities and cross sections for muon-pair production with capture of the negative muon into the K-shell of the target caused by the time-dependent electromagnetic fields generated in peripheral relativistic heavy-ion collisions. Our approach is nonperturbative in that we calculate probabilities by solving the time-dependent Dirac equation on a three-dimensional Cartesian lattice using the basis-spline collocation method. Use of the axial gauge for the electromagnetic potentials produces an interaction easier to implement on the lattice than the Lorentz gauge.

I. INTRODUCTION

Lepton-pair production has been widely discussed as a possible tool to help probe the formation and the decay of the quark-gluon plasma phase of matter in ultrarelativistic heavy-ion collisions [1]. Suggestions by several authors indicate that other sources of lepton pairs might possibly mask the leptonic signals from the plasma phase [2-4]. Electromagnetic lepton-pair production caused by highly stripped heavy ions in relativistic motion is estimated to be a major contribution to the physical background for the plasma pairs as the fields involved contain large Fourier components.

The electromagnetic production of lepton pairs with heavy ions is fundamentally different from the production mechanism with protons or electrons because the coupling constant is strongly enhanced; e.g., for very heavy systems ($Au + Au, U + U$) $Z\alpha \approx 0.5$. Applying low-order perturbation theory to this process in the high-energy regime results in probabilities which violate unitarity and cross sections which violate the Froissart bound [5, 6]. This evidence, along with nonperturbative model studies, clearly suggests that higher-order QED effects will be important at the Relativistic Heavy-Ion Collider (RHIC) [7].

In addition to these fundamental questions regarding lepton-pair production and nonperturbative QED, a valid description of electromagnetic lepton-pair production is important for both detector design and collider performance at experimental facilities such as the Relativistic Heavy-Ion Collider. Pair production with capture of the negative lepton into atomic bound states changes the charge state of the heavy ion leading to a decrease in the luminosity lifetime of the collider [2].

There is a long history of the use of perturbative methods in studying the electromagnetic production of lepton-pairs and a review article exits which discusses recent developments in this field [8]. However, as mentioned above, we desire to study nonperturbative effects in lepton-pair production. One nonperturbative approach discussed for use at mildly relativistic energies is a coupled-channels calculation which shows a five-fold increase in the cross section over perturbative predictions for electron capture into the K-shell of a participant in a fixed-target collision of $Pb + Pb$ at $1.2\ GeV/A$ [9].

In this paper, we discuss our nonperturbative approach which is applicable at ultrarelativistic energies. Beginning with the QED Lagrange density, we make reasonable assumptions about the nature of the lepton and radiation fields in peripheral relativistic heavy-ion collisions which simplify the equations of motion to the time-dependent Dirac equation and the classical Maxwell equations. In doing this, we maintain the field theoretical tools for calculating particle production. We implement the solution of the Dirac equation using the lattice basis-spline collocation method [10, 11]. In these preliminary calculations we limit ourselves to muon-pair creation with capture into the K-shell, neglecting, for now, the more probable and more nonperturbative electron capture process. We do this because the necessary $1s$ bound state of the target is easier to compute for the muonic atom than for the electronic atom with our lattice methods. We neglect free-pair production for now, as this process requires the evolution in time of many states, whereas the capture process requires time-evolution of only a few states.

In developing our approach to nonperturbative lepton-pair production, we have performed $1 + 1$ and $3 + 1$ dimensional model calculations [3, 10, 12]. In the $3 + 1$ dimensional calculation for the system $^{197}Au + ^{197}Au$ at collider energies of 0.2, 1.0 and $2.0\ GeV/A$, we estimated the probabilities for muon-pair production with capture into the K-shell to be between 1.0×10^{-3} and 7.0×10^{-2} at the grazing impact parameter of $b = 8.8\lambda_c$ [12]. This work gave the first clear computational indication for the nonperturbative nature of lepton-pair production in relativistic heavy-ion collisions, though it required small lattices and a screened projectile interaction to make the size of the calculation manageable. Here we improve upon these three-dimensional calculations by using the physical interaction, i.e. no screening. This is made possible by taking advantage of our freedom in choosing the gauge for the electromagnetic interaction.

II. THEORETICAL FRAMEWORK

Our formalism of nonperturbative lepton-pair production in relativistic heavy-ion collisions follows closely the derivation given in Ref. [12]. We begin with the standard QED Lagrange density operator ($\hbar = c = m_0 = 1$)

$$\mathcal{L}_{QED} = \hat{\psi}^\dagger \gamma^0 [\gamma^\mu i\partial_\mu - 1]\hat{\psi} - \frac{1}{4}\hat{F}_{\mu\nu}\hat{F}^{\mu\nu} - \hat{j}^\mu \hat{A}_\mu, \qquad (1)$$

where

$$\hat{j}^\mu = \hat{j}^\mu_{lept} + \hat{j}^\mu_{ext} = e\hat{\psi}^\dagger \gamma^0 \gamma^\mu \hat{\psi} + \hat{j}^\mu_{ext} \qquad (2)$$

denotes the total four-current density operator which consists of the lepton current and the conserved external current generated by the colliding heavy nuclei. By varying the action integral with respect to the field operators $\hat{\psi}$ and $\hat{A}_\mu$, we obtain the Euler-Lagrange equations of motion for the quantum fields

$$[\gamma^\mu(i\partial_\mu + e\hat{A}_\mu) - 1]\hat{\psi}(x) = 0, \quad (3)$$

$$\partial_\mu \hat{F}^{\mu\nu}(x) = \hat{j}^\nu(x). \quad (4)$$

These equations of motion are difficult to solve so we will now make the following simplifying assumptions. First, we neglect the leptonic current because the external heavy-ion current is much larger. Second, we shall assume that the radiation field may be treated classically. Hence, Eqs. (3) and (4) decouple. We now have the problem of solving the time-dependent Dirac equation for the lepton field $\hat{\psi}$ interacting only with an external, classical four-vector potential A_μ^{ext} which is determined independently by the Maxwell equations.

We study the electromagnetic production of lepton pairs in a reference frame in which one of the nuclei, henceforth referred to as the target, is at rest. The target nucleus and the muon interact via the static Coulomb field, A_T^0. The only time-dependent interaction, $A_P^\mu(t)$, arises from the classical motion of the projectile. Thus, it is natural to recast the Dirac equation in Schrödinger form

$$[H_F + H_P]\hat{\psi}(\vec{r},t) = i\frac{\partial}{\partial t}\hat{\psi}(\vec{r},t), \quad (5)$$

where the static Furry Hamiltonian, which describes the lepton field in the presence of the strong external Coulomb field of the target nucleus, is given by

$$H_F = -i\vec{\alpha}\cdot\nabla + \beta - eA_T^0, \quad (6)$$

and the time-dependent interaction between the lepton field and the projectile is

$$H_P(t) = e\vec{\alpha}\cdot\vec{A}_P(t) - eA_P^0(t). \quad (7)$$

Following the usual practice, we expand the field operator $\hat{\psi}(\vec{r},t)$, defined in Eq. (5), into complete orthonormal sets of single-particle basis states. First we consider the Furry basis $\{\chi_i(\vec{r})\}$, i.e. the stationary eigenstates of the Furry Hamiltonian H_F defined in Eq. (6)

$$H_F \chi_i(\vec{r}) = E_i \chi_i(\vec{r}), \quad (8)$$

which are also proper in- and out-states for asymptotic times $|t| \to \infty$, where the interaction $H_P(t)$ is zero

$$\hat{\psi}(\vec{r},t) = \sum_i \hat{a}_i(t)\chi_i(\vec{r})exp(-iE_i t). \quad (9)$$

From the anticommutation relations for the fermion field operators $\hat{\psi}$ and $\hat{\psi}^\dagger$, one readily shows that the $\hat{a}$'s are operators which describe the creation and annihilation of leptons in the static Coulomb field of the target nucleus. They define what one may call the mathematical vacuum $|0\rangle$, i.e., $\hat{a}_i|0\rangle = 0$. We also expand the fermion field operator in terms of the time-dependent basis $\{\phi_j(\vec{r},t)\}$ of solutions to the full Dirac Hamiltonian $H_F + H_P(t)$

$$[H_F + H_P(t)]\phi_j(\vec{r},t) = i\frac{\partial}{\partial t}\phi_j(\vec{r},t). \tag{10}$$

The operator-valued expansion coefficients, denoted by $\hat{\alpha}_j$, are quasi-particle destruction operators

$$\hat{\psi}(\vec{r},t) = \sum_j \hat{\alpha}_j \phi_j(\vec{r},t). \tag{11}$$

The QED ground state, in the presence of the undercritical external field of the target nucleus, $|\Phi_0\rangle$, is a many-lepton state of time-independent one-particle solutions of the Furry Hamiltonian where all states with energies less than $-m_0c^2$ are occupied, i.e. a single Slater determinate of the form

$$|\Phi_0(-\infty)\rangle \equiv |\Phi_0\rangle = \prod_{i<F}(\hat{a}_i^\dagger)|0\rangle, \tag{12}$$

where $i < F$ denotes states below the Fermi energy $E_f = -mc^2$. The time-evolved QED ground state is

$$|\Phi_0(t)\rangle = \hat{U}(t, t_0 \to -\infty)|\Phi_0\rangle \ , \ \hat{U}^\dagger(t,t_0)\hat{U}(t,t_0) = 1, \tag{13}$$

where $\hat{U}(t,t_0)$ is the time-evolution operator defined in the Schrödinger picture by

$$\hat{U}(t,t_0) = e^{-i(H_F+H_P(t))(t-t_0)}. \tag{14}$$

Equations (9), (11), and (12) are consistent with the particle-hole (Dirac sea) picture with both positive- and negative-energy states. However, we wish to work in the space-time picture with physical particles and antiparticles. Therefore, we now define antilepton creation and annihilation operators with respect to the QED ground state $|\Phi_0\rangle$ via

$$\hat{b}_i \equiv \hat{a}_i^\dagger, \ \hat{b}_i^\dagger \equiv \hat{a}_i, \ i < F. \tag{15}$$

Note that these new operators describe creation and annihilation of antiparticles in the static Coulomb field of the target nucleus. Also, the QED ground state behaves as the Furry-vacuum state for the creation of physical particles

$$\hat{a}_p^\dagger |\Phi_0\rangle = |\chi_p^{(+)}\rangle, \ \hat{a}_p|\Phi_0\rangle = 0, \ p > F, \tag{16}$$

where $|\chi_p^{(+)}\rangle$ is a Furry state with positive energy. We also define a new set of quasiparticle and quasiantiparticle creation operators with respect to the time-evolved Furry-vacuum $|\Phi_0(t)\rangle$

$$\hat{\beta}_i \equiv \hat{\alpha}_i^\dagger, \ \hat{\beta}_i^\dagger \equiv \hat{\alpha}_i, \ i < F. \tag{17}$$

We now equate the two representations of the field operator, Eqs. (9) and (11), using the representations defined in Eqs. (15) and (17) to establish the following connection between these two sets of Fock space operators

$$\hat{a}_p(t) = \sum_{s>F} \hat{\alpha}_s \langle \chi_p^{(+)} | \phi_s^{(+)}(t) \rangle e^{iE_p t} + \sum_{r<F} \hat{\beta}_r^\dagger \langle \chi_p^{(+)} | \phi_r^{(-)}(t) \rangle e^{iE_p t} \, , \quad p > F. \qquad (18)$$

Having derived all the necessary field theoretical tools, we now evaluate the probability for lepton-pair production with capture in relativistic heavy-ion collisions. First, using Eq. (18) and the anticommutation relations for the $\hat{\alpha}$'s, we calculate the expected number of leptons produced in state $p > F$ to be

$$\langle \hat{n}_p(t) \rangle = \langle \Phi_0(t) | \hat{a}_p^\dagger(t) \hat{a}_p(t) | \Phi_0(t) \rangle = \sum_{r<F} |\langle \chi_p^{(+)} | \phi_r^{(-)}(t) \rangle|^2, p > F. \qquad (19)$$

Since the expectation value $\langle \hat{n}_p(+\infty) \rangle$ in Eq. (19) is positive-definite and normalized to unity as a consequence of the Pauli principle and the unitarity of the time-evolution operator, $\langle \hat{n}_p(+\infty) \rangle$ is interpreted as the inclusive probability P_p that a lepton will be created in state $p > F$.

In many experimental situations, one is interested only in the probability for producing a lepton of a specific energy regardless of its angular momentum. Calculating this requires that one averages the probabilities for all leptons produced with a given energy. However, if one is interested in pair production with capture, there will be no need for such averaging over the magnitude of the total angular momentum because of the discrete nature of the spectrum. Only averaging over the momentum projections is required; e.g. the probability of creating a lepton with energy E_p is

$$P_{E_p} = \frac{1}{2\mu_p + 1} \sum_{\mu_p=-j_p}^{+j_p} P_p = \frac{1}{2\mu_p + 1} \sum_{\mu_p=-j_p}^{+j_p} \langle \hat{n}_p(+\infty) \rangle. \qquad (20)$$

From Eqs. (19) and (20) we see that to compute probabilities for lepton-pair production, we project time-evolved single-particle states onto static Furry states, i.e. compute single-particle transition amplitudes. Measurable probabilities are the asymptotic ($t \to \infty$) limit of the squares of these amplitudes. With this in mind, we identify these asymptotic transition amplitudes with matrix elements of the scattering operator, $\hat{S} \equiv \hat{U}(+\infty, -\infty)$, defined in the Furry basis as

$$\langle \chi_j | \hat{S} | \chi_i \rangle = S_{i \to j} = \langle \chi_j | \phi_i(\infty) \rangle. \qquad (21)$$

We use the time-reversal symmetry of the Dirac equation to reduce the effort needed to compute capture probabilities by applying to Eqn. (20) the principle of semi-detailed balance [13]

$$\sum_{\mu_a=-j_a}^{j_a} \sum_{\mu_b=-j_b}^{j_b} |S_{a \to b}| = \sum_{\mu_a=-j_a}^{j_a} \sum_{\mu_b=-j_b}^{j_b} |S_{b \to a}|. \qquad (22)$$

For example, to compute Eq. (20), we must solve for the entire time-evolved negative energy continuum. Rewriting Eq. (20), using Eqs. (19), (21), and (22), we obtain an expression where only $2j_p + 1$ time-dependent solutions of the Dirac equation are required

$$P_{E_p} = \frac{1}{2\mu_p + 1} \sum_{\mu_p=-j_p}^{+j_p} \sum_{r<F} |S_{p \to r}|^2, p > F. \qquad (23)$$

In the case of capture into the K-shell of the target, one needs to compute only two time-dependent Dirac states $|\phi_{1s}^{\pm 1/2}(t)\rangle$. Currently, we are forced to assume that both magnetic substates contribute equally to $P_{E_{1s}}$, as our method chosen to compute static bound states is unable to solve for degenerate eigenvectors.

III. NUMERICAL IMPLEMENTATION

In the course of Section II, we reduced the problem of electromagnetic pair production in relativistic heavy-ion collisions to that of computing single-particle excitation amplitudes in the Furry basis. We implement the solution of the Dirac equation (Eq. (8)), necessary for computing these excitation amplitudes, using the lattice basis-spline collocation method (BSCM) which is discussed in detail in Refs. [10, 11]. In this method, quantum-state vectors and coordinate-space operators are given by expansions in terms of basis-spline functions and represented on a spatial lattice. We write the BSCM lattice representation of the Dirac equation for the Furry Hamiltonian in matrix notation as

$$\mathbf{H_F} \chi_i = E_i \chi_i, \qquad (24)$$

where χ_i gives the value of the spinor only at the N^3 lattice points. In effect, the BSCM reduces the partial differential equation to a series of linear algebraic equations which may be solved (at least partially) using iterative techniques. This method is very efficient and allows us to avoid notorious pathologies, e.g. fermion doubling, and to maintain basic conservation laws on the lattice [10]. We use three-dimensional Cartesian coordinates which avoid the pathologies of rotating frames and the complicated metrics of spherical coordinate systems. The time-dependent electromagnetic fields exhibit no useful symmetry in any case.

The complete eigensolution of $\mathbf{H_F}$, providing its full spectrum of stationary states, currently exceeds the state-of-the-art in computational capabilities due to the rank of $\mathbf{H_F}$ being $N^3 \times 4 \times 2$ where the lattice size N can be as large as 100. For this reason, we compute the $1s$ state needed for K-shell capture cross sections by partial eigensolution of $\mathbf{H_F}$ using an efficient iterative Lanczos algorithm [14, 15]. The Lanczos algorithm has features which are attractive for our purposes; the memory requirements are small and the method approximates extremal eigenvalues in the spectrum very well. However, the algorithm has the limitation that it cannot determine the multiplicity of any eigenvalue, i.e. the solutions are not eigenstates of the projection of the total angular momentum. A description of this algorithm as we apply it can be found elsewhere [16].

We approximate the continuum states of $\mathbf{H_F}$ by requiring the eigenstates of the lattice representation of the free Dirac Hamiltonian, $H_0 = -i\vec{\alpha} \cdot \nabla + \beta$, to be orthogonal to the bound states of $\mathbf{H_F}$. Thus, we impose onto the free Dirac continuum the correct initial condition that transition S-matrix elements between the bound and continuum states are zero before switching on the interaction. Even though the free Dirac continuum states are known analytically, we cannot use these states in our lattice calculation by simply evaluating these functions at the lattice points, as such states would not be eigenstates of our lattice Hamiltonian. We must construct the eigenstates of $\mathbf{H_0}$ which we denote as $\xi_{a,\lambda,s}$

$$\mathbf{H_0}\boldsymbol{\xi}_{a,\lambda,s} = \lambda E_a \boldsymbol{\xi}_{a,\lambda,s}; \ \lambda = \pm 1, \qquad (25)$$

where s is the helicity [17]. Therefore, we define our Furry continuum states by modifying the states $\boldsymbol{\xi}_{a,\lambda,s}$

$$\bar{\boldsymbol{\chi}}_{a,\lambda,s} \equiv \boldsymbol{\xi}_{a,\lambda,s} - \sum_{b=1}^{n_b} \boldsymbol{\chi}_{E_b,\kappa_b}^{\mu_b} \left(\boldsymbol{\chi}_{E_b,\kappa_b}^{\dagger\mu_b} \boldsymbol{\xi}_{a,\lambda,s} \right). \qquad (26)$$

Currently, the sum above includes only the $1s$ bound state.

We approximate the scattering operator $\hat{S}$ by a Taylor series expansion of the lattice representation of the time-evolution operator $\mathbf{U}$. We begin and end time-evolution at finite times $\pm T$, where we may neglect the projectile's interaction with the lepton. Time is discretized in the sense that the interactions are taken as constant in each of a series of successive small intervals $(t, t+\tau)$. In each interval, the solution of the lattice representation of the time-dependent single-particle Dirac equation, Eq. (10), is obtained by

$$\phi(t+\tau) = \mathbf{U}(t+\tau,t)\phi(t) = \left[1 + \sum_{n=1}^{N} \frac{[-i\tau(\mathbf{H_F} + \mathbf{H_P}(t))]^n}{n!} \right] \phi(t). \qquad (27)$$

All numerical procedures discussed for implementing our lattice methods reduce to a series of *matrix* × *vector* operations which can be efficiently executed on vector or parallel supercomputers without explicitly storing the matrix in memory.

IV. ELECTROMAGNETIC INTERACTION

We saw in Eqs. (6) and (7) that the physics of lepton pair production is defined by the electromagnetic fields of two particles in relative motion, and that these fields enter the Hamiltonian via the dimensionless interaction energy between the lepton and the colliding nuclei $\bar{A}^\mu = -eA^\mu$. We assume a spherical and homogeneous charge density for both nuclei, as finite nuclear size effects are important in the $1s$ state for heavy leptons. We also neglect recoil effects. Therefore, in the fixed-target frame of reference, the projectile moves with constant velocity β_f along a straight-line trajectory in z-direction with impact parameter b.

Since the Dirac equation is covariant under a gauge transformation of the EM potentials, the gauge may be chosen for convenience in any problem. The most familiar gauge used in problems with electric sources is the Lorentz gauge, defined by the condition, $\partial_\mu A^\mu = 0$. Since we assume the projectile to move with constant velocity, the time-dependent EM interaction between the projectile and muon in the Lorentz gauge can be generated by a Lorentz-boost of the static Coulomb field. The Lorentz factors for the fixed target and collider frames are related by $\gamma_f = 2\gamma_c^2 - 1$. Figure 1 shows the lattice representation of the temporal component of the lepton's interaction energy with the target and projectile, $\bar{A}^0(\vec{r},t) = \bar{A}_P^0(\vec{r})+ \bar{A}_T^0(\vec{r},t)$, in a 100 GeV/A collision of $^{197}Au + {}^{197}Au$ in the collider frame. Features of the interaction in the Lorentz gauge cause difficulties for our lattice methods. First, the interaction is very large as it is proportional to γ_f, and sharply peaked in the boost direction $\hat{e}_3$. Its width is inversely proportional to γ_f. Also, the long-range nature of the Coulomb potential remains after the boost as the Lorentz gauge interaction behaves asymptotically as $1/z$. The large magnitude of the interaction requires that the time step be kept small so as to ensure a proper expansion of the time-evolution operator. Also, the long-range z dependence of the interaction requires that we start numerical calulations at large distances between projectile and target. Another major difficulty lies in representing spiked functions on a finite lattice. These features combine to make realistic, three-dimensional calculations too demanding computationally [12].

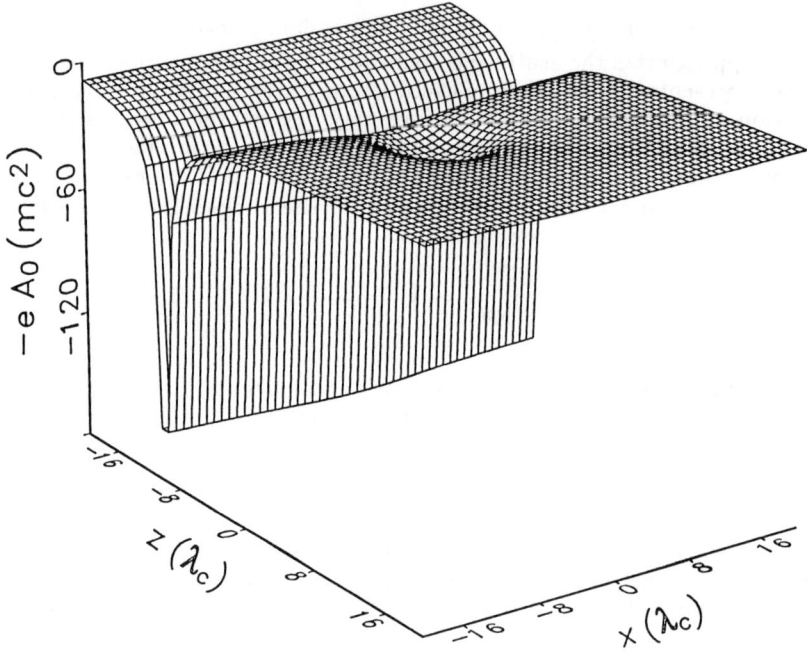

Figure 1. Depicted is a slice taken at $y = 0$ of the temporal component of the lepton's interaction energy with the target and projectile for the Lorentz gauge in a collision of $^{197}Au + {}^{197}Au$ at a collider energy of 100 GeV/A. The center of the projectile nucleus is located at the point $(0, 8.732\,\lambda_c, -10\,\lambda_c)$; the target is located at the origin. A square mesh with 51×51 points is used to represent the interactions. The narrow spike of the time-dependent interaction is cut off. Actually, the extrema of this interaction is $-141,210\,m_0 c^2$.

Another gauge choice exists which, for relativistic energies, produces an interaction that is better represented on the lattice [18]. The new gauge we use is the axial gauge in which the component of the vector potential in the direction of the projectile's motion is required to be zero, i.e.

$$\bar{A}_P^3[r'(t)] \to \tilde{A}_P^3[r'(t)] = \bar{A}_P^3[r'(t)] + \partial_z \Lambda[r'(t)] \equiv 0. \tag{28}$$

The new interaction is found by using the function Λ in a gauge transformation

$$\bar{A}_P^\mu[r'(t)] \to \tilde{A}_P^\mu[r'(t)] = \bar{A}_P^\mu[r'(t)] - \partial^\mu \Lambda[r'(t)]. \tag{29}$$

Figure 2 shows the x component of the axial gauge interaction of the muon with the projectile, $\tilde{A}_P^1[r'(t)]$, for the 100 GeV/A collision discussed in figure 1. The following features of the axial gauge interaction enable its easy representation on the lattice. First, the temporal component, $\tilde{A}_P^0[r'(t)]$, is very small as it is inversely proportional to γ_f. The dominate components in the interaction, $\tilde{A}_P^1[r'(t)]$ and $\tilde{A}_P^2[r'(t)]$, have a step function behavior in the large γ_f limit. Step functions require fewer lattice points than spiked functions to ensure a faithful representation on a mesh. Also, almost all of the energy dependence for the interaction is contained in the slope of the step in the x and y components which is proportional to γ_f; the extrema are independent of γ_f for large γ_f. As a result, the axial gauge interaction is largely independent of γ_f in the large γ_f limit. Therefore, all three of the numerical difficulties associated with the Lorentz gauge interactions are removed in the axial gauge.

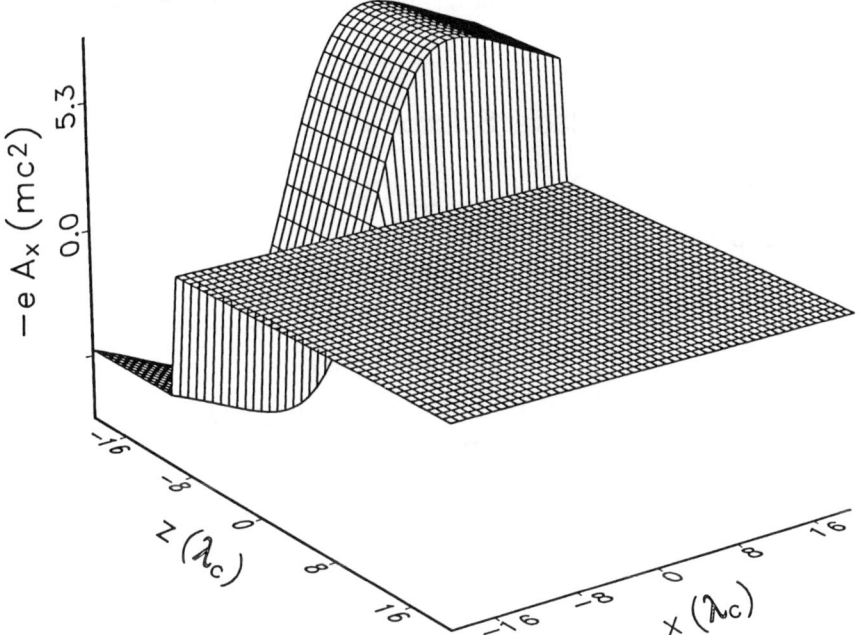

Figure 2. Depicted is a slice taken at $y = 0$ of the x component of the lepton's interaction energy with the projectile for the axial gauge in a 100 GeV/A collision of $^{197}Au + ^{197}Au$ in the collider frame. The center of the projectile nucleus is located at the point $(0, 8.732\,\lambda_c, -10\,\lambda_c)$. A square mesh with 51×51 points is used to represent the interaction.

V. RESULTS AND CONCLUSIONS

The first step in calculating the probability for muon-pair creation with the capture of the negative muon into the K-shell is the calculation of the 1s state of the Furry Hamiltonian. The static Coulomb interaction between the target nucleus and muon is evaluated for a homogeneous, spherical nuclear charge distribution with a radius $R = 1.40 \times A^{1/3}\, fm$ where A is the atomic number. Using the basis-spline collocation method and the iterative Lanczos algorithm referred to in Section III, we compute the muonic 1s state of ^{197}Au on a uniform cubic lattice within a box extending from $-20\,\lambda_c$ to $+20\,\lambda_c$ in the three Cartesian coordinates where λ_c is the muon Compton wavelength. We perform the calculation on uniform lattices varying in size from $N = 16$ to 39. We require for the convergence of each solution that the energy fluctuation $\eta = [<H^2> - <H>^2]^{1/2}$ be less than 10^{-6}. For the $N = 16$ lattice, the Lanczos algorithm required about 120 recursions to obtain a fluctuation of 3×10^{-7}. The Cray-2 CPU time required to perform these static calculations varies from 1 minute for $N = 16$ to 40 minutes for $N = 39$. The energy eigenvalues obtained for different lattices are shown in figure 3. The accuracy of our static solution is checked by performing an integration of the radial Dirac equations which gives an energy of $0.9127 m_0 c^2$ for the muonic 1s state in ^{197}Au. We observe a N^{-2} convergence of the static lattice solutions.

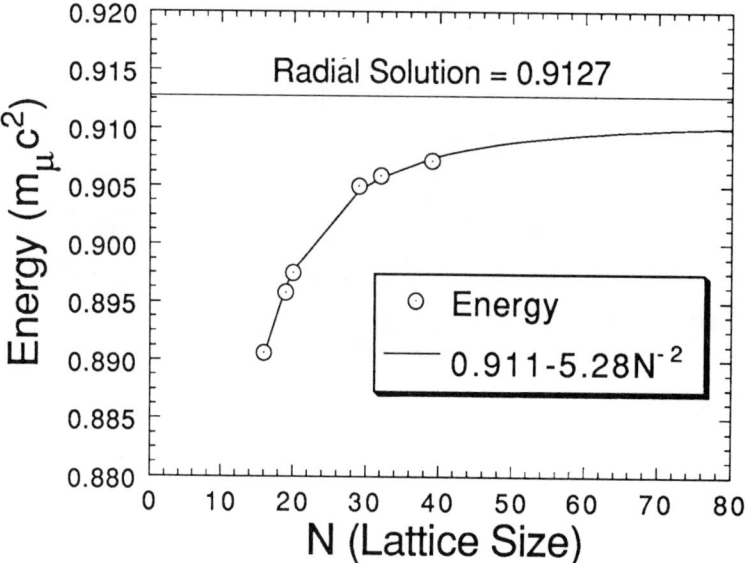

Figure 3. Energies for the 1s muonic state of ^{197}Au computed with the basis-spline collocation method as a function of the lattice size N. The energies approach a limiting value as the lattice size is increased.

We perform the time-development of the muonic 1s state of ^{197}Au, i.e. solve Eq. (10), on the 16^3 lattice under the influence of the time-dependent external fields produced in the collision of $^{197}Au + ^{197}Au$ at a collider energy of 100 GeV/A. A maximum number of 15 terms in the Taylor series expansion of the time-evolution operator is sufficient to preserve conservation of the norm of the wavefunction at the 1 part in 10^6 level. This is an important indicator for the numerical accuracy of the time-development of the Dirac spinor. We use the axial gauge for the interaction of the muon with the projectile. Initially, the projectile nucleus is positioned at $(0, b, -200\,\lambda_c)$ giving rise to a negligible interaction with the muon at the position of the target nucleus which is fixed at the center of the cubic lattice. We evolve the wavefunction in time for 2000 time steps with $\Delta t = 0.2/\beta_f$ in units of λ_c/c, stopping the evolution when the projectile is positioned at $(0, b, 200\,\lambda_c)$. One such time-dependent run for a given impact parameter requires 33 minutes of Cray-2 CPU time.

The dependence of the muon capture probability on the beam energy is shown in figure 4 by varying the energy between 0.1 and 100 GeV/A in the collider frame of reference for the grazing impact parameter $b = 8.732\,\lambda_c$. Notice that the capture probability is independent of the beam energy for $E_c \geq 16\ GeV/A$. The energy independence of the interaction of the muon with the projectile in the high-energy limit was discussed in Section IV in the context of the axial gauge.

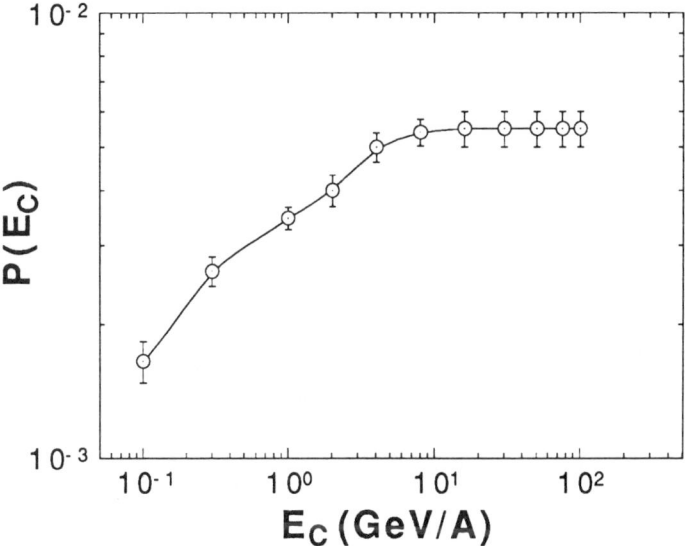

Figure 4. Muon capture probability into the K-shell in collisions of $^{197}Au + {}^{197}Au$ as a function of beam energy in the collider frame at the grazing impact parameter $b = 8.732\,\lambda_c$. The calculations are performed on a uniform cubic lattice with 16^3 collocation points. These results have yet to converge because of the use of small lattice sizes. We give a subjective estimate of the error in the probabilities from this limitation by the indicated error bars.

The dependence of the muon capture probability on impact parameter at a collider energy of 100 GeV/A is shown in figure 5. For peripheral impact parameters (i.e. $b \geq 8.732\,\lambda_c$), the capture probability decreases proportionally to b^{-2} as shown by the fit to the results in figure 5. We calculate the total cross section for muon capture into the K-shell of ^{197}Au by integrating the capture probabilities over b with the formula

$$\sigma = 2\pi \int_0^\infty bP(b)db. \tag{30}$$

We cut off the integral at $b = 700\,\lambda_c$ as the b^{-2} dependence of the probability results in a logarithmic divergence. Therefore, our estimate of the total cross section for muon-pair production with K-shell capture in collisions of $^{197}Au + ^{197}Au$ at 100 GeV/A is 0.57 b with 0.43 b coming from peripheral collisions.

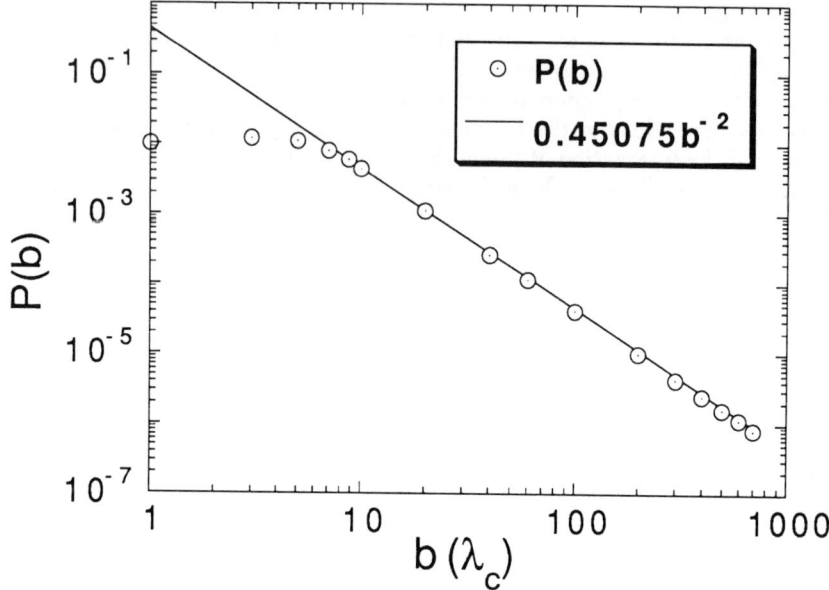

Figure 5. Probability for muon capture into the K-shell as a function of the impact parameter in units of λ_c for $^{197}Au + ^{197}Au$ collisions at a collider energy of 100 GeV/A. We show a b^{-2} fit to the probabilities for impact parameters greater than grazing. Error bars from the subjective error estimates are smaller than the data indicators because of the large logarithmic scale.

We stress that our nonperturbative calculations are preliminary and have not converged essentially because of the size of the collocation lattice dictated by the computational expense and, therefore, the results should be viewed only as an order-of-magnitude estimate. We estimate that realistic calculations for μ^- capture require 60 − 100 lattice points in each direction. We plan to meet the computational demands of such large lattices by implementing the calculation on the Intel iPSC/860 hypercube parallel supercomputer at the Oak Ridge National Laboratory.

VI. ACKNOWLEDGMENTS

This research was sponsored in part by the U.S. Department of Energy under contract No. DE-AC05-84OR21400 with Martin Marietta Energy Systems, Inc. and under contract No. DE-FG05-87ER40376 with Vanderbilt University. The numerical calculations were carried out on the CRAY-2 supercomputer at the National Center for Supercomputing Applications, Illinois.

REFERENCES

[1] K. Kajantie, M. Kataja, L. McLerran, and P.V. Ruuskanen, Phys. Rev. D **34**, 811 (1986); K. Kajantie, J. Kapusta, L. McLerran, and A. Mekjian, Phys. Rev. D **34**, 2746 (1986).

[2] H. Gould, in *Atomic Theory Workshop on Relativistic and QED Effects in Heavy Atoms*, Gaithersburg, Maryland, 1985, edited by Hugh P. Kelly and Yong-Ki Kim (AIP Conf. Proc. **136**), (AIP, New York, 1985).

[3] C. Bottcher and M.R. Strayer, in *Physics of Strong Fields*, proceedings of the International Advanced Course, Maratea, Italy, 1986, edited by W. Greiner, (NATO ASI Ser. B. Vol. 153) (Plenum, New York, 1987), page 629.

[4] E. Teller, in *Proceedings of the Ninth International Conference on the Application of Accelerators in Research and Industry*, edited by J. L. Duggan (North-Holland, New York, 1986).

[5] G. Baur, invited talk, presented at the workshop "Can RHIC be used to test QED?", Brookhaven National Laboratory, Upton, New York, April 1990, BNL 52247.

[6] C. Bottcher and M. R. Strayer, Phys. Rev. D **39**, 1330 (1989).

[7] C. Bottcher and M.R. Strayer, Nucl. Inst. and Meth. **B31**, 122 (1988); M. Fatyga, M. J. Rhoades-Brown, and M. J. Tannenbaum, Workshop Summary, "Can RHIC be used to test QED?", Brookhaven National Laboratory, Upton, New York, April 1990, BNL 52247.

[8] J. Eichler, Phys. Rep., **193**, 167 (1990).

[9] K. Rumrich, K. Momberger, G. Soff, W. Greiner, N. Grün, and W. Scheid, Phys. Rev. Lett. **66**, 2613, (1991).

[10] C. Bottcher and M.R. Strayer, Ann. Phys. (N.Y.) **175**, 64 (1987).

[11] A. S. Umar, J. Wu, M. R. Strayer, C. Bottcher, J. Comp. Phys., **93**, 426 (1991).

[12] M. R. Strayer, C. Bottcher, V. E. Oberacker, and A. S. Umar, Phys. Rev. **A41**, 1399 (1990).

[13] J.M. Jauch, F. Rohrlich, *The Theory of Photons and Electrons, 2nd Ed.*, (Springer-Verlag, New York, 1976) p.140ff; F. Coester, Phys. Rev. **84**, 1259 (1951).

[14] J. Cullum, R. Willoughby, *Lanczos Algorithms for Large Symmetric Eigenvalue Computations, Vol.I*, (Birkhäuser, Boston, 1985).

[15] B. Parlett, *The Symmetric Eigenvalue Problem*, (Prentice-Hall, Inc., Englewood Cliffs, N.J.).

[16] J. S. Wu, A.S. Umar, M.R. Strayer, C. Bottcher, J. C. Wells, J. Drake, and R. Flanery, manuscript in preparation.

[17] J. C. Wells, V. E. Oberacker, A. S. Umar, C. Bottcher, M. R. Strayer, J. S. Wu, G. Plunien, (manuscript in preparation) (1991).

[18] G. Katz, G. Batrouni, C. Davis, A. Kronfeld, P. Lepage, P. Rossi, B. Svetitsky, and K. Wilson, Phy. Rev. **D37**, 1589 (1988).

[19] C. Bottcher and M. R. Strayer, J. Phys. **G16**, 975 (1990).

Quark-Gluon Transport Theory: A Monte-Carlo Simulation

KLAUS GEIGER and BERNDT MÜLLER
Physics Department, Duke University, Durham, NC 27706

ABSTRACT

We present a QCD based relativistic kinetic model for high energy nuclear collisions, which is inspired by the parton picture of hadronic interactions. The nuclear dynamics is traced back to the microscopic level of quark and gluon interactions in the framework of perturbative QCD. The time evolution of the nuclear system is described in real time by a relativistic transport equation for the parton distributions that is solved by the Monte Carlo method.

INTRODUCTION

Ultrarelativistic nuclear collisions have attracted widespread interest because they are the unique tool for studying the physics of hadronic matter at extremely high energy densities [1]. At sufficiently high energies ($E_{CM} > 50$ GeV/u) it is expected that colliding nuclei penetrate each other, depositing a substantial fraction of their kinetic energy behind them [2], as illustrated in Fig. 1. In the space-time region between the two fast receding nuclear discs an energy density one or two orders of magnitude greater than that of ground state nuclear matter (0.15 GeV/fm^3) is expected shortly after the collision, thus duplicating conditions that prevailed in the very early universe until about 20 μs after the initial big bang.

It has usually been assumed that this matter quickly reaches conditions of approximate thermodynamic equilibrium, forming a dense plasma of quarks, antiquarks and gluons. The properties of this quark-gluon plasma have been extensively studied by theorists, and various signals for its detection have been proposed [1,3]. Its final dissolution into individual hadrons has also been widely modeled. On the other hand, the processes leading up to the formation of a thermally equilibrated QCD plasma have received only rather scant attention [4]. It is for this reason that we have begun to develop a comprehensive description for the dissipative processes occurring in the *pre-equilibrium* phase of a nuclear collision at very high energy.

As will become clear, our approach is not designed to yield a complete description of the nuclear reaction up to the prediction of details of hadronic spectra, at least in its present state of development. Various models and "event generators" have already been developed for this purpose [5,6]. Rather we wish to answer the question whether, and how, a local thermodynamic equilibrium will be established in the central rapidity region.

Since there is little experimental information to guide the theoretical analysis, our strategy has been to use as few modelling assumptions as possible,

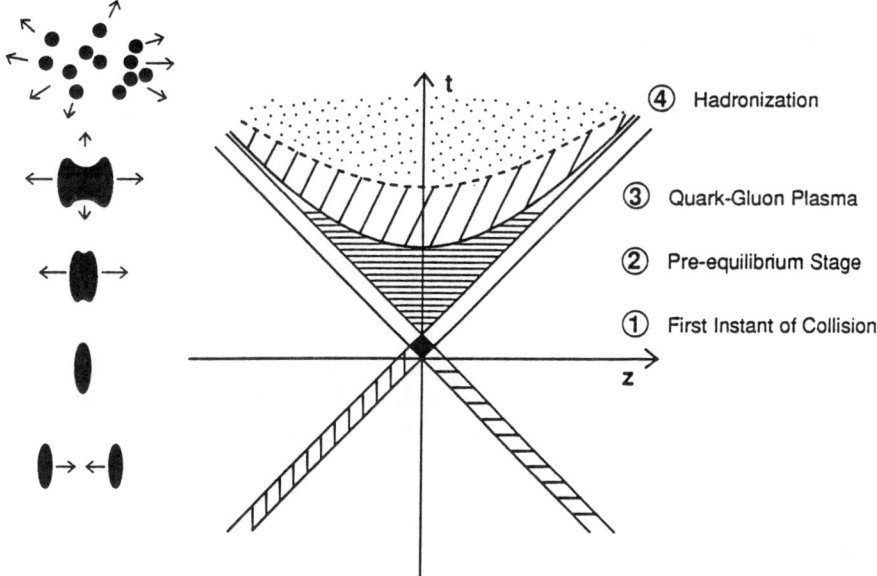

Fig. 1: Space-time of the (longitudinal) evolution of a relativistic nucleus-nucleus collision from the moment of collision, via pre-equilibrium and quark-gluon plasma to the final formation of hadrons.

and to base our analysis almost exclusively on the firmly established framework of perturbative QCD [7]. Wherever we reach the limits of its validity, we parametrize our lack of better knowledge by cut-off parameters, rather than trying to model nonperturbative QCD. We mention, however, one aspect, where we extend perturbative QCD clearly beyond its proven range of validity: by applying the parton model to the construction of the initial phase-space distribution of quarks and gluons in the colliding nuclei. This assumption is not covered by the celebrated factorization theorems of perturbative QCD but, lacking knowledge of the complete QCD wavefunction of the nucleon, there is simply no better approach available at the present time.

THE QCD TRANSPORT EQUATIONS

We start from a relativistic transport equation for the phase space density of partons which we can solve by means of Monte Carlo techniques, i.e. by simulating many collision events and taking their average. The basic idea is to embody the perturbative parton cascade picture into a relativistic version of the semiclassical kinetic theory. Our approach is related to, but differs from, the kinetic equations formulated by Elze and Heinz [8] and Mrowczynski [9]. In their framework, quarks and gluons interact via long-range color forces which give rise to a mean color field. The corresponding transport equations for quarks and gluons are generalized froms of the Vlasov equation, in which individual parton-parton interactions are neglected. In our approach, there is no color mean field, and all interactions are considered to be due to independent parton-parton collisions, which are calculable in the framework of perturbative QCD. We think that this is more appropriate for the preequilibrium phase of a hadronic reaction, since the initial hadrons are color singlets. Although models similar in spirit to ours have been studied earlier by Boal and by Mrwoczynski and Rafelski [9], we feel that this is the first attempt to apply the full knowledge of perturbative QCD to the description of the early stage of relativistic nuclear reactions.

We represent the partons as classical point particles of various "flavors" as specified by their internal degrees of freedom. The spin degrees of freedom are not explicitly taken into account. The state of a parton is characterized by its flavor $a = q_f, \bar{q}_f, g$ (quark, antiquark, or gluon), its momentum $\vec{p}$ and position $\vec{r}$. Its energy is determined by $E^2 = p^2 + m_a^2 + M^2$ where m_a is the rest mass and M a possible space- or time-like virtual mass. We define a Lorentz invariant single particle distribution function for the time-dependent phase space density of partons of species a [10]:

$$F_a(\vec{p}, \vec{r}, t) \, d^3p \, d^3r \tag{1}$$

The transport equation for the single particle distribution function (1), characterizing the time evolution of the system by balancing the various mechanisms by which partons can be gained or lost through interactions, has the manifestly covariant form

$$p^\mu \partial_\mu F_a = \sum_{\text{processes}} E_a C_a(\vec{p}, \vec{r}, t) = \sum_{\text{processes}} I_a \tag{2}$$

with Lorentz invariant collision integrals $I_a = E_a C_a$. The left-hand side of eq. (2) describes the free propagation of partons, here generally taken to be on mass shell in order to avoid violations of gauge invariance.

The collision term C_a, in general, describes all possible $m \to n$ interaction processes where at least one parton of type a is involved. All the information about these processes is contained in invariant matrix elements of the type

$$\mathcal{M}_{m \to n} = \langle 1', 2', \ldots, n | M | 1, 2, \ldots m \rangle, \tag{3}$$

which give the amplitudes for processes with m particles in the initial state and n particles in the final state. Since it is impractical to include all possible ($m \to n$) processes in the collision term of the transport equation, we restrict ourselves to a subset of processes that give the major contributions in perturbative QCD.

In accordance with the intuitive picture of the evolution of a nuclear collision, we consider the important class of asymptotically (in energy) dominant

$(2 \to n)$ processes that can be obtained from elementary $(2 \to 2)$ scatterings including associated space-like and time-like branchings [11,12]. For this set of $(2 \to n)$ processes, we formally write the invariant amplitudes squared as

$$\begin{aligned}|\mathcal{M}|^2_{2\to n} &= |\langle a_1 a_2 \ldots b_1 b_2 \ldots c_1 c_2 \ldots d_1 d_2 \ldots cd|M|ab\rangle|^2 \\ &= [S_a(x_a, Q^2, Q_0^2) S_b(x_b, Q^2, Q_0^2)] \\ &\quad \times |\mathcal{M}_{ab\to cd}(\hat{s}, \hat{t}, \hat{u}, Q^2)|^2 [T_c(Q^2, \mu_0^2) \cdot T_d(Q^2, \mu_0^2)],\end{aligned} \quad (4)$$

where a_i, b_i are the partons radiated from a, b before the elementary scattering

$$a + b \longrightarrow c + d,$$

and c_i, d_i are the partons radiated from the final state partons c, d. Eq. (4) expresses that the total squared amplitude can be factorized into three components, graphically illustrated in Fig. 2:

1. the squared matrix element $|\mathcal{M}_{ab\to cd}|^2$ for the elementary $(2 \to 2)$ subprocess $a + b \to c + d$ at a scale of interaction Q^2,
2. the Sudakov form factors $S_{a,b}$ determining the probabilities $P^{(S)}_{a,b} = 1 - S_{ab}$ that the scattering partons a, b have emitted a number of partons before the scattering, $a \to a + a_1 + \ldots, b \to b + b_1 \ldots$ (space-like branchings),
3. the Sudakov form factors $T_{c,d}$ giving the probabilities $P^{(T)}_{c,d} = 1 - T_{c,d}$ that the scattered partons c, d have initiated a sequence of branchings after the scattering, $c \to c + c_1 + \ldots, d \to d + d_1 + \ldots$ (time-like branchings).

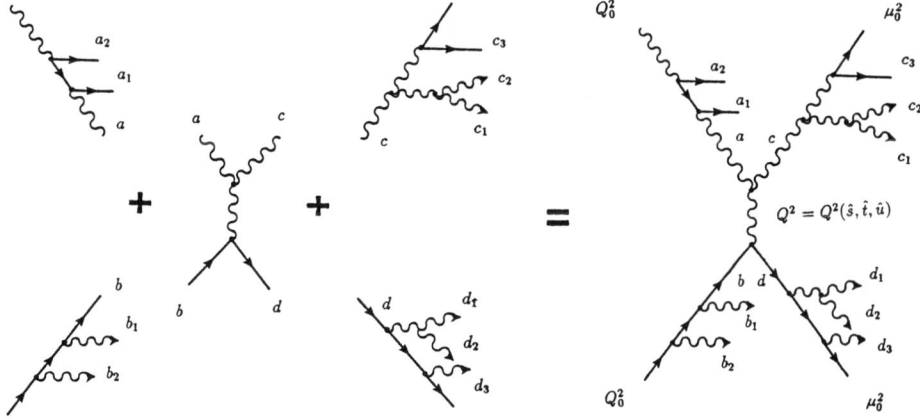

Fig. 2: Schematical illustration of a $(2 \to n)$ interaction process, factorizing in a $(2 \to 2)$ scattering $ab \to cd$ and associated space-like branchings $a \to a_1 a_2 \ldots, b \to b_2 b_2 \ldots$ and time-like branchings $c \to c_1 c_2 \ldots, d \to d_1 d_2 \ldots$.

Initial Parton Distributions

The initial distribution functions $F_a^0(\vec{p},\vec{r},t_0)$ of the partons in the incoming nuclei are represented in the factorized form

$$F_a^0(\vec{p},\vec{r},t_0) = \sum_{i=1} P_a^{N_i}(\vec{p},\vec{P},\mathcal{P}) \cdot R_a^{N_i}(\vec{r},\vec{R},\vec{\mathcal{R}})\Big|_{t=t_0}. \tag{5}$$

Here $P_a^{N_i}$ and $R_a^{N_i}$ give the initial momentum and spatial distributions, respectively, for partons of type a in each individual nucleon N_i, and the sum runs over all $N_A + N_B$ nucleons in nuclei A and B. The vectors $\vec{p},\vec{P},\vec{\mathcal{P}}$ denote the momenta of partons, nucleons and nuclei, respectively, and similarly do the positions $\vec{r},\vec{R}$ and $\vec{\mathcal{R}}$. Factorizing the parton phase space distribution function F_a^0 into a momentum part and a spatial part means that we neglect any correlation between flavor and momentum composition and the positions of the partons within the nucleons. This is in accord with the interpretation of $F_a^0(\vec{p},\vec{r},t)$ as single-parton distribution function, averaging over all correlations in the many-parton wavefunction of a single nucleon, which is the object of the Boltzmann equation.

The distribution of the parton flavors a inside the nucleons N_i in the variable $x = p_a/P_{N_i}$ is given by the nucleon structure functions $f_a^{N_i}(x,Q_0^2)$, where Q_0^2 denotes the scale at which the nucleon is probed. In the c.m. frame of the colliding nuclei the nucleons carry equal fractions of the total c.m. momentum P_{CM}. That is, neglecting their Fermi motion, the nucleons only have nonvanishing longitudinal momentum along the beam axis. In this frame, the variable x specifies the longitudinal momentum $p_\parallel = p_z$ fraction of a nucleon carried by the parton.

Our ansatz for the initial momentum distribution $P_a^{N_i}$ of parton flavors a in each nucleon N_i is

$$P_a^{N_i}(\vec{p},\vec{P},\vec{\mathcal{P}}) = \frac{x}{x_R} \cdot f_a^{N_i}(x,Q_0^2) \cdot g(\vec{p}_\perp) \tag{6}$$

where x specifies the longitudinal momentum fraction, and the radial variable x_R denotes the energy fraction carried by the parton [14]:

$$x_R^2 = x^2 + \frac{p_\perp^2 + m_a^2}{(P_{CM}/N_{\text{nuc}})^2}. \tag{7}$$

The factor x/x_R is included to form the invariant volume element $d^3p/2E(2\pi)^3$ of the distribution $P_a^{N_i}$ which is a product of the x-dependent structure functions f and the x-independent transverse momentum distribution g.

The scale Q_0^2, at which the structure functions $f_a^{N_i}(x,Q_0^2)$ must be evaluated, is in principle arbitrary as long as Q_0^2 is smaller than the scale Q^2 of any primary scattering process which involves at least one of the initial partons (we choose $Q_0^2 = 1 \text{GeV}^2$). Because we include initial (space-like) radiation processes for each such parton, the precise scale Q^2 entering into the structure functions is computed event by event on a statistical basis by the Q^2-evolution of the structure functions in the branching cascade, as explained further below. We have employed the set of structure functions from the higher order parameterization of Glück, Reya, and Vogt [15] with allowed variable range $0.2 \text{ GeV}^2 < Q_0^2 < 10^6$ GeV^2; $10^{-4} < x < 1$.

We take into account the transverse momentum $\vec{p}_\perp$ of the initial partons which results from their confinement within the radius of a nucleon (of the order 0.8 fm). According to the uncertainty relation, the partons must have a momentum spread of at least $\Delta p \sim 0.25$ GeV transverse to the nucleon's direction of motion. This intrinsic $\vec{p}_\perp$ can be observed in the Drell-Yan process, where the $\vec{p}_\perp$-distribution is found to be approximately Gaussian, roughly independent of s and Q^2 [13]:

$$g(\vec{p}_\perp) = \frac{1}{2\pi p_0^2} \cdot \exp[-|\vec{p}_\perp|^2/p_0^2] \qquad (8)$$

with $p_0 \simeq 0.35 - 0.45$ GeV. We do not treat p_0 as a parameter, but rather fix it to satisfy the kinematical constraint that the invariant mass of the total parton distribution of a nucleon equals the nucleon rest mass.

With the 3-momenta $\vec{p} = (\vec{p}_\perp, p_z)$ determined, each valence quark is put on mass shell, which fixes its energy $\varepsilon = \sqrt{p^2 + m_a^2}$, where $a = u_v, d_v$. Each sea quark or gluon, on the other hand, is assigned an energy $\varepsilon = \beta p_z$, where β is the velocity of the parent nucleon in the c.m. frame of the collision. This determines the initial space-like virtuality of the sea quark or gluon $q^2 = \varepsilon^2 - (p^2 + m_a^2) < 0$. Thus we obtain the 4-momenta p_v and $p_{s,g}$ for the valence quarks, respectively, the sea quarks and gluons:

$$p_v = \left(\sqrt{(xP)^2 + p_\perp^2 + m_a^2},\ \vec{p}_\perp, xP\right)$$
$$p_{s,g} = (\beta(xP), \vec{p}_\perp, xP). \qquad (9)$$

The sampling of the flavor and momentum distribution is carried out independently for each nucleon subject to the requirement the combined invariant mass of all partons belonging to the same nucleon is equal to the nucleon mass M_N. This constrains the width of the Gaussian distribution of transverse momenta. For nuclear collisions at 100 GeV/u c.m. energy (the final RHIC energy, for which we will present results below) the width p_0 in eq. (8) turns out to be 0.42 GeV/c.

The spatial distribution of partons $R_a^{N_i}(\vec{r}, \vec{R}, \vec{\mathcal{R}})$ is generated as follows. The individual nucleons are assigned positions randomly throughout the volume of the colliding nuclei A and B, in the rest frame of each nucleus. Next, the parton coordinates are uniformly distributed in this reference frame within a sphere of $R_{\text{nuc}} = 0.8$ fm around the nucleon centers. Finally, the positions of nucleons and their valence quarks are boosted into the c.m. frame, while the sea quarks and gluons are smeared out longitudinally by an amount $(\Delta z) \approx \hbar/p_\parallel < 2R_{\text{nuc}}$ around the valence quarks. This procedure is called *distributed Lorentz contraction* and yields a "fuzzy" pancake shape of the incoming nuclei [16].

ELEMENTARY TWO-PARTON INTERACTIONS

We now outline how the elements of the collision term C_a can be calculated in the framework of perturbative QCD. The $2 \to 2$ scattering cross sections are expressed in terms of the spin and color averaged squared amplitudes $|M|^2$:

$$\frac{d\hat{\sigma}(ab \to cd)}{d\hat{t}} = \frac{1}{16\pi \hat{s}^2} \cdot \langle |M_{(ab \to cd)}(\hat{s},\hat{t},\hat{u})|^2 \rangle. \tag{10}$$

where the "hatted" variables are the Mandelstam variables $\hat{s} = (p_a + p_b)^2$, $\hat{t} = (p_a - p_c)^2$, $\hat{u} = (p_a - p_d)^2$ pertaining to the subsystem of the two scattering partons. We include all elementary interactions between quarks and gluons, i.e. the following processes:

$$gg \to gg, \quad gg \to q\bar{q},$$
$$qg \to qg, \quad qq \to qq,$$
$$q\bar{q} \to q'\bar{q}', \quad q\bar{q} \to gg.$$

The scale at which the scattering occurs is determined by a variable Q^2 which determines the strength of the QCD coupling constant $\alpha_s(Q^2)$ associated with the vertex. The functional dependence of Q^2 on $\hat{s},\hat{t},\hat{u}$ is process dependent and not unambiguous. Here we choose the form

$$Q^2 = \frac{2\hat{s}\hat{t}\hat{u}}{\hat{s}^2 + \hat{t}^2 + \hat{u}^2}. \tag{11}$$

FORM FACTORS FOR SPACE-LIKE BRANCHINGS

A parton that directly originates from the initial nuclear parton clouds with small space-like virtuality $q^2 \gtrsim -Q_0^2$ and encounters its first hard scattering at a scale $Q^2(\hat{s},\hat{t},\hat{u}) > Q_0^2$ must be evolved from the initial scale Q_0^2 up to Q^2, the scale of the interaction vertex. The initial parton undergoes successive space-like branchings and increases its virtuality $|q^2|$ up to a value that is implied by the Q^2-scale of the scattering. The amplitude for such a sequence of consecutive branchings is calculable in perturbative QCD.

Usually, in the treatment of nucleon-nucleon collisions, the Q^2- evolution is included in the initial state parton distributions, by evolving the QCD structure functions $f_a(x,Q^2)$ to the required value of Q^2 with the help of the Altarelli-Parisi (AP) equations [17]. This is impossible here, because the initial phase-space distribution of partons must be set up once and for all as input for the

Monte-Carlo simulation. It turns out that the space-like branchings before the $2 \to 2$ scattering describe precisely the same physics, i.e. they account for the AP evolution of the initial parton distribution on an event-by-event basis.

Three different branchings are possible: $q \to qg$, $g \to gg$ and $g \to q+\bar{q}$. The probability for a single branching process is given by the so-called AP splitting functions, where $z, 1-z$ are the fractions of momentum carried by the daughter partons:

$$P_{q \to qg}(z) = \frac{4}{3} \cdot \frac{1+z^2}{1-z}$$

$$P_{g \to gg}(z) = 6 \cdot \frac{(1-z(1-z))^2}{z(1-z)}$$

$$P_{g \to q\bar{q}}(z) = \frac{1}{2} \cdot (z^2 + (1-z)^2). \quad (12)$$

In the leading-log approximation (LLA) the probability of a sequence of n branchings becomes the product of probabilities for each single branching. During a change dQ^2 a parton b may be resolved into a parton a. The relative probability dP_b for this to occur is given by df_b/f_b. Summing up the cumulative effect of small changes dQ^2, the probability for no branching to occur exponentiates. Therefore one may define the *Sudakov form factor* for space-like branchings:

$$S_b(x_b, Q^2_{\max}, Q^2)$$
$$= \exp\left\{-\int_{Q^2}^{Q^2_{\max}} \frac{dQ'^2}{Q'^2} \frac{\alpha_s(Q'^2)}{2\pi} \sum_a \int dz P_{a \to bc}(z) \frac{x_a f(x_a, Q'^2)}{x_b f(x_b, Q'^2)}\right\} \quad (13)$$

giving the probability that a parton b remains at x_b from $Q^2_{\max}$ to $Q^2 < Q^2_{\max}$. It is important to realize, however, that in our parton cascade picture of a nuclear collision S is only of relevance for the initial space-like partons that directly originate from the incoming nuclei and have their first interaction. For all other partons, those who have emerged either from a scattering or have been produced in a time-like branching process, the form factor S is equal to unity.

FORM FACTORS FOR TIME-LIKE BRANCHINGS

If the momentum transfer in an elementary $(2 \to 2)$ scattering provides the scattered partons with more energy than is necessary to be on mass shell $p^2 = m^2$, they may gain a time-like virtual mass $M^2 > 0$ and initiate a sequence of time-like branchings $q \to qg$, $g \to gg$, or $g \to q+\bar{q}$. This process of successive radiative emissions (bremsstrahlung) in the final state can be described in a way similar to the space-like branchings discussed above. However, now the evolution of the parton virtualities starts at the Q^2-scale of the hard scattering, decreasing in a strongly ordered manner, down to some final value μ_0^2.

The probability for such a time-like n-branching process is (in the LLA) given by the product of probabilities for individual branchings $a \to bc$, where a parton of time-like mass M_a^2 decays into partons b, c with mass $M_{b,c}^2 < M_a^2$, carrying fractions z and $z_1(1-z)$, respectively, of the momentum p_a of the parent parton [12,18]. Thus a sequence of time-like final state branchings can be represented, similar to the case of space-like initial state branchings, as a tree of successive single branchings connected to the scattering vertex. However, the two processes differ in that the virtualities and momentum fractions evolve in the opposite direction.

The time-like branchings are again associated with the AP splitting functions of eq. (12). For a branching $a \to bc$ their behavior with varying M_a^2 is guided by perturbative QCD according to

$$dP_{a \to bc} = \frac{\alpha_s(Q^2)}{2\pi} \cdot \frac{dM_a^2}{M_a^2} P_{a \to bc}(z) dz. \tag{14}$$

The probability that a parton a does not branch between some initial maximal mass $M_{\max}^2$ and a mass $M^2 < M_{\max}^2$ is again given by the exponentiation of (14). This yields the so-called *Sudakov form factor* for time-like branchings [19]:

$$T_a(M_{\max}^2, M^2) = \exp\left\{-\int_{M^2}^{M_{\max}^2} \frac{dM'^2}{M'^2} \frac{\alpha_s(M'^2)}{2\pi} \int dz P_{a \to bc}(z)\right\}. \tag{15}$$

It is useful to compare the form factor T_a with the corresponding expression S_b for space-like branchings, eq. (13). The most obvious difference between S_b and T_a is the explicit appearance of structure functions in the former. This expresses the fact that the probability for a parton b to originate from a branching of a parton a is proportional to the number of partons of type a present in the nucleus of the incoming nuclei. In contrast to that, the probability for a given parton a to branch, once it is present, is independent of this number.

The Monte-Carlo Algorithm

The partons propagate along classical trajectories until they scatter. To decide whether a scattering between a pair of partons occurs, the total cross-section $\hat{\sigma}_{ab}$ for this particular pair is used to define an effective area of interaction. Next the distance of closest approach for the two partons is computed, which is defined as the point where the scalar product of relative velocity and the separation of the pair is zero, which represents a Lorentz invariant. As is customary in the cascade approach, a collision is assumed to occur, if the distance of closest approach between two partons is within the area of interaction specified by $\hat{\sigma}_{ab}$. The kinematics of the elementary scattering $ab \to cd$ are evaluated in the c.m. frame of a and b with their momenta along the z-axis. The scattering angle

$\cos\hat{\theta}_{CM}$ is determined by generating the value of $\hat{t}$ according to the probability distribution

$$\Xi(\hat{t}) = \frac{1}{\hat{\sigma}_{ab\to cd}} \int_{t_{\min}}^{\hat{t}} \left(\frac{d\hat{\sigma}}{d\hat{t}}\right)_{ab\to cd} d\hat{t}, \qquad (16)$$

the differential QCD cross section is given by eq. (10).

For the initial state branchings we use a "backward" evolution scheme [19], which is based on the form factor $S_a(x_a, Q^2, Q_0^2)$ with the already determined Q^2 of the interaction as starting point. The evolution is then reconstructed step by step in falling q^2 sequence back towards the original $q^2 = Q_0^2$, for which the parton distributions have been initialized. We generate this backward evolution with the help of an algorithm described by Bengtsson and Sjöstrand [20].

As explained before the probabilities for bremsstrahlung processes after the scattering are determined by the no-branching probabilites $T_{c,d}(Q^2, \mu_0^2)$ given in eq. (15). The Monte Carlo algorithm in our program that generates such sequences of time-like branchings follows the procedure of Sjöstrand [21]. The procedure works as follows: First the virtuality M^2 of parton c is sampled from the distribution $T_c(Q^2, M^2) = T_c(Q^2, \mu_0^2)/T_c(M^2, \mu_0^2)$. If the virtuality M^2 is less than the cut-off μ_0^2, then the branching is forbidden. If the branching does occur, the second step is to select a branching channel, i.e. the flavor of the daughters c_1 and c_2, from the relative probabilities, given by the integrals of the splitting functions $\int dz P_{c\to c_1 c_2}(z)$.

With $M_c^2 = M^2$ and c_1, c_2 determined, the value of the splitting variable z is generated according to the probability distribution given by $P_{c\to c_1 c_2}(z)dz$. Finally, the four-momenta of the daughter partons c_1 and c_2 are constructed and the procedure is repeated with Q^2, the scale of the scattering $ab \to cd$, replaced by M_c^2, the virtuality of the first branching, and so on. The branching cascade is complete when all selected virtualities in the branching sequence have fallen below the cut-off scale μ_0^2. In our computer simulations we have chosen $\mu_0 = m_\rho$, the rho-meson mass, because the cut-off determines the scale where nonperturbative QCD begins to dominate the description of hadronic structure.

RESULTS

With the Monte Carlo computer program, described above, we have simulated and analyzed the evolution of energetic collisions at RHIC (Relativistic Heavy Ion Collider) energies $\sqrt{s}/A = 100$ GeV/u. The space-time evolution of the nuclear collisions in our approach is controlled by the parameters Λ, $p_\perp^{\min}$, and μ_0. We chose a value of $\Lambda = 0.16$ GeV/c, corresponding to the structure function parametrization of ref. [15] employed in our calculations. Furthermore we chose $p_\perp^{\min} = 1$ GeV/c and $\mu_0 = m_\rho = 0.77$ GeV. For more details we refer to our extended manuscript [22].

The choices of μ_0 does not have a very strong motivation at this point, since the required value depends primarily on the model to be used for hadronization, which we do not consider here. The value of $p_\perp^{min}$, which guides the strength and the number of parton-parton scatterings, has been taken from an educated guess as to the borderline between perturbative and nonperturbative QCD. It may be far too low for QCD purists, but the coupling $\alpha_s(p_\perp^{min}) = 0.375$ is still sufficiently small to make a perturbative expansion credible. A more reliable determination of the optimal value of $p_\perp^{min}$ may be derived from the total inelastic nucleon-nucleon cross section; this study is presently in progress.

We have studied four symmetric collision systems: $p+p$, $\alpha+\alpha$, $^{16}O+^{16}O$, and $^{32}S+^{32}S$. The simulations were performed in the c.m. reference frame. Because of the purely geometric impact parameter dependence in our model,

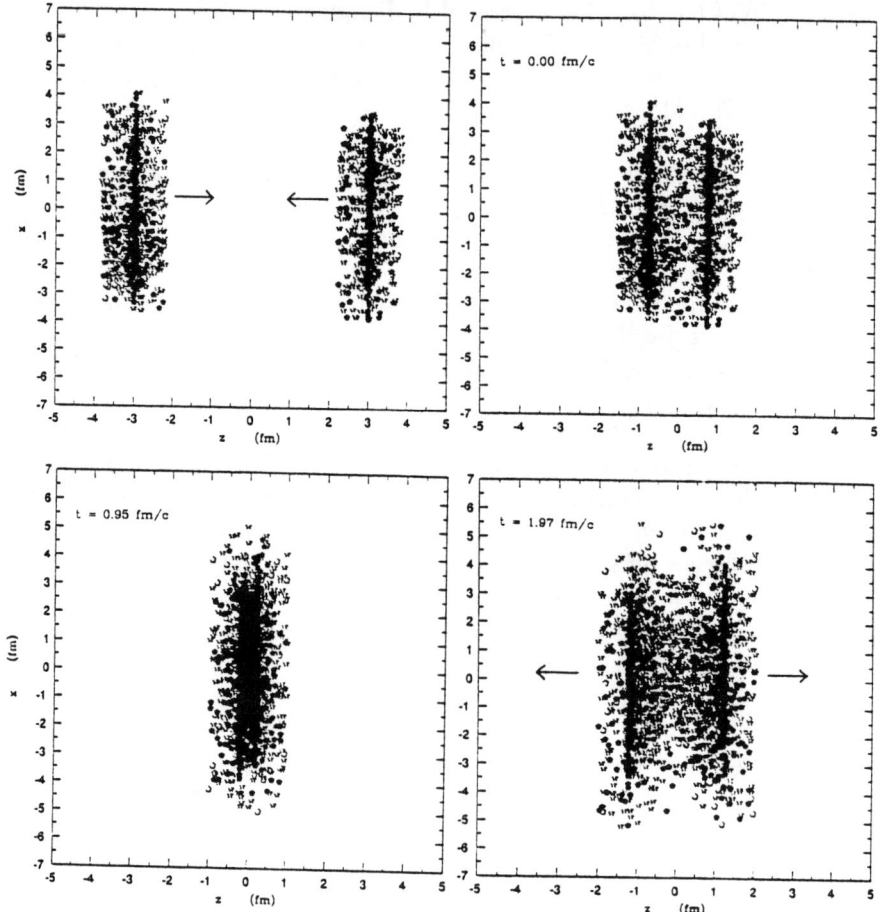

Fig. 3: Space-time evolution of an $^{32}S + ^{32}S$ collision at $E_{CM} = 100$ GeV/u in position space viewed from the center of mass frame of the colliding nuclei. The pictures are projections of the parton coordinates onto a plane through the collision axis for different times during the collision.

we considered only central collisions. The size of the time steps was taken to be $\Delta t = 0.01$ fm/c. The value of Q_0, the scale used in the structure functions for initialization and space-like branchings was initially chosen as 1 GeV, and was updated after each event. The final accumulated average values for Q_0 were 2.12 GeV for $p + p$, 2.09 GeV for $\alpha + \alpha$, 1.95 GeV for $^{16}O + ^{16}O$ and 1.94 GeV for $^{32}S + ^{32}S$. This decrease with the mass number A of the nuclei reflects the increased importance of secondary interactions between partons for larger nuclei, which occur at a reduced partonic c.m. energy $\hat{s}$. The reactions were stopped after $t_f = 1$ fm/c for $p+p$, $\alpha+\alpha$ and $t_f = 2$ fm/c for $^{16}O+^{16}O$, $^{32}S+^{32}S$. After this moment no further scatterings were found to occur, because the remaining "hard" partons have been spatially separated. The evolution of the parton distribution in space is illustrated in Fig. 3 for the $S + S$ system. The fuzziness due to distributed contraction is clearly seen.

The results shown below are based on 1000 event simulations for $p + p$ and $\alpha + \alpha$, 100 events for $^{16}O + ^{16}O$ and 40 events for $^{32}S + ^{32}S$. The calculations were performed on a Silicon Graphics R-4D220S computer (2 CPU's rated at 3.5 M flops each) and on the Cray Y-MP computer of the North Carolina Supercomputing Center. The required CPU time per event is 9 seconds, respectively 5 seconds, for $p + p$ collisions and grows quadratically with the mass number A of the beam particles.

A measure of the violence of the nuclear collisions is the change of the parton momentum distributions during the reaction. Commonly used representations are projections of the momentum distributions on the rapidity variable

$$y = \tfrac{1}{2} \ln \left(\frac{E - p_z}{E + p_z} \right) \tag{17}$$

where E and p_z are the energy and the longitudinal momentum (along the nuclear collision axis) of a parton and on the transverse momentum. Fig. 4a shows the evolution of the number of partons per rapidity interval, dN/dy, for $^{32}S + ^{32}S$ collisions, starting from the initial distribution of partons in the incoming beam particles. Only those partons that are on mass shell are included in the spectra. That is, the initial state reflects only the distribution of valence quarks, whereas at later times more and more of the initially virtual gluons and sea quarks gain enough energy in scatterings to become "real". The two broad bumps correspond to the distributions of valence quarks in the initial nuclei at rapidities $y_N = \pm 5.36$. Clearly, for the majority of quarks, $|y| < |y_N|$, corresponding to small momentum fractions x, while only few are hard quarks with $|y| > |y_N|$. With progressing time the number of partons that have scattered and consequently the number of produced partons increases, giving rise to a strong enhancement in the central rapidity region around $y = 0$, which is dominated by gluons.

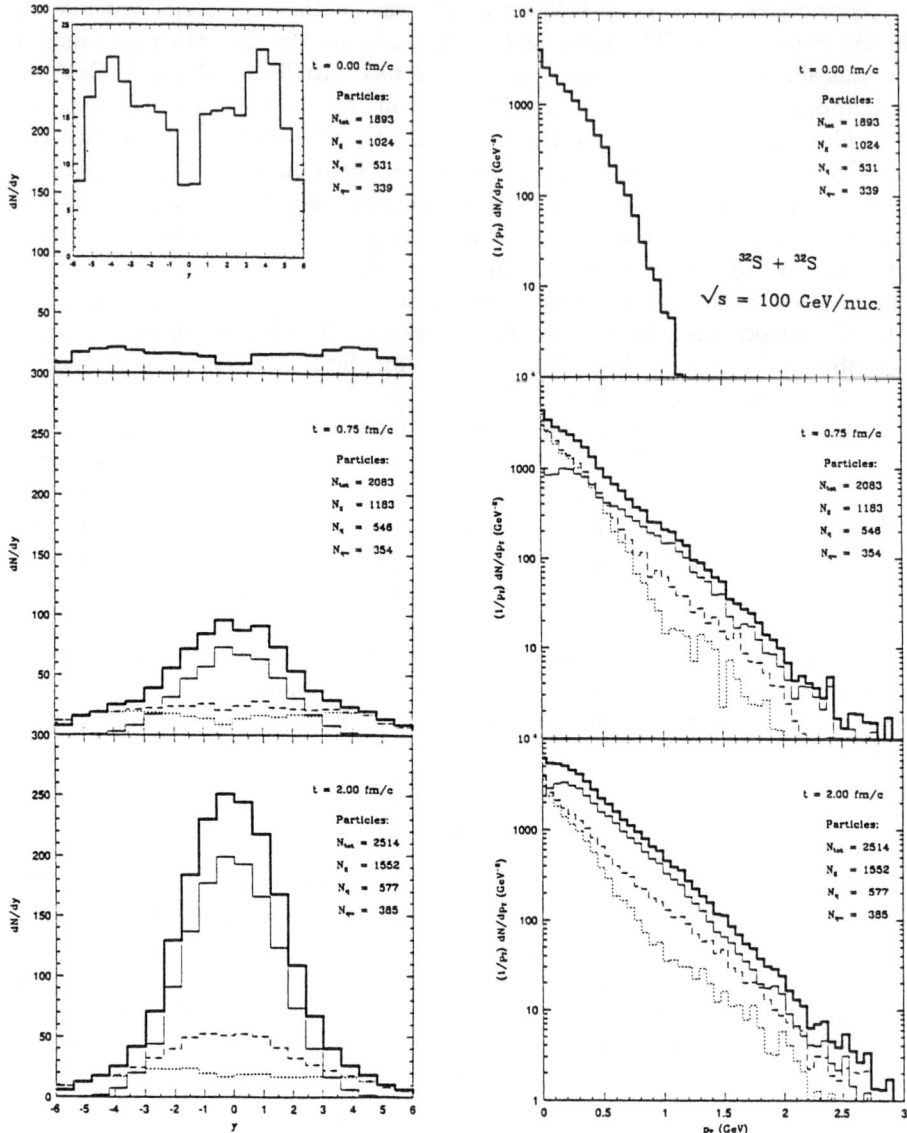

Fig. 4: (a) The dN/dy distribution (left) and (b) the transverse momentum distribution $1/p_\perp dN/dp_\perp$ of partons (right) at different times during an $^{32}S + {}^{32}S$ collision ($E_{CM} = 100$ GeV/u). The inlet shows the initial distribution in a larger scale. Bold lines represent all partons, full, dashed and dotted lines are the distributions of gluons, quarks + antiquarks and quarks - antiquarks, respectively.

The change of the transverse parton momenta is exhibited in Fig. 4b, where the time-evolution of $(1/p_\perp)dN/dp_\perp$ in $^{32}S+^{32}S$ collisions is shown. The first picture represents the initial $p_\perp$-distribution which has a Gaussian shape of width $p_0 = 0.42$ GeV/c. During the collision the distribution bends out to larger $p_\perp$ and becomes more exponential with final slope $p_0 \simeq 0.34$ GeV/c. The exponential nature of the transverse parton momentum spectrum indicates a high degree of (transverse) thermalization of incident energy. However, one must remain cautious with this claim until the local three-dimensional momentum distribution has been investigated carefully.

The transverse energy ($E_\perp$) produced in nuclear collisions is a direct measure of how much of the beam energy has been redirected in interactions and characterizes the violence of the collision. It may also serve as an indicator of the energy density achieved during the collision [23]. The transverse energy is defined as the sum of the energies of particles weighted with the sine of their angle to the beam axis, $E_\perp = \sum_i E_i \sin\theta_i$. In Fig. 5 the dependence of $E_\perp$ on the mass number A of the beam particles is displayed. It is found to increase approximately with $A^{\frac{1}{3}}$, i.e. it roughly scales with the nuclear radius, which is close to what is observed in nuclear collisions at lower energies [24]. A certain saturation is seen to occur around $A = 100$.

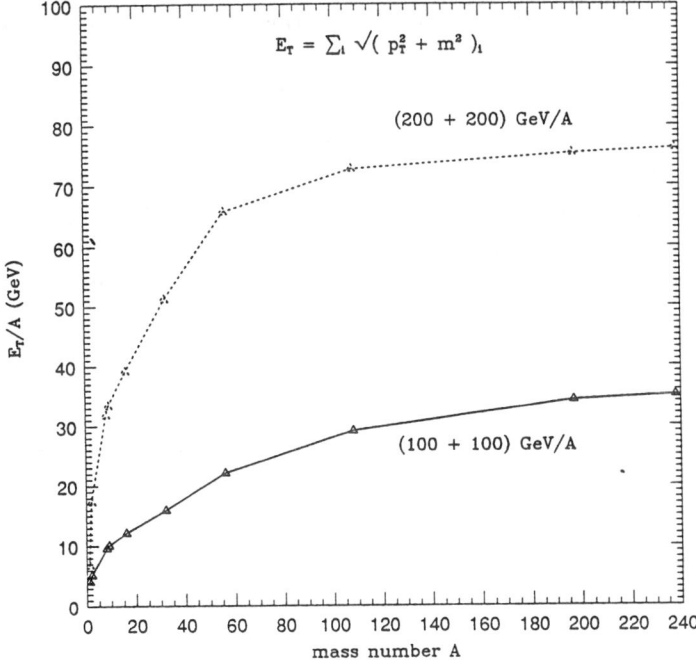

Fig. 5: Production of transverse energy $E_\perp$ as function of mass number A at $E_{CM} = 100$ GeV/u per collision event. For $p+p$ the impact parameter was averaged. All the other collisions were purely central.

Summary

We have presented a relativistic space-time approach for the simulation of relativistic nuclear collisions on the microscopic level of quark and gluon interactions. Although at this point it does not include a model for the formation of final hadronic states, it already provides a new and interesting opportunity to study the formative phase of highly relativistic nuclear collisions, which can serve as a basis for future investigations.

Clearly, our approach has its limitations and, as it stands so far, cannot be expected to describe all the features of high energy nuclear collisions equally well. In this work we focused on the microscopic dynamics of parton cascades and their consequences for the macroscopic space-time development of the collision system. The results that we presented are to be understood as preliminary. The question, how well experimental data can be described and how accurate quantitative predictions are, requires the implementation of an appropriate hadronization model.

Many possibilities for improvements to the current version of our computer code come to mind, some relatively simple to incorporate and some requiring major extensions. The most prominent on our list are: (1) a dynamical hadronization scheme based on fragmentation and cluster formation of partons; (2) corrections to the initial parton distributions due to nuclear shadowing, EMC-type effects, Fermi motion of nucleons and correlations between the flavor-momentum and the spatial distributions of partons; (3) the effect of a time delay in the branching processes of partons; (4) the incorporation of electromagnetic processes, such as $gq \to \gamma^* q$, $q\bar{q} \to \gamma^* g$, and $q\bar{q} \to \ell^+\ell^-$. In particular, the latter would allow for a systematic study of the influence of pre-equilibrium contributions to characteristic signatures of a possible quark gluon plasma. We also plan to study the formation time scale, energy density and entropy flow in the collision, as a basis for the construction of more reliable models of the reaction dynamics.

Acknowledgments

We would like to acknowledge the many motivating discussions and critical comments that we experienced with numerous colleagues during the course of this work. Especially we thank L. Frankfurt, M. Strikman, E. Shuryak and T. Sjöstrand. This work was supported in part by the U.S. Department of Energy (Grant DE-FG05-90ER40592). The calculations were partly performed on the Cray Y-MP of the North Carolina Supercomputing Center (NCSC).

REFERENCES

[1] B. Müller, *The Physics of the Quark-Gluon Plasma*, Lecture Notes in Physics, vol. **225** (Springer, Heidelberg, 1985).
[2] R. Anishetty, P. Koehler, and L. McLerran, *Phys. Rev.* **D22**, 2793 (1980).
[3] K. Kajantie and L. McLerran, *Ann. Rev, Nucl. Part. Science* , (1987).
[4] J. P. Blaizot and A. H. Mueller, *Nucl. Phys.* **B289**, 847 (1987).
[5] K. Werner, *Z. Phys.* **C42**, 85 (1989).
[6] H. Sorge, H. Stöcker, and W. Greiner, *Nuc. Phys.* **A498**, 567c (1989); *Ann. Phys.* **192**, 266 (1989).
[7] see e.g.: A. Buras, *Rev. Mod. Phys.* **52**, 199 (1980).
[8] H.-T. Elze and U. Heinz, *Phys. Rep* **183** (1989), 81.
S. Mrówczyński, *Phys. Rev.* **D39** (1989), 1940.
[9] D. H. Boal, *Phys. Rev.* **C33**, 2206 (1986);
S. Mrówczyński and J. Rafelski, *Phys. Rev.* **C40**, 1077 (1989).
[10] J. J. Duderstadt and W. R. Martin, *Transport Theory* (Wiley, New York, 1979). S. de Groot, W. A. van Leuwen and C. G. van Weert, *Relativistic Kinetic Theory* (North-Holland, Amsterdam, 1980).
[11] R. D. Field, *Applications of Perturbative QCD*, (Addison-Wesley, Redwood City, 1989).
[12] B. R. Webber, *Ann. Rev. Nucl. Part. Science* **36**, 253 (1986).
[13] D. Antreasyan et al., *Phys. Rev. Lett.* **48**, 302 (1983).
[14] R. P. Feynman and R. D. Field, *Phys. Rev.* **D15**, 2590 (1977); *Nucl. Phys.* **B136**, 1 (1978).
[15] M. Glück, E. Reya, and A. Vogt, *Z. Phys.* **C48**, 471 (1990).
[16] R. C. Hwa and K. Kajantie, *Phys. Rev. Lett.* **56**, 696 (1986).
[17] G. Altarelli and G. Parisi, *Nucl. Phys.* **B126**, 298 (1977).
[18] A. Bassetto, M. Ciafaloni, and G. Marchesini, *Phys. Rep.* **100**, 203 (1983).
[19] G. Marchesini and B. R. Webber, *Nucl. Phys.* **B238**, 1 (1984).
[20] M. Bengtsson, T. Sjöstrand, and M. van Zijl, *Z. Phys.* **C32**, 67 (1986); T. Sjöstrand, *Phys. Lett.* **157B** (1985), 321.
[21] M. Bengtsson and T. Sjöstrand, *Nucl. Phys.* **B289**, 810 (1987).
[22] K. Geiger and B. Müller, *Parton Cascades in Relativistic Nuclear Collisions*, preprint DUKE-TH-90-15.
[23] J. D. Bjorken, *Phys. Rev.* **D27**, 140 (1983).
[24] see e.g.: M. Tannenbaum, *Int. Journ. Mod. Phys.* **A14**, 3377 (1983).

RENORMALIZATION GROUP APPROACH TO THERMAL QUANTUM FIELD THEORY

Eric Braaten
Department of Physics and Astronomy
Northwestern University
Evanston, IL 60208

ABSTRACT

A new approach to thermal quantum field theories based on renormalization group ideas is developed. Integrating out thermal loop corrections involving high energy particles results in a nonlocal effective lagrangian for low energy particles. The approach is illustrated using thermal QED, and applications to thermal QCD are discussed.

INTRODUCTION

The Renormalization Group (RG) is a misnomer for a general strategy developed by Ken Wilson[1] for dealing with problems involving different length or energy scales. It is not a group and does not necessarily have anything to do with renormalization, which began as a mysterious process in QED by which infinities were thrown away, leaving finite results which agreed with experiments to amazing accuracy. Understanding the renormalization of divergences in quantum field theories was indeed part of Wilson's original motivation, but the RG is much more than that. It has become the conceptual framework within which modern quantum field theory is understood, and it is also a valuable computation tool for carrying out explicit calculations.

I believe the RG approach can be equally valuable for thermal quantum field theories, i.e., theories at nonzero temperature and/or particle density. It can provide a conceptual framework for understanding features of thermal quantum field theories which are otherwise very puzzling. It can also be used as a computational tool for carrying out calculations which have not been feasible in the past. The implementation of the RG approach in thermal quantum field theories is very different from at zero temperature and density, since it involves nonlocal effective field theories instead of the local effective field theories encountered in conventional RG applications. To illustrate the RG approach, I will use thermal QED, which is the simplest field theory which exhibits the full complexity of the RG approach. For simplicity, I will consider only the case of 0 electron density. I will also take the limit of high temperature T, so that the electron mass can be ignored.

VIRTUAL AND THERMAL CORRECTIONS

In order to set the stage for the discussion of the RG approach to thermal QED, I first describe some of the virtual and thermal loop corrections that arise in QED. Fig. 1 shows the free propagator for photons, the 1-loop virtual correction, and a typical 1-loop thermal correction. The virtual correction in Fig. 1b represents the photon of 4-momentum k^μ splitting into a virtual electron-positron pair, which recombine to give a photon. The amplitude for Fig. 1b must be integrated $\int d^4p$ over the 4-momentum p^μ of the virtual electron. The thermal correction in Fig. 1c represents the elastic scattering of the photon off a thermal electron of 3-momentum $\vec{p}$ (represented by the open circle) in such a way that the electron is scattered back into the same momentum state (represented by the closed circle). The net effect is just the propagation of the photon. The amplitude for Fig. 1c must be multiplied by the Fermi distribution for the thermal electron and integrated over its 3-momentum: $\int d^3p(e^{\beta E} + 1)^{-1}$, where $\beta = 1/T$ is the inverse temperature and $E = |\vec{p}|$ is the energy of the on-shell thermal electron.

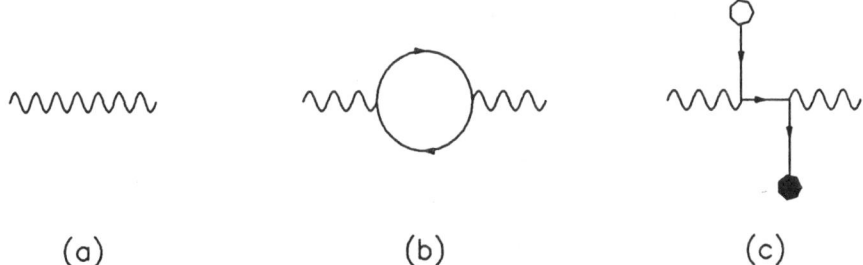

Fig. 1 Photon propagator: (a) free propagator, (b) virtual correction, and (c) thermal correction.

Not only are the propagators of photons and electrons modified by thermal corrections, but their interactions are modified as well. In Fig. 2, we show the fundamental QED vertex for the electron-photon interaction, the 1-loop virtual correction, and a typical 1-loop thermal correction. The virtual correction in Fig. 2b involves a loop of virtual electrons and photons, and must be integrated over all values of the loop momentum p^μ. In the thermal correction of Fig. 2c, the incoming electron scatters off a thermal photon of 3-momentum $\vec{p}$ in such a way that a photon is scattered back into the same state, while a second photon and an electron are emitted. The net effect is a nonlocal correction to the electron-photon interaction. The amplitude for Figure 1c must be multiplied by the Bose distribution for the thermal photon, and integrated over its 3-momentum: $\int d^3p(e^{\beta E} -1)^{-1}$, where $E = |\vec{p}|$ is the photon energy.

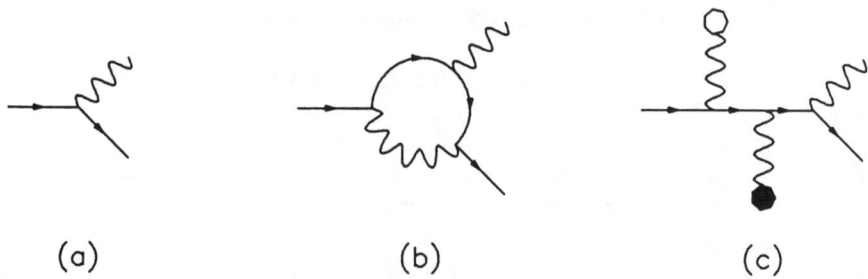

Fig. 2 Electron-photon interaction: (a) fundamental vertex, (b) virtual correction, and (c) thermal correction.

RG APPROACH TO VIRTUAL CORRECTIONS

Virtual corrections involve integrals $\int d^4 p$ over loop momenta. These integrals may require an ultraviolet cutoff $p^2 < \Lambda^2$ to regularize ultraviolet divergences. One can always introduce a scale μ and split the integral up into 2 regions, depending on whether the invariant mass p^2 is smaller or larger than μ^2:

$$\int d^4 p = \int_{p^2 < \mu^2} d^4 p + \int_{p^2 > \mu^2} d^4 p . \tag{1}$$

In QED, which has Lorentz and gauge symmetries, it is convenient to implement the splitting of loop momenta in a more sophisticated way that preserves these symmetries, but this is only a technical complication. The RG approach is based on the observation that for small external momenta satisfying $|k^\mu| \ll \mu$, the effects of virtual corrections from $p^2 > \mu^2$ can be well approximated by local interactions whose form is independent of μ. For example, the photon propagator correction in Fig. 1b after integrating over $p^2 > \mu^2$ reduces to the free propagator in Fig. 1a multiplied by a constant proportional to $e^2 \log(\Lambda/\mu)$. The vertex in Fig. 2b after integrating over $p^2 > \mu^2$ reduces to the fundamental vertex in Fig. 2a multiplied by a constant proportional to $e^2 \log(\Lambda/\mu)$.

The RG strategy for computing low energy observables involving momenta with $|k^\mu| \ll \mu$ is to separate the calculation into two steps. In the first step, one integrates over virtual particles with large invariant mass $p^2 > \mu^2$. The results of this first step can be conveniently summarized by a local effective lagrangian $\mathcal{L}_{eff}(\mu)$ with μ-dependent coefficients. Because QED is a renormalizable theory, $\mathcal{L}_{eff}$ reduces, after allowing for field renormalizations, to the usual QED lagrangian $\mathcal{L}_{QED}(e)$, with the coupling constant e replaced by a running coupling constant $e(\mu)$:

$$\mathcal{L}_{eff}(\mu) = \mathcal{L}_{QED}(e(\mu)) + O(1/\mu^2) . \tag{2}$$

The corrections, which fall like powers of μ, involve higher dimension local interactions such as the 4-electron interaction $\bar{\psi}\gamma^\mu\psi\bar{\psi}\gamma_\mu\psi$.

The second step in the RG strategy is to compute the low energy observable using $\mathcal{L}_{eff}(\mu)$, while imposing a cutoff $p^2 < \mu^2$ on the virtual corrections. If both steps could be carried out exactly, the dependence on μ would cancel exactly between the μ-dependence of $\mathcal{L}_{eff}$ and the μ-dependence due to the cutoff. In practice one has to resort to some approximation, such as perturbation theory to a given order in e. In this case, the cancellation is not exact, but the dependence on μ is higher order in e. The incomplete cancellation of the dependence on μ can be exploited by adjusting μ so as to minimize the size of the uncalculated higher order corrections.

RG APPROACH TO VIRTUAL CORRECTIONS

We now discuss how the same strategy can be applied to thermal corrections. These corrections involve integrals $\int d^3p$ over 3-momenta weighted by Bose and Fermi distributions. One can always introduce a scale μ and split up such an integral into 2 regions, depending on whether the energy E is greater or less than μ:

$$\int d^3p \frac{1}{e^{\beta E} \mp 1} = \left(\int_{E<\mu} d^3p + \int_{E>\mu} d^3p \right) \frac{1}{e^{\beta E} \mp 1} . \tag{3}$$

The RG approach is based on the observation that for external momenta satisfying $|k^\mu| \ll \mu$, the effects of thermal corrections from energies $E > \mu$ can be approximated by nonlocal interactions whose form is independent of E. For example, the thermal correction to the photon propagator in Fig. 1c, when expanded in powers of $1/E$, includes a term $(e^2/E)/(k_0 - \hat{p}\cdot\vec{k})$, where $(k_0, \vec{k})$ is the 4-momentum of the photon and $\hat{p}$ is a unit vector in the direction of the momentum of the thermal electron. Multiplying by the Fermi distribution of the electron and integrating over energies $E > \mu$, we obtain

$$\int_{E>\mu} \frac{d^3p}{(2\pi)^3} \frac{1}{e^{\beta E}+1} \frac{e^2}{E(k_0 - \hat{p}\cdot\vec{k})} = \left(\frac{e^2}{2\pi^2} \int_\mu^\infty \frac{EdE}{e^{\beta E}+1} \right) \int \frac{d\Omega}{4\pi} \frac{1}{k_0 - \hat{p}\cdot\vec{k}} . \tag{4}$$

The result is a correction to the photon propagator whose only dependence on μ is in the overall coefficient. This and all other thermal corrections proportional to 1/E can be summarized compactly by a nonlocal effective lagrangian that will be presented later.

The RG strategy for computing low energy observables involving momenta $|k^\mu| \ll \mu$ separates the calculation into two steps. In the first step, one integrates out the thermal loop corrections due to thermal particles with energies $E > \mu$. The result of this step can be conveniently summarized by a nonlocal effective lagrangian $\mathcal{L}_{eff}(\mu)$ with μ-dependent coefficients. The second step of the RG strategy is to compute the low energy observable using $\mathcal{L}_{eff}(\mu)$, while imposing a cutoff $E < \mu$ on the thermal corrections. If both steps could be carried out exactly, the dependence on μ would cancel exactly between the μ-dependence of $\mathcal{L}_{eff}$ and the μ-dependence due to the cutoff. In practice, we have to resort to some approximation, such as perturbation theory to a given order in e. In this case, the cancellation is not exact, but the dependence on μ is higher order in e. It should be possible to exploit this incomplete cancellation of the dependence on μ by adjusting μ so as to minimize the size of the uncalculated higher order corrections.

EFFECTIVE LAGRANGIAN FOR THERMAL QED

The RG approach is a general strategy that can be applied to any problem involving different energy scales, but it is not always an effective strategy. The result of the first step could be an effective lagrangian so complicated that it would be impossible to use it for explicit calculations. Fortunately, this is not the case for thermal quantum field theories. I present below the terms in $\mathcal{L}_{eff}$ for QED that arise from thermal corrections proportional to 1/E, where E is the energy of the thermal particle. These thermal corrections have been labelled "hard thermal loops"[2]. The thermal corrections from higher order in the expansion in powers of 1/E have not yet been analyzed.

The effective lagrangian due to the hard thermal loops in thermal QED consists of the fundamental QED lagrangian and two thermal correction terms:

$$\mathcal{L}_{eff}(\mu) = \mathcal{L}_{QED} + \mathcal{L}_\gamma(\mu) + \mathcal{L}_e(\mu). \tag{5}$$

These terms were first deduced by Braaten and Pisarski[2] and were summarized in the form of an effective lagrangian by Taylor and Wong.[3] The simplified form of the effective lagrangian presented below is due to Braaten and Pisarski. The terms $\mathcal{L}_\gamma$ in (5) involve only the photon field A_μ:

$$\mathcal{L}_\gamma(\mu) = \frac{3}{2}m_\gamma(\mu,T)^2 \left(A_0^2 - \int \frac{d\Omega}{4\pi}(A_0 - \hat{p}\cdot\vec{A})\frac{\partial_0}{\partial_0 - \hat{p}\cdot\vec{\nabla}}(A_0 - \hat{p}\cdot\vec{A}) \right), \quad (6)$$

where $\int d\Omega$ represents integration over the directions of the unit vector $\hat{p}$. The expression (6) is manifestly nonlocal, because it contains the inverse of the differential operation $\partial_0 - \hat{p}\cdot\vec{\nabla}$. It is gauge invariant, although the invariance is not manifest. Under the gauge transformation $A_\mu \to A_\mu + \partial_\mu\omega$, the integrand of the angular integral changes by $2(A_0 - \hat{p}\cdot\vec{A})\partial_0\omega$, which integrates to $2A_0\partial_0\omega$. This is precisely what is needed to cancel the gauge transform of the A_0^2 term.

The only dependence on the scale μ in (6) is in the overall coefficient, which is given by

$$m_\gamma(\mu,T)^2 = \frac{4e^2}{3\pi^2} \int_\mu^\infty \frac{E\,dE}{e^{\beta E}+1}. \quad (7)$$

The notation m_γ is intended to be suggestive, because adding (6) to the lagrangian of QED modifies the propagation of photons so that they have rest mass m_γ. For $\mu \ll T$, this mass reduces to $m_\gamma = eT/3$, which is the classic result for the thermal mass of a photon in an ultrarelativistic plasma.[4] Finally note that (6) is quadratic in the photon field, so its only effect is to modify the photon propagator.

The term $\mathcal{L}_e$ in the effective lagrangian (5) involves the electron field ψ as well as the photon field:

$$\mathcal{L}_e(\mu) = i\, m_e(\mu,T)^2\, \bar{\psi}\left(\int \frac{d\Omega}{4\pi}(\gamma_0 - \hat{p}\cdot\vec{\gamma})\frac{1}{D_0 - \hat{p}\cdot\vec{D}} \right)\psi. \quad (8)$$

This is manifestly gauge invariant, since it is expressed in terms of the gauge-covariant derivative $D_\mu = \partial_\mu - ieA_\mu$. It is nonlocal, since the differential operator $D_0 - \hat{p}\cdot\vec{D}$ appears in the denominator. The only dependence on the scale μ is in the overall coefficient:

$$m_e(\mu,T)^2 = \frac{e^2}{2\pi^2}\left(\int_\mu^\infty \frac{E\,dE}{e^{\beta E}+1} + \int_\mu^\infty \frac{E\,dE}{e^{\beta E}-1} \right). \quad (9)$$

The notation m_e is intended to be suggestive, because adding (8) to the Lagrangian of QED changes the propagation of an electron, so that it has a rest mass m_e. If the rest mass $m_0 = 0.511$ MeV for an electron propagating through the vacuum is not ignored, its rest mass in the thermal environment is $m_0/2 + (m_e(\mu,T)^2 + m_0^2/4)^{1/2}$. For $\mu \ll T$, the mass given by (8) reduces to

$m_e = eT/\sqrt{8}$, which is the classic result for the thermal mass of an electron in an ultrarelativistic plasma.[7]

The term (8) in the effective lagrangian for thermal QED not only gives a correction to the electron propagator, but it also gives corrections to the interactions of electrons and photons. The inverse of the differential operator $D_0 - \hat{p} \cdot \vec{D}$ in (8) can be expanded in powers of the photon field A_μ:

$$\frac{1}{D_0 - \hat{p} \cdot \vec{D}} = \frac{1}{\partial_0 - \hat{p} \cdot \nabla} + i e \frac{1}{\partial_0 - \hat{p} \cdot \nabla} \left(A_0 - \hat{p} \cdot \vec{A}\right) \frac{1}{\partial_0 - \hat{p} \cdot \nabla} + \ldots \quad (10)$$

The term of first order in A_μ gives a thermal correction to the fundamental QED vertex in Fig. 2a. The terms of higher order in A_μ give rise to thermally induced nonlocal interactions between an electron and N photons for any N. These effective interactions were first derived by Braaten and Pisarski.[2] Note that these interactions are related to the thermal electron propagator correction by gauge invariance. If the thermal propagator correction is taken into account in a calculation, it may also be necessary to take into account the thermal interactions in order to get a gauge invariant result.

For non-gauge theories, the effective lagrangian summarizing the effects of hard thermal loops is much simpler. For a scalar field with a $g^2\phi^4/4!$ interaction, the thermal term in $\mathcal{L}_{eff}$ is a local mass term $-g^2T^2\phi^2/24$. For a scalar and a spinor with a Yukawa interaction, there is no hard thermal loop in the 1-loop scalar propagator correction. In the spinor propagator correction, there is a cancellation of the hard thermal loop for $\mu << T$ between the thermal corrections due to the scalar and the spinor.

The case of nonabelian gauge theories is only a little more complicated than QED. For thermal QCD, hard thermal loops also give rise to two terms in the effective lagrangian:

$$\mathcal{L}_{eff}(\mu) = \mathcal{L}_{QCD} + \mathcal{L}_g(\mu) + \mathcal{L}_q(\mu). \quad (11)$$

The quark term $\mathcal{L}_q(\mu)$ has the same form as (8), except that the quark field ψ^i has a color index and D_μ is the nonabelian covariant derivative $D_\mu = \partial_\mu + i g_s A_\mu^a T^a$. The coefficient $m_q(\mu,T)^2$ differs from (9) only in that e^2 is replaced by $4g_s^2/3$. For $\mu << T$, $m_q(\mu,T)$ reduces to $m_q = g_s T/\sqrt{6}$, which is the classic result for the thermal quark mass in the quark-gluon plasma.[5] The gluon term $\mathcal{L}_g(\mu)$ in (11) is the nonabelianization of (6). The terms quadratic in the gluon field A_μ^a have the same form as (6) except for the presence of the color index. The coefficient is $(3/2) m_g(\mu,T)^2$, which reduces for $\mu << T$ to $m_g = (g_s T/\sqrt{3})(1 + n_f/6)^{1/2}$, where n_f is the number of light quark flavors. This is the classic result for the thermal mass of a gluon in the quark-gluon plasma.[6,7] The nonabelian gauge symmetry requires that the terms quadratic in A_μ^a be supplemented by an infinite series of terms of increasing order in the gluon field. Thus the effective lagrangian for thermal QCD includes nonlocal N-gluon interactions for any N which were first derived by Braaten and Pisarski.[2] While closed form expressions for the gluon term $\mathcal{L}_g(\mu)$ have been written down,[3,8] they are not as elegant as (6).

PLASMON PROBLEM

Sometimes problems that appear very difficult can become simple when viewed within the appropriate conceptual framework. An example in thermal quantum field theory is the "plasmon problem", which remained unsolved for 10 years, but becomes simple when viewed from the perspective of the RG approach. The problem, which was first discussed in 1980,[6] is to calculate the damping rate γ_g for a zero momentum gluon in the quark-gluon plasma. In the absence of quarks, the thermal gluon mass to leading order in g_s is $m_g = g_s T/\sqrt{3}$. Higher order corrections give the mass an imaginary part, which is called the damping rate and can be written

$$\gamma_g = a \frac{g_s^2 T}{8\pi}, \qquad (12)$$

where a is a numerical coefficient. The plasmon problem is that 1-loop calculations of γ_g give gauge-dependent answers. The 1-loop calculations are equivalent to calculating the decay rate of a thermal gluon at rest (E = m_g, p = 0) into two massless gluons ($E_1 = p_1 = m_g/2$, $E_2 = p_2 = m_g/2$) via the fundamental 3-gluon interaction of QCD: g → gg. In Coulomb and time-like axial gauge, the 1-loop calculation gives a = 1 for the coefficient in (12). In covariant gauges, a depends quadratically on the gauge parameter, but is always negative: a ≤ -11/4. There were many attempts to avoid the gauge-dependence by using a gauge-invariant formulation of the problem, but different gauge-invariant methods give different gauge-invariant answers. These methods succeeded only in masking the symptoms without curing the disease.[9]

From the RG point of view, the source of the gauge dependence is obvious. The thermal mass of the initial gluon is derived from the effective lagrangian $\mathcal{L}_{eff}$ for thermal QCD given in (11). The decay amplitude into massless gluons is derived from the fundamental QCD lagrangian $\mathcal{L}_{QCD}$. Since $\mathcal{L}_{eff}$ and $\mathcal{L}_{QCD}$ are both gauge-invariant, they should give gauge-invariant results for physical quantities provided the same lagrangian is used throughout a calculation. The plasmon problem arises because $\mathcal{L}_{eff}$ was used for the inital states, while $\mathcal{L}_{QCD}$ was used for the final states and for calculating the transition amplitude.

The solution to the plasmon problem is also obvious from the RG point of view. The damping rate should be calculated using the effective Lagrangian $\mathcal{L}_{eff}$ throughout. In particular, all gluons should be given the thermal mass m_g. The decay mechanism g → gg is then kinematically forbidden. The actual physical mechanisms are 2 → 2 scattering process (gg → gg) and 3 → 2 scattering process (ggg → gg), where the extra gluons in the initial state are thermal gluons with energies on the order of T. The calculation of the damping rate involves not only the effective gluon propagator, but also the nonlocal effective 3-gluon and 4-gluon vertices. While they are considerably more complicated than the fundamental vertices of QCD, the calculation of the damping rate is in fact tractable. The final result[10] for the coefficient in (12) is a = 6.63538.

Further evidence for the tractability of the RG approach is provided by calculations of the production rate of soft dileptons (i.e. with energies on the order of $g_s T$) from the quark gluon plasma.[11] These calculations involve not only the effective quark propagator derived from the QCD analog of (8), but also an effective photon-quark vertex derived by integrating out thermal quarks and gluons. Another explicit calculation is the neutrino emissivity due to decay of thermal photons in the ultrarelativistic QED plasma that is generated in the collapse of a massive star. A calculation[12] using the correct photon dispersion relations derived from (6) reveals significant discrepancies with the formulae[13] that have been used in the astrophysics literature for 24 years.

SCREENING EFFECTS

For many of the fundamental properties of a plasma, tree level calculations suffer from logarithmic infrared divergences. Examples are the transport coefficients of the plasma, the energy loss of a charged particle propagating through the plasma, and the production rate of weakly interacting particles from the plasma. The divergences are due to long range gauge interactions, which are screened by the plasma, but that screening is not taken into account in tree level calculations. The effects of screening can however be calculated within the RG framework. The photon term (6) in the effective lagrangian for thermal QED provides Debye screening of the static Coulomb interaction with inverse screening length $\sqrt{3}\, m_\gamma$. There is also dynamical screening of the long-range magnetic interaction[14], with inverse screening length that vanishes in the static limit, but sufficiently slowly that logarithmic infrared divergences are screened.[15]

A method for calculating the effects of screening to leading order in the coupling constant has been developed by Braaten and Yuan.[16] The effective lagrangian from hard thermal loops is used to calculate the contributions from small momentum transfer where screening is important. This calculation is then matched consistently on to a tree level calculation of the contribution from large momentum transfer. This method was applied to calculate the emissivity due to axion emission via the Primakoff process from a high temperature QED plasma. It has also been applied to calculate the energy loss dE/dx for a massive lepton in an ultrarelativistic QED plasma and for a heavy quark in the quark-gluon plasma.[17]

An ultrarelativistic plasma not only screens the long range gauge interactions, but also screens the mass singularities due to other particles in the plasma. For example, a tree level calculation of the production rate for hard photons from a quark-gluon plasma has a logarithmic sensitivity to the mass of a light quark. When the screening of the plasma is taken into account, the logarithm is cutoff at the scale of the thermal quark mass $m_q = g_s T/\sqrt{6}$. Calculations of the production rate for hard photons that correctly take into account the screening of the quark mass singularity have recently been carried out.

CONCLUSIONS

I have outlined a renormalization group approach to thermal quantum field theories. Thermal loop corrections from thermal particles with energies above some scale μ are integrated out to give a nonlocal effective lagrangian that describes the propagation and interaction of low energy particles. Observables involving low energy particles are calculated using the effective lagrangian with an upper cutoff μ imposed on thermal loop corrections. This approach provides a conceptual framework within which thermal quantum field theories can be understood. The resolution of the plasmon problem in particular is easily understood within this framework. The renormalization group approach is also a useful tool for carrying out explicit calculations in thermal quantum field theories. These calculations are tractable in spite of the nonlocal propagators and interactions, as evidenced by a number of explicit calculations involving soft particles or screening of long range interactions. The full power of the renormalization group approach, such as the exploitation of the μ-independence of physical quantities, has yet to be brought to bear on computations in thermal quantum field theory.

REFERENCES

1. K.G. Wilson, Rev. Mod. Phys. $\underline{55}$, 583 (1983).
2. E. Braaten and R.D. Pisarski, Nucl. Phys. $\underline{B337}$, 569 (1990); $\underline{B339}$, 310 (1990).
3. J.C. Taylor and S.M.H. Wong, Nucl. Phys. $\underline{B346}$, 115 (1990).
4. V.P. Silin, Sov. Phys. JETP $\underline{11}$, 1136 (1960).
5. V.V. Klimov, Sov. J. Nucl. Phys. $\underline{33}$, 934 (1981); H.A. Weldon, Phys. Rev. $\underline{D26}$, 2789 (1982).
6. O.K. Kalishnikov and V.V. Klimov, Sov. J. Nucl. Phys. $\underline{31}$, 699 (1980).
7. V.V. Klimov, Sov. Phys. JETP $\underline{55}$, 199 (1982); H.A. Weldon, Phys. Rev. $\underline{D26}$, 1394 (1982).
8. R.D. Pisarski, "A Shortcut to Hard Thermal Loops", Brookhaven Preprint BNL-46013 (1991).
9. E. Braaten, "Diagnosis and Treatment of the Plasmon Problem of Hot QCD", Northwestern Preprint NUHEP-TH-90-31 (October, 1990), to appear in the proceedings of the QCD '90 meeting, Montpellier, France (July, 1990).
10. E. Braaten and R.D. Pisarski, Phys. Rev. $\underline{D42}$, 2156 (1990).
11. E. Braaten, R.D. Pisarski, and T.C. Yuan, Phys. Rev. Lett. $\underline{64}$, 2242 (1990); S.M.H. Wong, Cambridge preprint DAMTP 91-18 (1990).
12. E. Braaten, Phys. Rev. Lett. $\underline{66}$, 1655 (1991).
13. G. Baudet, V. Petrosian, and E.E. Salpeter, Astrophys. J. $\underline{150}$, 979 (1967).
14. H.A. Weldon, Phys. Rev. $\underline{D26}$, 1394 (1982).
15. G. Baym, H. Monien, C. Pethick, and D.G. Ravenhall, Phys. Rev. Lett. $\underline{64}$, 1867 (1990).
16. E. Braaten and T.C. Yuan, Phys. Rev. Lett. $\underline{66}$, 2183 (1991).
17. E. Braaten and M. Thoma, Lawrence Berkeley preprint LBL-30303 (February, 1991), to be published in Phys. Rev. $\underline{D44}$; Lawrence Berkeley preprint LBL-30998 (July, 1991).
18. J. Kapusta, P. Lichard, and D. Seibert, Minnesota preprint (1991); R. Baier, H. Nakkagawa, A. Niegawa, and K. Redlich, Bielefeld preprint BI-TP 91/15 (May, 1991).

ELECTROMAGNETIC LEPTON–PAIR PRODUCTION IN RELATIVISTIC COLLISIONS

C. J. Albert and D. J. Ernst
Physics Department and Center for Theoretical Physics
Texas A&M University, College Station, TX 77843-4242

M. R. Strayer and C. Bottcher
Physics Division
Oak Ridge National Laboratory, Oak Ridge, TN 37831

ABSTRACT

Electromagnetic lepton–pair production in relativistic collisions is studied in an *ab initio* approach with no free parameters. After a semi–classical approximation to the relative motion of the two incident particles is made, the resulting second-order diagram is calculated using a Monte Carlo technique to evaluate the resulting seven–dimensional integral. We examine the case of electron–positron pair production in $\pi^- p$ collisions at $p_\pi = 17$ GeV. We find that a significant fraction of the measured pairs in this reaction are produced via the magnetic spin–flip current of the proton. Approaches, such as the equivalent photon approximation, which neglect this part of the current predict much too small a cross section. This feature is traced to the cuts imposed in taking the experimental data. Lepton–pair production in the scattering of ^{3}He,^{4}He and ^{4}He,^{4}He is proposed as a clean way of experimentally separating the spin–flip and non–flip processes; predictions are made for these systems.

INTRODUCTION

Electromagnetic lepton–pair production by the fields of two colliding particles is a problem that has been studied[1] since the 1930's. A vast majority of the work, with few exceptions[2–4], has followed Landau and Lifshitz[5] and uses the "equivalent photon" or "Weizsäcker–Williams" approximation. Recent work[3,4] has begun to clarify the limitations of this approximation.

In Table 1, we give the value of the effective electromagnetic coupling constant for pair production, $Z_1^2 Z_2^2 \alpha^4 \gamma^4$, which can be achieved at contemporary accelerators. We see that this effective coupling constant can be quite large indicating that pair production may not be properly treated in lowest order perturbation theory. This also provides a motivation to perform the lowest order calculation without making the equivalent photon approximation. We examine here electromagnetic lepton–pair production from the collision of two heavy ions. We make a semi–classical approximation to the motion of the two ions but do *not* make the equivalent photon approximation. In particular, we investigate the role of magnetic spin–flip currents which are neglected in the equivalent photon approximation. The formalism for calculating cross sections for electromagnetic lepton pair production can be extended to look at the electromagnetic production of more exotic particles, such as the Higgs boson[6], $W^\pm$–pairs[7], or particles of high spin[8].

TABLE 1. Accelerator information.

ACCELERATOR	C.M. ENERGY (GeV/A)	$Z_1^2 Z_2^2 \alpha^4 \gamma^4$
AGS	1.6	6.85†
GSI	1.0	12.5
CERN COLLIDER	9.0	1,620†
FERMILAB	1000	2840
RHIC	100	11,000,000
CERN (LHC)	8000	11,600,000
SSC	20,000	458,000,000

† Assumes a target nucleus of ^{92}Pb

FORMALISM

The Feynman diagram for the two–photon process is given in Fig. 1. We take a semi–classical limit on the relative motion of the two ions, treat their motion as a classical straight–line trajectory, and then calculate exactly the resulting Feynman diagram. To do this, we require the field for an ion with impact parameter $\vec{b}/2$ travelling in the z direction

$$A^\mu(q) = \frac{-8\pi^2 Z e \, e^{i\vec{q}\cdot\frac{\vec{b}}{2}}}{q^2}\delta(p_0 - p_0' - q_0)J^\mu(p,q) \ . \tag{1}$$

The semi-classical S–matrix amplitude is

$$\langle \vec{k}\,s'\,|\,S_{\vec{b}}\,|\,\vec{q}\,s\rangle = (-ie)\sqrt{\frac{m}{E_f}}\bar{u}_{s'}(k)\int d^4p\big[\not{A}_1(k-p)\,S_F(p)\,\not{A}_2(p+q) \\ + \not{A}_2(k-p)\,S_F(p)\,\not{A}_1(p+q)\big](-ie)\sqrt{\frac{m}{E_i}}v_s(q) \ , \tag{2}$$

where $S_F(p)$ is the Feynman propagator.

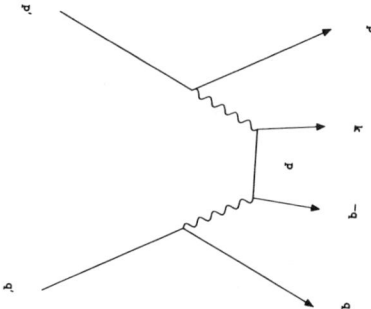

Fig. 1. The Feynman diagrams for the two–photon process for lepton pair production.

The current matrix element in Eq. 1 for a spin 1/2 particle with an anomalous magnetic momement κ is given by, for the non–spin–flip term, $J^\mu(p,q) = \bar{u}_+(p-q)\Gamma^\mu(p,q)u_+(p)$ and for the spin-flip currents $J^\mu(p,q) = \bar{u}_-(p-q)\Gamma^\mu(p,q) \times u_+(p)$ where the current operator $\Gamma^\mu(p,q)$ is

$$\Gamma^\mu(p,q) = \gamma^\mu F_1(q^2) + i\sigma^{\mu\nu} + \kappa F_2(q^2)\frac{i\sigma^{\mu\nu}q_\nu}{2M} \ . \tag{3}$$

Explicit calculation of the current matrix elements give, for the non–spin–

flip current,

$$J^0 = \frac{2M}{N}\left\{\left[\frac{\gamma - \frac{q_0}{2M}}{1 + \frac{q^2}{4M^2}}\right]\left[1 + \gamma + \frac{q_0 + \gamma q_0 - \gamma\beta q_3}{2M}\right]G_E - \left[\left(\frac{\gamma + \frac{q_0}{2M}}{1 + \frac{q^2}{4M^2}}\right)\right.\right.$$
$$\left.\left.\times\left(1 + \gamma + \frac{q_0 + \gamma q_0 - \gamma\beta q_3}{2M}\right) + \gamma + \gamma^2 + \frac{q_0 + \gamma q_0 + \gamma\beta q_3}{2M}\right]G_M\right\}$$

$$J^1 = \frac{1}{N}\left\{q_1(G_M - G_E)\left[\frac{1 + \gamma + \frac{q_0+\gamma q_0 - \gamma\beta q_3}{2M}}{1 + \frac{q^2}{4M^2}}\right] + G_M(1+\gamma)(q_1 - iq_2)\right\}$$

$$J^2 = \frac{1}{N}\left\{q_2(G_M - G_E)\left[\frac{1 + \gamma + \frac{q_0+\gamma q_0 - \gamma\beta q_3}{2M}}{1 + \frac{q^2}{4M^2}}\right] + G_M(1+\gamma)(iq_1 + q_2)\right\} \quad (4)$$

$$J^3 = \frac{2M}{N}\left[\frac{\beta\gamma - \frac{q_3}{2M}}{1 + \frac{q^2}{4M^2}}\right]\left(1 + \gamma + \frac{q_0 + \gamma q_0 - \gamma\beta q_3}{2M}\right)(G_E - G_M)$$
$$+ \frac{2M}{N}\left(\beta\gamma + \beta\gamma^2 + \frac{q_3 + \gamma q_3 + \beta\gamma q_0}{2M}\right)G_M$$

and for the spin-flip current,

$$J^0 = -\frac{\gamma\beta}{N}(q_1 - iq_2)\left\{G_M - \left[\frac{\gamma - \frac{q_0}{2M}}{1 + \frac{q^2}{4M^2}}\right]\right\}(G_M - G_E)$$

$$J^1 = \frac{1}{N}\left\{\gamma\beta\frac{q_1}{2M}(G_M - G_E)\frac{q_1 - iq_2}{1 + \frac{q^2}{4m^2}} + G_M(-\beta\gamma q_0 + q_3 + \gamma q_3)\right\}$$

$$J^2 = \frac{1}{N}\left\{\gamma\beta\frac{q_2}{2M}(G_M - G_E)\frac{q_1 - iq_2}{1 + \frac{q^2}{4m^2}} + G_M(-i)(-\beta\gamma q_0 + q_3 + \gamma q_3)\right\} \quad (5)$$

$$J^3 = -\frac{1}{N}(q_1 - iq_2)\left\{(1+\gamma)G_M + \beta\gamma\left[\frac{\beta\gamma - \frac{q_3}{2M}}{1 + \frac{q^2}{4M^2}}\right](G_M - G_E)\right\},$$

where we have changed from the form factors $F_1(q^2)$ and $F_2(q^2)$ to the electric and magnetic form factors $G_E(q^2)$ and $G_M(q^2)$, with $G_E(q^2) = F_1(q^2) - (q^2/2M)F_2(q^2)$ and $G_M(q^2) = F_1(q^2) + 2M F_2(q^2)$. The normalization factor N is given by

$$N = \frac{2\sqrt{(1+\gamma)(1+\gamma-\frac{q_0}{M})}}{(2\pi)^3 E_p E_{p'}}. \quad (6)$$

In order to make the connection to our earlier work and to explicitly provide the formulas which are calculated numerically, we change notation. The Feynman propagator is expanded into its particle and antiparticle parts

$$\frac{\not{p}+m}{p^2 - m^2} = \frac{1}{p_0 - E_p}\frac{m}{E_p}\sum_s u_s(\vec{p})\bar{u}_s(\vec{p}) + \frac{1}{p_0 + E_p}\frac{m}{E_p}v_s(\vec{p})\bar{v}_s(\vec{p}). \quad (7)$$

The covariant spinors are replaced by non-invariantly normed spinors $|\chi_p^s\rangle = (m/E)^{1/2}u_s(\vec{p})$ for $s = 1, 2$ and $|\chi_p^s\rangle = (m/E)^{1/2}v_s(\vec{p})$ for $s = 3, 4$, and $E_p^s = E_p$ for $s = 1, 2$ and $E_p^s = -E_p$ for $s = 3, 4$, and $V(q) = \gamma_o Ze\,A(q)$. These replacements translate Eq. 2 into

$$\langle \vec{k}s'|S_b|\vec{q}s\rangle = \frac{-ie^2 m}{E_k E_q} \int \frac{d^2 p}{(2\pi)^2} \sum_{s''} \frac{m}{p_o - E_p^s} e^{ib(p+k-q)}$$
$$\{\langle\chi_k^{s'}|V_1(k-p)|\chi_p^{s''}\rangle\langle\chi_p^{s''}|V_2(p+q) + V_2(k-p)|\chi_p^{s''}\rangle\langle\chi_p^{s''}|V_1(p+q)\}|\chi_q^s\rangle . \tag{8}$$

Written in this form the structure of the amplitude is that of a standard second-order quantum mechanics calculation.

The square modulus of the S–matrix is the cross section of the interaction for one particular point in phase space

$$|S|^2 = \frac{e^4 m^2}{E_i E_f} \langle\chi_k^{s+}| \int \frac{d^2 p}{(2\pi)^2} \sum_{s_p} \sum_{s_{p'}} \frac{1}{p_0 + E_p}$$
$$[V_1(k-p)|\chi_p^s\rangle\langle\chi_p^s|V_2(p+q) + V_2(k-p)|\chi_p^s\rangle\langle\chi_p^s|V_1(p+q)]|\chi_q^{s-}\rangle \tag{9}$$
$$\times \langle\chi_q^{s-}| \int \frac{d^2 p'}{(2\pi)^2} \frac{1}{p'_0 + E_{p'}} [V_2(k-p')|\chi_{p'}^s\rangle$$
$$\times \langle\chi_{p'}^s|V_1(p'+q) + V_1(k-p')|\chi_{p'}^s\rangle\langle\chi_{p'}^s|V_2(p'+q)]\,|\chi_k^{s+}\rangle .$$

The total cross section for a specified final spin s' is found by integrating this over the final state momenta, averaging over initial spins, integrating over impact parameter and dividing by the incident current flux (for our normalization this is 2β)

$$\sigma = \frac{1}{2}\sum_s \int d^2 b \int \frac{d^3 k}{(2\pi)^3} \frac{d^3 q}{(2\pi)^3} \frac{|S|^2}{2\beta} . \tag{10}$$

We use the integral over $d^2 b$ to produce a delta function in p and p':

$$\int d^2 b\, e^{i\vec{b}(\vec{p}'+\frac{\vec{k}-\vec{q}}{2})} e^{-i\vec{b}(\vec{p}+\frac{\vec{k}-\vec{q}}{2})} = \int d^2 b\, e^{i\vec{b}(\vec{p}'-\vec{p})}$$
$$= (2\pi)^2 \delta^2(p' - p) .$$

Thus equation (9) becomes

$$|S|^2 = \frac{e^4 m^2}{E_i E_f} \sum_{s,s'=1}^{4} \int \frac{d^2 p}{(2\pi)^2} \{\langle \chi_k^{s+}|$$

$$[V_1(k-p)\frac{|\chi_p^s\rangle\langle\chi_p^s|}{p_0 + E_p^s}V_2(p+q) + V_2(k-p)\frac{|\chi_p^s\rangle\langle\chi_p^s|}{p_0 + E_p^s}V_1(p+q)]|\chi_q^{s-}\rangle$$

$$\times \langle\chi_q^{s-}|[V_2(k-p)\frac{|\chi_p^s\rangle\langle\chi_{p'}^s|}{p_0 + E_p^{s'}}V_1(p+q) + V_1(k-p)\frac{|\chi_p^s\rangle\langle\chi_p^s|}{p_0 + E_p^{s'}}V_2(p+q)]|\chi_k^{s+}\rangle\} \; .$$

(11)

and so equation (10) becomes

$$\sigma = \frac{e^4 m^2}{2\beta} \sum_{s_+,s_-} \frac{1}{2} \sum_{s,s'} \int \frac{d^2 p \, d^3 k \, d^3 q}{(2\pi)^8 E_k E_q} \{\langle \chi_k^{s+}|$$

$$[V_1(k-p)\frac{|\chi_p^s\rangle\langle\chi_p^s|}{p_0 + E_p^s}V_2(p+q) + V_2(k-p)\frac{|\chi_p^s\rangle\langle\chi_p^s|}{p_0 + E_p^s}V_1(p+q)]|\chi_q^{s-}\rangle$$

$$\times \langle\chi_q^{s-}|[V_2(k-p)\frac{|\chi_p^s\rangle\langle\chi_{p'}^s|}{p_0 + E_p^{s'}}V_1(p+q) + V_1(k-p)\frac{|\chi_p^s\rangle\langle\chi_p^s|}{p_0 + E_p^{s'}}V_2(p+q)]|\chi_k^{s+}\rangle\} \; .$$

(12)

Equation (12) is the total cross section for electromagnetic pair production, to lowest order, in the semi-classical formulation. This is the expression which was programmed. Note that (12) is still a general equation insofar as the potentials V_1 and V_2 have not been specified; they are as yet general matrix elements. The equation is therefore valid for both spin–flip and non–spin–flip processes for scalar or Dirac particles.

THE COMPUTATION

The FORTRAN program that evaluates (12), YEW, is written to handle the production of a pair of any spin-$\frac{1}{2}$ point–particles via the collision of either spin-$\frac{1}{2}$ or spin–0 particles. In addition, the program calculates the spin–flip and non–spin–flip process for an incident spin-$\frac{1}{2}$ particle, or the proton–delta transition for an incident proton. (However, YEW will not yet yield accurate results for the proton–delta transition until difficulties with the modeling of the transition form factor is resolved).

The evaluation of (12) amounts to an integration of a positive function over a seven-dimensional phase space; of the eight variables

$$x_1 \ldots x_8 = p_1, p_2, q_1, q_2, q_3, k_1, k_2, k_3 \; ,$$

one is expressed as a function of the others by symmetry. Equation (12) becomes

$$\sigma = F_0 \int f(\vec{x}) d\vec{x} \; , \qquad (13)$$

with F_0 chosen so that $f < 1$. A mapping is found so that the seven remaining integration variables x_i are re-written as $x_i = y_i X_i$ where $0 < y_i < 1$.

The particular variation of the Monte Carlo method used here is thus: a random "throw" is made in an eight-dimensional space y_i, and if $f(\vec{y} \cdot \vec{X}) > y_8$, the throw is called "successful". After T throws and S successes, the cross section (13) is given by

$$\sigma = F_0 \frac{S}{T} \prod_{i=1}^{7} X_i \; ,$$

with a proportional error in σ/F_0 of $\sqrt{S^{-1} + T^{-1}}$.

RESULTS

In a previous work[4], we examined the case of electron–positron pair production in 17 Gev/c pion–nucleon collisions — a reaction for which there exists data[9]. Figures 2a - 2c show the results of our calculation. In all three differential cross sections we see that the theoretical predictions pass through the error bars at the low end of the scales, in clear contrast to the behavior of the equivalent photon prediction, which is everywhere a few orders of magnitude low. Moreover, it is the spin–flip cross section arising from the anomalous magnetic moment coupling which is dominant. This situation arises because a cut on the invariant mass of the lepton pair has been imposed in the experiment (and in the calculation). In this kinematic region, the non–spin–flip electric term is small and is reasonably approximated by the equivalent photon approach.

At the higher mass, transverse momentum, or Feynman X end of Figures 2, our curves are too low by an order of magnitude. This is in contrast to our earlier[4]

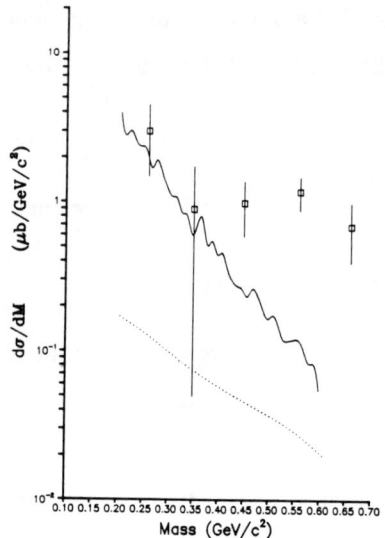

Fig. 2a. The differential cross section $d\sigma/dM$ for electron–positron production in 17 GeV/c $\pi^- p$ collisions. M is the invariant pair mass, $M^2 = (E_+ + E_-)^2 - (\vec{p}_+ + \vec{p}_-)^2$. The data are from Ref. 9.

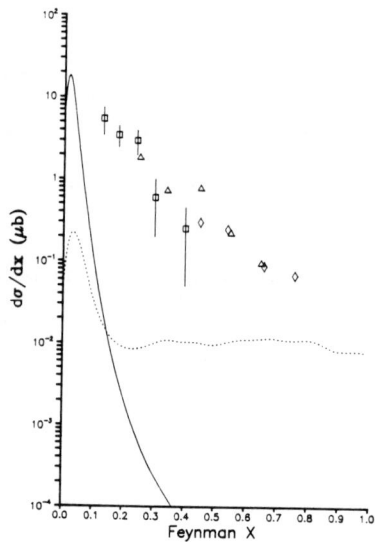

Fig. 2b. The same as Fig. 2a except the cross section is $d\sigma/dx$ with x equal to Feynman x, the total longitudinal momentum of the pair divided by maximum longitudinal momentum kinematically possible.

results which used a boosted potential model rather than the more exact current matrix elements, Eqs. 5 and 6, used here. Clearly our lowest order exclusive calculation does not adequately reflect the inclusive data.

The important and interesting point is that there are regions where the magnetic spin–flip piece can be dominate the electric non–spin–flip piece. This point can be addressed directly by examining the collision between He isotopes. We find that in helium-helium collisions, the spin-flip term dominates the non-flip term, provided the same cut on invariant mass as was used in obtaining the data of Ref. 8 is imposed. Figure 3a illustrates this in the case of ^{3}He,^{4}He collisions, and Figure 3b shows that the non-flip case of ^{3}He,^{4}He closely resembles the ^{4}He,^{4}He collision (which has no spin flip channel). Thus the difference between lepton pair production in ^{3}He,^{4}He collisions and ^{4}He,^{4}He collisions is a measurement of the magnetic spin–flip contribution to the reaction.

For completeness, we present in Fig. 4 the results for proton–proton collisions. Again, if we impose a cut on the invariant pair–mass, we find a dominance by the magnetic spin–flip current.

CONCLUSION

We have found the rather surprising result that the spin–flip piece of the effective potential is what dominates lepton–pair production cross sections for the 17 Gev pion–proton data of Ref. 8. We have found this to be the result of the experimental cut imposed on the invariant pair mass.

This surprising result is not true in general: the spin–flip term does not usually dominate the non–spin–flip term in QED processes. The spin–flip current brings in an extra factor of q/m which usually renders the magnetic terms smaller than the electric terms. If the cut on the invariant pair mass is removed, we find that the electric non–spin–flip current does indeed dominate and that the equivalent photon approximation and our previous boosted potential approximation do well at predicting total cross sections.

We have used this formalism to calculate pair production from ^{3}He-^{4}He and ^{4}He-^{4}He collisions, and explained why these systems will provide a clean signature

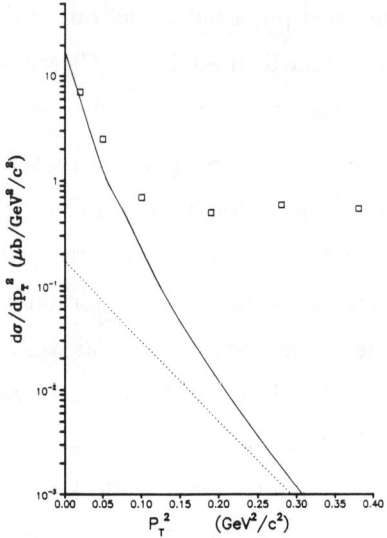

Fig. 2c. The same as Fig. 2a except the cross section is $d\sigma/dp_{\perp^2}$, where $p_\perp$ is the total transverse momentum of the pair.

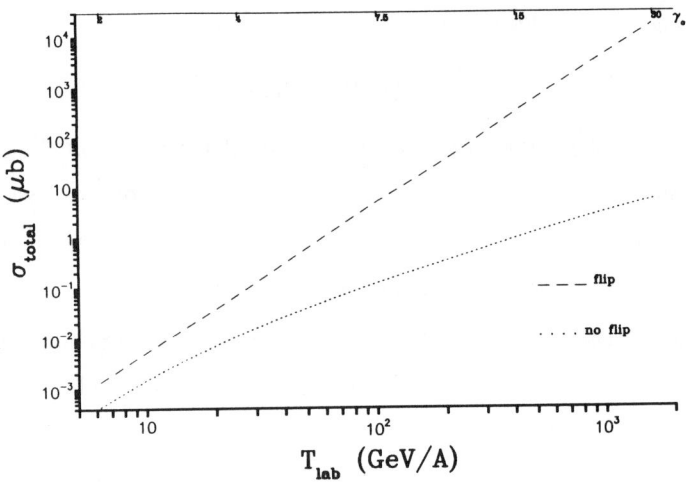

Fig. 3a. The total cross section within the experimental cuts made in Ref. 9 for electron–positron production in ^{3}He,^{4}He collisions verses T_{lab}. The dashed curve is the cross section produced by the spin–flip current; the dotted curve is the cross section produced by the non–spin–flip current.

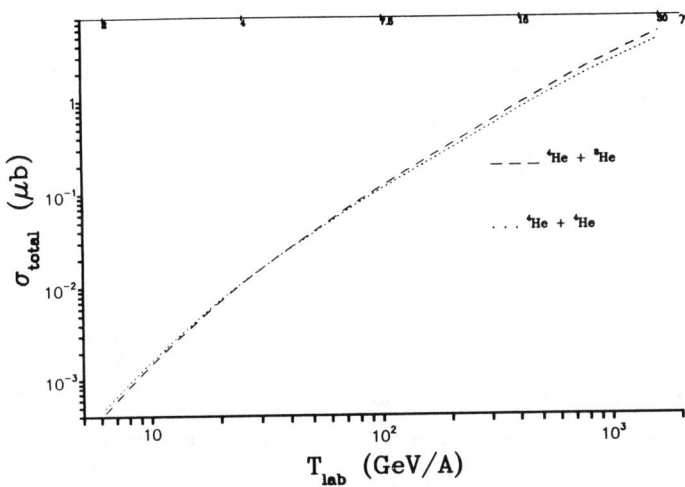

Fig. 3b. The total cross section within the experimental cuts made in Ref. 9 for electron–positron production for ^{3}He,^{4}He collisions (dashed curve) and for ^{4}He,^{4}He collisions (dotted curve). Both are non–spin–flip cross sections.

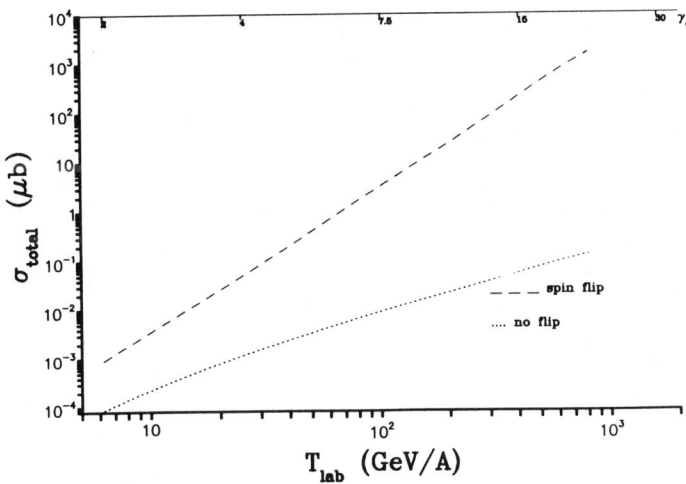

Fig. 4. The same as Fig. 3a except the reaction is proton on proton.

for the spin–flip mechanism.

The work presented here is an *ab initio* calculation which involved no free parameters. The Monte Carlo technique used proved an efficient and convergent approach to calculating these cross sections.

BIBLIOGRAPHY

1. W. H. Furry and J. F. Carlson, Phys. Rev. 44, (1933) 237; W. Heitler and L. Nordheim, J. de Physique 5, (1934) 449; J. R. Oppenheimer, Phys. Rev. 47, (1935) 146; H. J. Bhabha, Proc. Roy. Soc. A152, (1935) 1532; C. A. Bertulani and G. Baur, Phys. Rep. 163, (1988) 299.
2. S. J. Brodsky, T. Kinoshita, and H. Terazawa, Phys. Rev. D 4, (1971) 1532.
3. M. R. Strayer and C. Bottcher, Phys. Rev. C 39, (1989) 1330;
4. M. R. Strayer, C. Bottcher, C. J. Albert, and D. J. Ernst, Phys. Lett. B (1990) 175; C. J. Albert, D. J. Ernst, C. Bottcher and M. R. Strayer, in *Progress in Nuclear Physics*, ed. W–Y. P. Hwang, S–C. Lee, C–E. Lee and D. J. Ernst (North–Holland, New York, 1990), p. 570.
5. L. Landau and E. Lifshitz, Physik Z. Sowjet. 6, (1934) 244; following C. v. Weizsäcker, Z. Phys. 88 612 (1934) and E. Williams, Phys. Rev. 45 (1934) 730.
6. J. S. Wu, C. Bottcher, M. R. Strayer, and A. K. Kerman, to appear in Ann. Phys. (N.Y.).
7. J. S. Wu, C. Bottcher, M. R. Strayer, D. J. Ernst, and A. K. Kerman, in *Progress in Nuclear Physics*, ed. W–Y. P. Hwang, S–C. Lee, C–E. Lee and D. J. Ernst (North–Holland, New York, 1990), p. 552.
8. D. V. Ahluwalia and D. J. Ernst, in preparation.
9. R. Stroynowski, D. Blockus, W. Dunwoodie, D. W. G. S. Leith, M. Marshall, C. L. Woody, B. Barnett, C. Y. Chien, T. Fiefuth, M. Gilchriese, D. Hutchinson, W. B. Johnson, P. Kunz, T. Lasinski, L. Madansky, W. T. Meyer, A. Pevsner, B. Ratcliff, P. Schacht, J. Scheid, S. Shapiro and S. Williams, Phys. Lett. 97B, 315 (1980); D. Blockus, W. Dunwoodie, D. W. G. S. Leith, M. Marshall, R. Stroynowski, C. L. Woody, B. Barnett, C. Y. Chien, T. Fieguth, M. Gilchriese, D. Hutchinson, W. B. Johnson, P. Kunz, T. Lasinski, L. Mandansky, W. T. Meyer, A. Pevsner, B. Ratcliff, P. Schacht, J. Sheid, S. Shapiro and S. Williams, Nucl. Phys. B201, (1982) 205; J. Stekas, G. Abshire, M. R. Adams, C. Brown, L. Cormell, E. Crandall, G. J. Donaldson, J. Goldberger, H. A. Gordon, P. D. Grannis, B. T. Meadows, G. R. Morris and P. Rehak, Phys. Rev. Lett. 47, 1686 (1981); M. R. Adams, G. Abshire, C. Brown, E. S. Crandall, J. Goldberger, P. D. Grannis, B. T. Meadows, J. Stekas, G. J. Donaldson, H. A. Gordon, G. R. Morris, P. Rehak and L. Cormell, Phys. Rev. D. 27, (1983) 1977.

Laser-Assisted Molecular Dissociation and Recombination of Diatomic Molecules

Zi-Min Lu, Michel Vallières, and Jian-Min Yuan

*Department of Physics and Atmospheric Science, Drexel University,
Philadelphia, PA 19104-9984*

August 23, 1991

Abstract

This talk reviews recent classical and quantum calculations of laser assisted dissociation and recombination of diatomic molecules. We treat the interaction of the laser field with the vibrational degree of freedom of the molecule. We emphasize the comparison of the classical and quantum mechanical solutions.

1 Introduction

We review in this talk recent classical and quantum calculations of photo-dissociation and recombination processes of diatomic molecules under the influence of a laser field. This is a rich topic where the comparison of classical and quantum results can be of significance. In particular, the dissociation of diatomic molecules under an EM field is predicted to exhibit chaotic behavior [1, 2], with a fractal dissociation threshold in the parameter space of the laser field. This multi-photon dissociation process has unfortunately never been observed cleanly in any experiments. A very similar system, the photo-ionization of hydrogen atoms, has on the other hand become a landmark experiment in the comparison of classical chaotic behavior and quantum mechanical approach [3]. We believe that the molecular dissociation will shortly provide another such fundamental comparison, due to advances in overtone spectroscopy.

We will also consider the complementary physical processes to photo-dissociation, namely collisions leading to molecular formation of two atoms under the influence of a laser field. This process should also shortly be of primary importance in analyzing the classical-quantum correspondence in chaotic systems [4]. We will show below how this system exhibits classical chaotic scattering and then show some preliminary quantum mechanical calculations of the same process. Classical chaotic scattering is a subject which has attracted much interest recently[5]. It provides for new advancements in non-linear dynamics. However, the systems so far studied were of multi-dimensions (at least two) so has to provide for chaotic solutions. The molecular system studied here is a one-and-a-half degrees of freedom system, a simplification over the higher dimensional systems. Another interesting aspect of this system is the existence of a dissociation threshold, an uncommon features in non-linear dynamics, but not so for quantum systems. The quantum solutions of chaotic scattering systems has been obtained for very few simple systems [6]. We will report here on preliminary results of such a study.

We will describe the model we use in section 2. Section 3 will deal with molecular dissociation, while section 4 will deal with chaotic atomic scattering and molecular recombination. We will conclude in section 5.

2 Model

We consider in this work diatomic molecules embedded in a laser field and consider multiphoton excitation of the molecule. Recent advances in overtone spectroscopy have revealed that the energy of highly excited molecules tends to be localized in single bonds so that a local-mode description of the molecule is better suited than its normal-mode counter part. The Morse oscillator is the time honored way to model the molecule in a local-mode approximation. We further consider only the vibrational degree of freedom since the vibrational motion induced by the E field will tend to take place in the direction parallel to the field.

The Hamiltonian describing such a non-rotating diatomic molecule embedded in an external laser field consists of two parts,

$$H = H_0 + H_I \tag{1}$$

where

$$H_0 = \frac{p^2}{2} + \frac{1}{2}\left(1 - e^{-x}\right)^2 \tag{2}$$

is the free Morse Hamiltonian describing the interaction between the two atoms and H_I is the interaction term treated in the dipole approximation,

$$H_I = -\frac{A\Omega}{2}\mu(x)cos(\Omega t) \tag{3}$$

between the field and the induced dipole moment of the molecule. The variables are all dimensionless scaled variables [1, 2]; x is the scaled distance between the two atoms. The laser field is characterized by two parameters, its frequency Ω and amplitude A.

The dipole moment [7] is taken as

$$\mu(x) = (x+a)e^{-\frac{x+a}{b}} \tag{4}$$

It will be approximated by a linear form in our dissociation studies since we do not need to track the atomic motion to large separation before deciding that the molecule is dissociated. This approximation neglects the possibility of recombination, expected to be a small effect. We will use the full dipole moment form in our scattering studies since the atoms originate from a large separation distance in atomic beam experiments.

A useful approximation to understand the physics of the model at a given frequency consists in replacing the continuously driving field by a one sided periodic string of delta functions (Dirac comb) [1].

$$cos(\Omega t) \longrightarrow \frac{2}{\Omega} \sum_{n=-\infty}^{+\infty} \delta(t - nT); \tag{5}$$

$\frac{2}{\Omega}$ arises by demanding the average values of the driving terms, over half period, to be the same. Therefore

$$H_{in} = -A\mu(x) \sum_{n=-\infty}^{+\infty} \delta(t - nT) \tag{6}$$

This provides for a *map* solution of the model which can be solved piecewise analytically, thereby providing for a better understanding of the physics. We will describe below results for both the impulsively and continuously driven systems.

3 Dissociation

An interesting aspect of this system concerns its response to the influence of the external

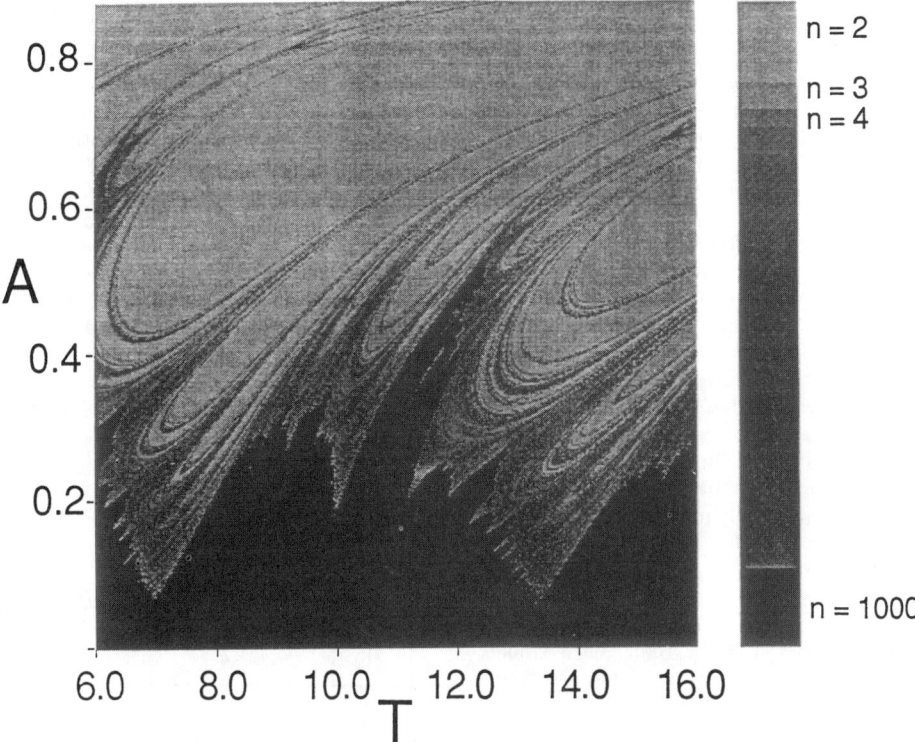

Figure 1: Lifetime of the diatomic molecule as a function of the period T and amplitude A of the external field.

electromagnetic field which tends to induce the dissociation of the molecule. The system exhibits chaotic behavior and the study of the features in phase space are interesting in their own right; but perhaps the more experimentally relevant characteristic behavior is to be found in the features in parameter space, Ω and A, the frequency and amplitude of the laser field.

Classically, the problem is described in the following terms: let the system be at the bottom of the Morse potential, $(x,p)=(0,0)$, initially. Follow the time evolution of the system by solving Hamilton's equations and record the dissociation time as a function of Ω and A, i.e., the time when the total energy of the two atoms crosses $E=0.5$, the dissociation threshold in scaled units. In the case of an impulsively driven system, Hamilton's equations can be solved analytically in between the impulses [1, 4]. The delta function interaction terms provide for an impulse, a sudden change in momentum proportional to $\frac{d\mu}{dx}$, with no change in position.

There exist closed Kolmogorov-Arnold-Moser (KAM) curves surrounding the origin in phase space; these KAM curves acquire ever increasing complicated shapes as they break up as we change the EM field parameters [1]. The molecule can dissociate if these KAM curves open up, allowing for paths for the trajectories to reach the $E=0.5$ separatrix. Due to the very complicated features in phase space, the response of the system to the external field can also be very complex [1]. Fig 1 shows the lifetime of the molecule as a function of the period $T = \frac{2\pi}{\Omega}$ and amplitude A of the laser field. The boundary between the dissociative and non-dissociative regions in (A,T) space is fractal, in that enlargements of scale of the figure reveal that ever more complicated structures emerge.

The question of the response of the system as a function of the external field parameters is still very relevant in a quantum solution. It is difficult to relate any classical features in phase space to features in a quantum solution since the classical phase space variables, x and p, lose their meaning quantum mechanically; yet, T and A remain experimental control parameters in the quantum solution. Further, the dissociation threshold is a well defined quantum mechanical concept. Therefore, the problem can be phrased in the following way: let the system be in the ground state of the Morse potential at the initial time. Follow the time evolution of the wave function and record the dissociation time, namely when the overlap of the wavepacket with the Morse bound state reaches below a specific value (typically 95% in what follows) [2].

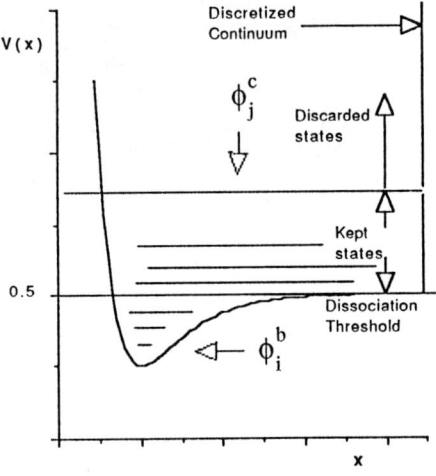

Figure 2: Box discretization of the continuum.

The time evolution of the wave function can be obtained via the use of the *Floquet* formalism [2] which takes advantage of the periodical time dependence of the Hamiltonian. The *Floquet* operator is defined as the time evolution operator for one period of the external field, namely

$$\hat{U}(T) = \hat{U}_{free}\hat{U}_{impulse} = e^{\frac{-i\hat{B}_0 T}{\hbar}} e^{\frac{iAx}{\hbar}}. \quad (7)$$

which leads to a simple form for the wavefunction after N periods

$$|\psi_N> = \left(\hat{U}(T)\right)^N |\psi_0> \quad (8)$$

where $|\psi_0>$ is the initial state of the system and $|\psi_N>$ the wave packet at the Nth impulse.

In order to calculate the *Floquet* operator we discretized the continuum by putting a hard wall at a large distance which leads to an approximation of the continuum states by bound states [2]. The validity of this approximation requires a good description of the density of states for energy close to the dissociation threshold, $E=0.5$. The quality of the solution can be easily verified by moving the position of the wall and requiring stability of the results. We also used an energy upperbound above which the states were not included in the calculation; this again can be checked for accuracy by varying the cutoff position and requiring stability of the results. We typically kept around 200 states in our calculations.

The quantum solutions depends on a new parameter, $\hbar$, which is a measure of the number of discrete bound states in the Morse potential. The result of a quantum dissociation calculation are shown in Fig. 3 for $\hbar=0.04$, or 25 bound states. The figure shows a much coarser sampling as a function of T and A than in Fig. 1; this is due to the prohibitive (for us) length of the calculation. The major characteristics of the classical results are nevertheless reproduced here, i.e., the existence of "finger" structures, the tilting of these fingers... However, the very fine structures leading to the fractal dissociation boundary are absent, due to the finiteness of $\hbar$.

The large dissociation probability occurs at values of the field parameters for which the overlap between some Floquet state and both the ground state and states in the continuum is large. These constitute *"favored channels"* through which the molecule mostly dissociates [2]. This quantum mechanical interpretation of the large dissociation probability is to be contrasted with the classical interpretation in terms of the breaking of the KAM curves.

4 Scattering

One of the most interesting recent applications of non-linear dynamics techniques resides in the studying of scattering systems. Physical systems often exhibit chaotic scattering; this behavior is characterized by the scattering functions, i.e., differential cross section, final energy, scattering time..., becoming discontinuous over a fractal set of initial conditions, i.e., initial energy, momentum, angular variables... Many physical systems have been observed to exhibit chaotic scattering: among others, Rankin and Miller [8], as early as 1971, found great sensitivity in the reactive and inelastic scattering of $H+Cl_2$; the Ericson scattering [9] is also an example in nuclear physics.

Great physical insights into the mechanisms of chaotic scattering and scenarios leading to chaotic scattering have been obtained in simple time independent model systems [10], usually in two dimensional model space. The currently accepted explanation of the chaotic scattering behavior is that a trapping mechanism results when the scattering orbits get close to the stable manifolds of the saddle orbits.

The system studied in the previous sections constitutes an excellent physical system which can be studied for chaotic scattering. This can be done with minimum extension of the model; for example, we consider the scatter-

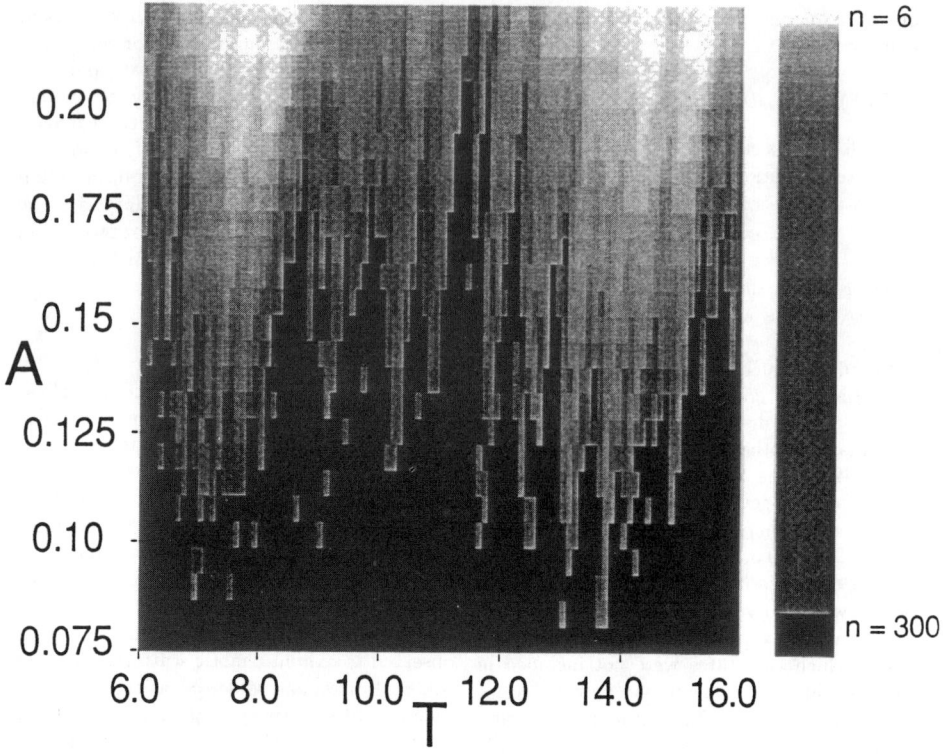

Figure 3: Quantum solution for the lifetime of the diatomic molecule as a function of the period T and amplitude A of the external field.

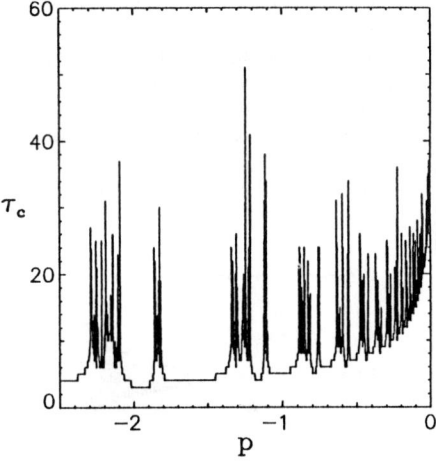 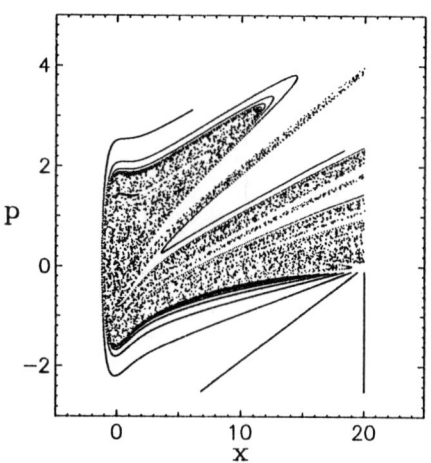

Figure 4: Scattering times for trajectories initiating at $x=20$ and initial momenta in the range $p=\{-2.0,-0.2\}$.

Figure 5: Stroboscopic surface of section for 2000 trajectories initiating at $x=20$ and initial momenta in the range $p=\{-2.0,-0.2\}$.

ing of two non-identical atoms in a typical atomic beam experiment set-up. We assume that the scattering region is exposed to a laser field, and study the sensitivity of the scattering functions as a function of the initial relative momentum of the atoms for various frequency Ω and amplitude A of the laser field. This system is simpler than models based on two-dimensional phase space, yet very rich in its dynamics because of the driving field, and realistic since it models atomic collisions.

The specific model [4] we use is given by the Hamiltonian in Eq. (1), the interaction with the field being taken in the dipole approximation with dipole moment given in Eq. (3). The latter realistically gives that $\mu(x) \longrightarrow 0$ for large x. The initial condition is given by a point in phase space well in the asymptotic region of both the Morse potential and dipole moment; typically, $x=20$ and p in the range $\{-2.0,-0.01\}$ are used. We then follow the time evolution of the ensuing trajectory by solving Hamilton's equation. If the time dependence is impulsive, this solution is analytic; note however that energy is exchanged with the $\mathcal{EM}$ field and that care must be exercised after each impulse to decide if the subsequent time evolution is in the bounded or unbounded region. If the time dependence is continuous, eq. 3, we use the Runge-Kutta algorithm of 4th order to solve numerically Hamilton differential equations [11].

The model possesses only one degree of freedom; therefore the characteristic function of the scattering is the *scattering time*, namely the time taken by the atom for a journey in from and back out into the asymptotic region. This scattering time is shown in Fig. 4 for an impulsively driven system. Note that the scattering time seems to be a very sensitive function of the initial momentum. The structures persist and repeat themselves under expansion of scale of this figure. The scattering time is discontinuous over a set of fractal initial momenta. Fig. 5 shows the stroboscopic surface of section, with the points on the trajectory being recorded just prior to the impulse, for the same initial conditions as in Fig. 4.

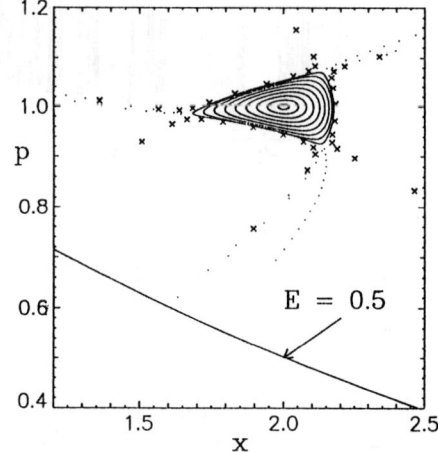

Figure 7: Period one orbit surrounded by period three orbit; the $E=0.5$ separatrix is shown to indicate that these structures are above the dissociation threshold. The crosses in the figure denote scattering points.

The chaoticity of the scattering is attributed to *trapping* by various unstable periodic orbits embedded in phase space. The stable manifolds of the unstable periodic orbits reach the asymptotic region. The intersections of these manifolds with the initial condition line gives the locations for which the scattering time becomes singular. In our impulsively driven atomic collision model, we can find the stable and unstable closed trajectories via the "*sym-*

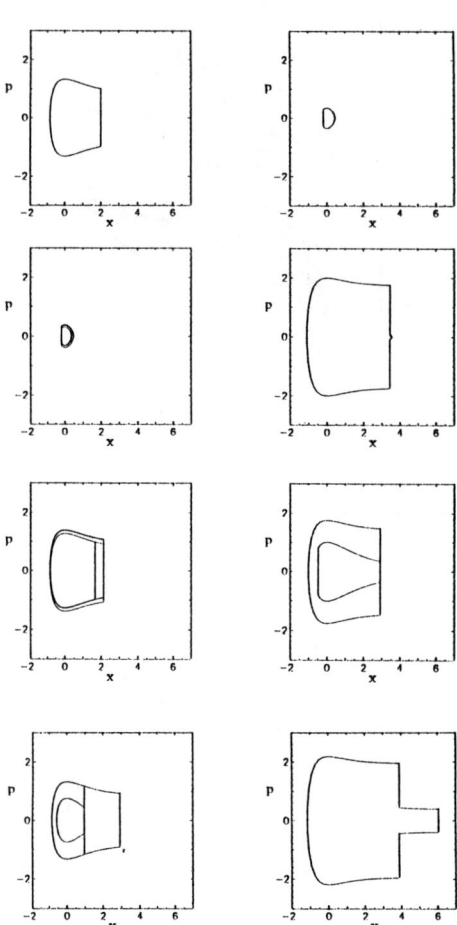

Figure 6: Example of closed trajectories of various periodicities.

metry lines" construct; this approach is well explained in Ref [4]. Examples of closed orbits are shown in Fig. 5. There are saddle-center bifurcations followed by period doubling sequences taking place as the parameters change leading to chaotic behavior.

An important feature in phase space which influences the scattering behavior is the stable period one orbit existing above dissociation threshold, which is surrounded by a period three saddle orbit. This structure is illustrated in Fig. 6. That this structure influences the scattering times is evidenced in Fig 7, where we show the correlation of the stable manifolds of the period three orbit to the peaks in the scattering times. Note however that higher periodicity orbits also contribute to the chaoticity of the scattering function. We are currently studying the symbolic dynamics of this system to understand better the transition to chaotic behavior.

The scattering trajectories do not penetrate the closed KAM curves in phase space, hence the latter forming a *"repeller"*[5]; this is illustrated in Fig. 5. A *dressed* molecule will be formed if the scattering trajectories could penetrate the stable island in phase space. The control procedures [12] proposed so far, usually based on a feedback mechanism, is difficult to apply for these fast molecular systems. To control the chaotic atomic scattering, we propose to exploit the switching on of the laser field so that the closed KAM curves will trap permanently the atoms which are within this region in phase space; namely, we switch on the field at the moment the incoming trajectory are within the neighborhood of the period one orbit. Fig. 8 illustrates this possibility of *controlling the chaotic scattering to enhance the molecular formation*.

The continuously driven system exhibits the same generic features as the impulsively driven one. The scattering time is discontinuous over a set of fractal initial momenta. The difference

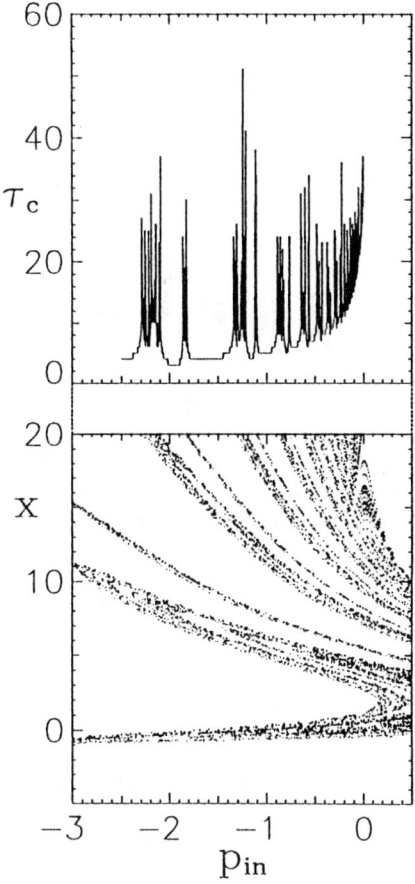

Figure 8: The unstable manifolds of the period three orbit asymptotically (at $x=20$) coincide in momentum with the location of the initial conditions which will produce large collision times.

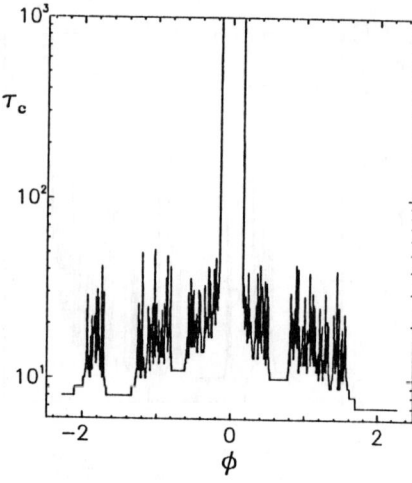
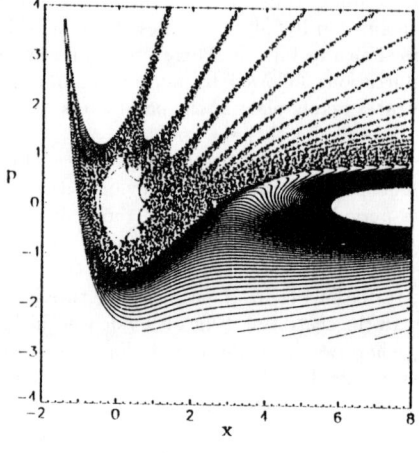

Figure 9: Control of the chaotic scattering via the switching-on of the laser field.

Figure 10: Same as Fig. 4 but for continuously driven molecules.

is that the phase space structures that influence the scattering are first a period one saddle along the x axis [11] and its homoclinic tangles. High periodicity orbits surrounding the origin which lie partially in the field-free continuum [11] can be used for permanent trapping. Therefore the control mechanism proposed earlier to enhance dressed-molecular formation works in this case also [11].

The solution of the scattering problem quantum mechanically is difficult. We cannot use easily the "*Floquet*" formalism or the "*box discretization*" since states high in energy in the continuum are important in a scattering situation. We resolve to use a direct approach in solving numerically Schrödinger Equation. We use a standard implicit numerical method [13] summarized by writing down the time evolution over a time step

$$\Psi^{n+1} = \frac{1 - i\frac{1}{2}H\delta t}{1 + i\frac{1}{2}H\delta t}\Psi^n \qquad (9)$$

This method maintains unitarity; we use a three step form for the second differential operator for efficiency. We have also compared the solution obtained by this approach to the solution obtained by the Feit-Fleck spectral method and found good agreement.

The first interesting question we should address is the existence of "*trapping*" in the quantum mechanical solution. Namely, we conjecture that the triangular structure surrounding the period one orbit will influence the quantum solution provided it has an area large enough compared to $\hbar$.

To prove this point, we launched a Gaussian wave-packet of minimum spread commensurate with the size of $\hbar$ from within the triangular structure,

$$\psi(x, t=0) = Ce^{\frac{i}{\hbar}p_0 x - \frac{x-x_0^2}{4\sigma^2}} \qquad (10)$$

and observe its time evolution. Fig. 10 shows the results of such calculations. The phase space portrait was obtained by plotting the momentum expected value $<p>$ versus the position expected value $<x>$. We recognize that the equivalence of the period one orbit is well reproduced, at least until the vavepacket leaks out. Quantum mechanics allows for a spreading of the wave-packet; Fig. 10 shows that the widening of the wave-packet is prevalent. The diffusion of the wave packet is a function of $\hbar$, occurring faster for larger $\hbar$. The classical calculation is a limiting case of the quantum calculation, when $\hbar$ is small.

The time scale for the wave packet spreading in the regular region and chaotic regions are quite different [14]. The spreading is much faster in the classical chaotic region than in the regular region since both quantum dispersion and classical stochasticity are present in this regime. This difference in time scale would favor our control mechanism. Before turning on the laser field, the classical phase space is regular; after switching on the laser field, the Hamiltonian within the interior region of the resonance island is only infinitesimally different from its normal form which is integrable. A wave packet with the right initial condition which is caught by the resonance structure will have a spreading through tunneling out of the resonance island much slower than the spreading of a similar wave packet with slightly different initial conditions which might end up in the classical chaotic region after the field is turned on. This indicates that the control mechanism we proposed earlier ought to work well for the quantum system as well.

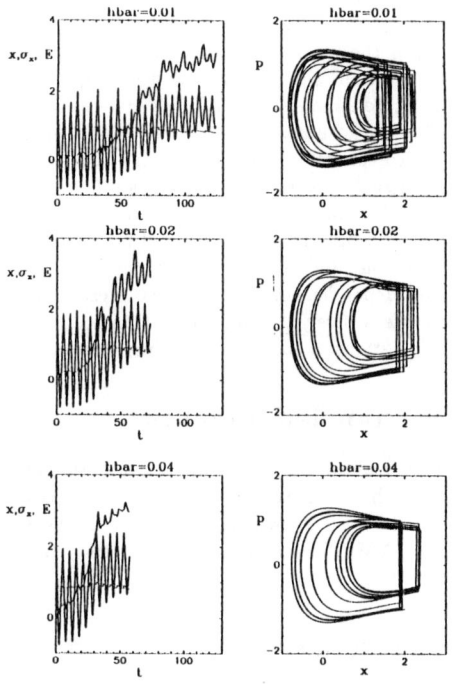

Figure 11: Quantum time evolution of a wavepacket originally centered within the classical triangular structure. The RHS figures show a *"phase space portrait"* in terms of expected values. The LHS shows $<x>$, the spread in x, and the energy of the solution for various $\hbar$.

5 Conclusion

The molecular calculations described here represent potential landmark experiments to be pursued in order to further our understanding of the classical-quantum correspondence. In addition, they describe a useful approach in photo-enhancement of chemical reactions. They make use of a potentially useful strategy to control chaotic scattering. The study of chaotic scattering is still in its infancy and yet it has already been related to many physical systems. These studies will no doubt enhance our understanding of classical chaos of time-dependent systems. At the present time we are actively pursuing quantum scattering calculations in the chaotic regime.

Acknowledgements

This work was partially supported by the NSF through Grant No. Phy90-04582 and Phy90-06186. We acknowledge using time on NCSA Cray YMP for our quantum scattering calculations.

References

[1] J. F. Heagy and J. M. Yuan, Phys. Rev. A **41**, 571 (1990); J. F. Heagy, Ph. D. dissertation, Drexel University, 1989.

[2] Zi-Min Lu, James F. Heagy, Michel Vallières, and Jian-Min Yuan, Phys. Rev. A **43**, 1118 (1991).

[3] R.V. Jensen, S.M. Susskind, and M.M. Sanders, Phys. Rev. Lett. **62** 1476 (1989); J.E. Bayfield and D.W. Sokol, Phys. Rev. Lett. **61** 2007 (1988); P.M. Koch in *Electronic and Atomic Collisions*, ed. by H.B. Gilbody, W.R. Newell, F.H. Read, and A.C.H. Smith, (Elsevier Science Publishers, 1988).

[4] *Controlling Chaotic Scattering: Stimulated Molecular Recombination*, Zi-Min Lu, Michel Vallières, and Jian-Min Yuan, Phys. Rev. Lett. (*submitted*); *Controlling Chaotic Scattering: I - Impulsively Driven Morse Oscillator*, Zi-Min Lu, Michel Vallières, and Jian-Min Yuan, Phys. Rev. A (*submitted*).

[5] B. Eckhardt and C. Jung, J. Phys. A **19**, L829 (1986); P. Gaspard and S. Rice, J. of Chem. Phys. **90**, 2225 (1988). G. Troll and U. Smilansky, Physica D **35** (1989) 34; R. Blumel and U. Smilansky, Phys. Rev. Lett. **60** 477 (1988); Physica **36 D**, (1989) 111.

[6] Quantum Chaotic Scattering by R. Blumel in *Directions in Chaos*, Vol. 4, ed. by B.L. Hao, D.H. Feng, and J.M. Yuan, (World Scientific, Singapore, to appear).

[7] Gerhard Herzberg, *Molecular Spectra and Molecular Structure I. Spectra of Diatomic Molecules*, (D. Van Nostrand Company, Inc., Toronto, 1950); R. Heather and H. Metiu, J. Chem. Phys. **86**, 5496 (1987).

[8] C.C. Rankin and W.H. Miller J. Chem. Phys, **55**, 3150 (1971).

[9] T. Ericson, T. Mayer-Kuckuk, Ann. Rev. Nuc. Sci. **16** 183 (1966).

[10] S. Bleher, E. Ott, and C. Grebogi, Phys. Rev. Lett. **63**, 919 (1989); S. Bleher, C. Grebogi, and E. Ott, Physica D, to appear; *Massive Bifurcation of Chaotic Scattering*, M. Ding, C. Grebogi, and E. Ott, Univ. of Maryland Preprint, 1990; M. Ding, C. Grebogi, and E. Ott, *Transition to Chaotic Scattering*, University of Maryland Preprint, 1990.

[11] Zi-Min Lu, Michel Vallières, and Jian-Min Yuan, *Controlling Chaotic Scattering: II - Continuously Driven Morse Oscillator*, Drexel University preprint, 1991.

[12] E. Ott, C. Grebogi, and J. A. Yorke, Phys. Rev. Lett. **64** 1196 (1990); in *CHAOS: Soviet-American Perspectives on Nonlinear Science*, Ed. by D. K. Campbell (American Institute of Physics, New York, 1990) 153; W. L. Ditto, S. N. Rauseo, and M. L. Spano, Phys. Rev. Lett. **65**, 3211 (1990).

[13] Computational Physics, S.E. Koonin and D.C. Meredith, (Adisson-Wesley Publishing, 1990).

[14] G.P. Berman, V. Yu Rubaev, and G.M. Zaslavsky, Nonlinearity **3** 1 (1991).

Real-time evolution in lattice gauge theory

A.Trayanov[*,†] and B.Müller[*]

[*] Duke University, Physics Department, Durham, NC 27706
[†] Research Institute, North Carolina Supercomputing Center,
Research Triangle Park, NC 27709-2889

Abstract

A new approach to study the real-time behavior of non-abelian lattice gauge theories off thermal equilibrium, based on Hamiltonian formalism is presented. The equations of motion are solved numerically in the semi-classical limit. Several examples of the SU(2) lattice gauge dynamics are discussed: single plaquette system, the existence of periodic solution, instability of the the color fields and chaotic behavior.

I. Introduction

In the recent years, computer simulations of lattice gauge theories have become one of the most active areas in computational physics.[1] So far, these studies were mostly concerned with ground state properties and equilibrium systems. As a result many aspects of the complex dynamics of the gauge field were neglected. Here we present an extension of the existing computational methods which is capable of overcoming this drawback and may shed light of these aspects. It may be relevant in connection with the quark-gluon dynamics in relativistic nuclear collisions, which are becoming a major testing ground for nonperturbative SU(3) gauge theory. In particular, we hope that our studies will eventually provide insight into mechanisms responsible for energy dissipation in QCD and give information on relaxation times of field configurations far off equilibrium. Studies of the long-time behavior of gauge field dynamics on the lattice may also be helpful in better understanding the possible chaoticity of gauge field dynamics, which has been conjectured to be a source of the confinement properties of SU(N) gauge fields.[2-4] We have been motivated by related work in the fields of quantum optics and spin systems by R. G. Brown and M. Ciftan, who studied the interplay between dynamical evolution and external thermalization of quantum systems.[5,6] Finally, we mention that a similar analysis has recently been performed to understand dynamical properties of topological excitations in gauge field - Higgs systems.[7] An excellent and very detailed review of the current progress in lattice QCD was presented by Dr. C. DeTar at this Conference.[8]

In this paper we present the initial stage of the project, studying the dynamics of Hamiltonian SU(2) lattice gauge theory in the classical limit. In the next section we describe our model and derive the equations of motion. Then in the result section we demonstrate that the system displays chaotic behavior and show that the static electromagnetic fields are unstable and their average values tend to zero.

II. Equations of motion

The standard approach to lattice gauge computer simulations is through Lagrangian formulation where the space-time is discretized on a four dimensional lattice. The discretized time naturally limits the simulations to very short time-intervals and therefore this approach is not well suited for studying the dynamics of the gauge fields.

A better alternative is to use the Hamiltonian formulation of the lattice gauge theory. Working in the temporal gauge makes the time continuous variable and only the space is discrete.[9] The d - dimensional space is divided by a simple cubic lattice with L lattice sites and lattice spacing a. In the case presented here $d = 3$. On each link which connects two neighboring sites of the lattice a triplet of dynamical variables, A_l^a, is defined. In SU(2) lattice gauge theory, each directed link is associated with an element of SU(2) by

$$U_{\pm l} = \exp(\mp i \frac{1}{2} \tau^a A_l^a). \tag{1}$$

Here the plus or minus sign in U specifies the direction of the link, τ^a are the Pauli matrices and the summation convention has been assumed. These group elements are the fundamental variables and they as well as their time derivatives describe completely the system. The SU(2) Hamiltonian[1,9,10] is a function of the space components of the gauge fields and it is given by

$$H = \frac{g^2}{a} \left\{ \sum_l \frac{1}{2} E_l^a E_l^a + \lambda \sum_p [1 - \cos(\frac{1}{2} B_p)] \right\}, \tag{2}$$

where g is the coupling constant, $\lambda = 4/g^4$, and the magnetic field B_p is defined as

$$\cos(\frac{1}{2} B_p) = \frac{1}{2} \mathrm{tr} U_p. \tag{3}$$

Here U_p is the product of the four U's along the links of the plaquette p (see Figure 1). The electric field operators, E_l^a, satisfy the following commutation relations

$$[E_l^a, U_{l'}] = \frac{1}{2} \tau^a U_l \delta_{ll'} \tag{4}$$

$$[E_l^a, U_{l'}^{-1}] = -U_l \frac{1}{2} \tau^a \delta_{ll'} \tag{5}$$

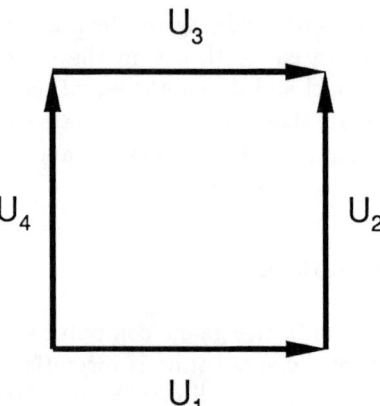

Figure 1: Relative positions of the SU(2) group elements on a plaquette.

and also satisfy the Lie algebra of SU(2)

$$[E_l^a, E_l^b] = -i\epsilon_{abc}E_l^c. \tag{6}$$

To derive the equations of motion we use the Heisenberg equation (with $\hbar = 1$) for the expectation values of the U_l operators

$$i\dot{U}_l = [U_l, H] \tag{7}$$

and the commutation relations (4, 6). After some algebra we arrive at

$$\dot{U}_l U_l^{-1} = i\frac{1}{2}\tau^a E_l^a. \tag{8}$$

This equation relates the expectation values of the color components of the electric color gauge field to the fundamental variables. Similarly, for the time derivative of electric field operator we have

$$\dot{E}_l^a = i\lambda \frac{g^2}{a} \sum_p [E_l^a, \cos\frac{1}{2}B_p]. \tag{9}$$

The expression for the last commutator has a very simple form[10] if we use the quaternion representation for the plaquette product

$$U_p = x_p^0 - i\tau^a x_p^a. \tag{10}$$

Then it becomes

$$[E_l^a, \cos\frac{1}{2}B_p] = \begin{cases} -i\frac{1}{2}x_{p,l}^a & \text{if } l \in p, \\ 0 & \text{otherwise} \end{cases}. \tag{11}$$

Here the second index, l, refers to the relative position of the link on the plaquette p (see Figure 1)

$$x^a_{p,1} = -x^a_{p,4} = x^a_p \tag{12}$$

$$x^a_{p,2} = R^{ab}(U_1)x^b_p \tag{13}$$

$$x^a_{p,3} = -R^{ab}(U_4)x^b_p. \tag{14}$$

Here

$$R^{ab}(U)y^b = (x^0 x^0 - x^c x^c)y^a + 2(x^0 \epsilon^{abc} y^b x^c + x^c y^c x^a). \tag{15}$$

Finally we get

$$\dot{E}^a_l = \lambda \frac{g^2}{2a} \sum_{p \ni l} x^a_{p,l}. \tag{16}$$

Differentiating eq.(8) with respect to time and combining it with the last equation yields the equations of motion:

$$\ddot{U}_l = W_l U_l \tag{17}$$

where we introduce a new quantity

$$W_l \equiv W^0_l - i\tau^a W^a_l \tag{18}$$

and the W's are defined as

$$W^0_l = -(\dot{U}^0_l \dot{U}^0_l + \dot{U}^a_l \dot{U}^a_l) \tag{19}$$

and

$$W^a_l = -\lambda \left(\frac{g^2}{2a}\right)^2 \sum_{p \ni l} x^a_{p,l}. \tag{20}$$

The equations of motions are derived from the principles of quantum mechanics. However, in order to solve them one needs to know the whole set of the link variables U_l as well as their time derivatives at any instant of time. The SU(2) group elements, U_l, are analogous to positions of particle, while their counterparts $\dot{U}_l$ act as momentum. It is clear that quantum mechanically, because of the uncertainty relation, this requirement is impossible to be satisfied. In this paper we employ the semi-classical approach, i.e. solve the quantum equations with classical (Newtonian) methods, as a first step toward understanding the full, complex dynamics of the quantum gauge fields. Moreover, since the lattice represents a regulator of the ultraviolet divergencies, ultimately one wants to remove this cutoff and to consider the continuum limit where the observable quantities tend to their physical values. From renormalization theory it is known that in the lattice gauge theory this corresponds to weak coupling limit. In this physically most interesting case and/or in the high temperature regime we believe that the classical approach yields reasonably adequate picture of the lattice gauge theory dynamics.

In the semi-classical approach, the lattice gauge theory is similar to molecular dynamics of system with four-body interactions. If we specify the initial conditions, i.e. the initial values of the U_l and their time derivatives, the dynamics of the system is completely deterministic. An extension of this approach to include the non-coherent thermal, stochastic bath interactions[5] is currently in progress.[14]

There are several different constants of motion and therefore not all of the Ld link variables are independent. In particular, for each lattice site s the Hamiltonian commutes with the operator

$$[(\nabla \cdot E)_s, H] = 0, \qquad (21)$$

where the lattice divergence of the electric field is the directed sum over all links emanating from s

$$(\nabla \cdot E)_s = \sum_{l \in s} E_l^a. \qquad (22)$$

Therefore

$$\frac{d}{dt}(\nabla \cdot E)_s = 0. \qquad (23)$$

This constraint for each of the lattice sites expresses the conservation of the color charge and it is often referred to as Gauss law. In the present study we are interested in the time evolution of *color-charge-free* gauge fields.

III. Computational aspects

There are different representations for an SU(2) group element. It can be viewed as a point on the 4-dimensional unit sphere and thus can be specified by 3 angles. It also can be represented as quaternion

$$U = x^0 - i\tau^a x^a \qquad (24)$$

where the four components are subject to additional condition

$$(x^0)^2 + (x^a)^2 = 1. \qquad (25)$$

We chose the last representation because the time consuming trigonometric calculations are avoided and also the equations of motion are simpler in this representation. The trade off is that it requires more memory to store the dynamical variables. If the lattice is a cube with N lattice points in each direction the number of all link variables is $4dN^d$.

The equations of motion are set of second-order differential equations. They can be rewritten as twice as many first-order coupled differential equations by introducing a new variable:

$$\frac{dU_l}{dt} = V_l. \qquad (26)$$

We integrate the set of equations numerically by the fourth-order Runge-Kutta method. Several criteria can be applied to test the accuracy of the integration. When the time step was kept of the order of 0.02 dimensionless units the total energy was conserved to better than 8 significant digits for the whole simulation. Another criterion is that the length of each link variable should not be time-dependent and remain 1 (see eq.25). For an interval less then 1 dimensionless time-unit the conservation was better than 12 significant figures. However, due to mainly accumulation of cut-off and rounding errors the precision was getting progressively worse. A special care is taken by rescaling to maintain fixed length of each dynamical variable since the subsequent integrations are very sensitive to the quaternion length preservation. This method allows to integrate the equations of motion with bigger time-step. The validity of the Gauss law (eq.23) is also an indication for accurate integration. The color charge was conserved to better than 10 significant figures.

For faster calculations of all sums in the equations of motion we built a list of the numbers of all plaquettes which contain a given link as well as list of the numbers of all 4 links which form given plaquette. The most time consuming part of the code is the calculation of W_l in eq.(17). This includes several matrix rotations and multiplications as well as calculation of the product of the group elements on each plaquette. However, this part of the code is fully vectorizable and runs at about 160 Mflops on a single Cray-Y-MP processor. Typically a single time-step integration of the set of equations of motion for $N = 10$ takes about 30 ms CPU time on the same processor.

Varying the initial conditions for the set of link variables $\{U_l\}$ and their time derivatives $\{\dot{U}_l\}$ different initial color field configurations can be modeled. However, the Gauss law, eq.(23) limits the possible initial conditions and often it is very hard to find a set of $\{U_l\}$ and $\{\dot{U}_l\}$ to yield source-free fields.

In the next section we present the results from computer simulation of the time evolution of a lattice gauge system ranging in size from a single plaquette to a $10 \times 10 \times 10$ lattice with periodic boundary conditions.

IV. Results

A. Single plaquette

The reason we are interested in a single plaquette system is two-fold. Its ground state is known exactly and thus it offers a testing ground for our model. In addition this small system is relatively simple and allows to better understand the dynamics in lattice gauge theory. The exact ground-state energy of a single plaquette is given by[10]:

$$\epsilon_0 \frac{a}{g^2} = \begin{cases} \lambda - \frac{1}{6}\lambda^2 + \frac{5}{432}\lambda^4 + \cdots & \lambda \ll 1 \\ \frac{3}{2}\lambda^{1/2} - \frac{21}{32} + \cdots & \lambda \gg 1 \end{cases} \qquad (27)$$

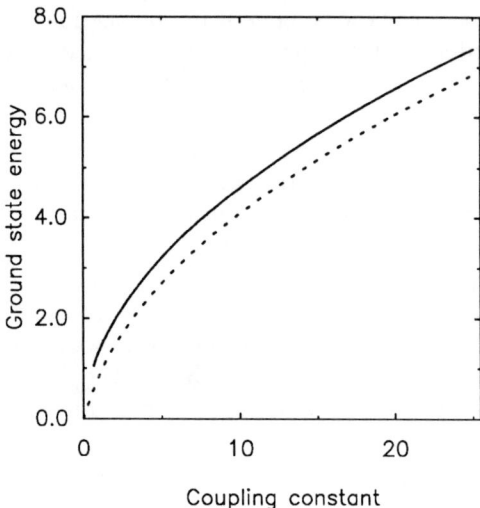

Figure 2: Comparison between the semi-classical (solid line) and quantum-mechanical (dashed line) ground state energies.

Semi-classically the ground state energy can be estimated by using the WKB relation

$$J = \frac{1}{2\pi} \oint p\,dq = (n + \frac{d}{2})\hbar \qquad (28)$$

with $n = 0$. Here J is the classical action, while p and q are the generalized momentum and coordinate.

Figure 2 presents a comparison between the two ground state energies. In the weak coupling limit the agreement is very good. This is not surprising – in this limit the effective potential (the magnetic energy) is almost quadratic and the system is like coupled harmonic oscillators where the WKB energies are exact.

We found that both electric and magnetic fields are always periodic in time. However, the frequency of the oscillations depends strongly on the total energy and resembles closely the behavior of a gravitational pendulum (Figure 3). As we will see in the next subsection this energy dependence is the major factor for the chaotic behavior in the system.

B. Periodic solutions

In the last decade a lot of effort was spent to study the classical solutions of the Yang-Mills field equations. A special class of periodic solutions was found recently.[11] However, all periodic gauge fields were spatially homogeneous. In

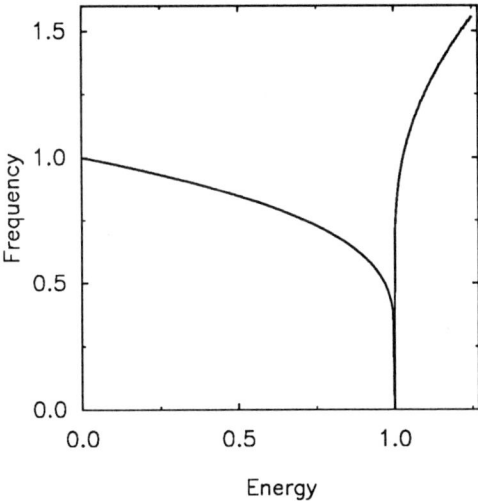

Figure 3: Plot of the frequency of oscillations as a function of total energy.

lattice gauge theory which is generalization of the the Yang-Mills equations we found that some special initial conditions lead to periodic solutions. Moreover, we found periodic solutions in which the fields are spatially inhomogeneous. An example for initial conditions which yields periodic solutions is:

$$\dot{U}_l = 0 \tag{29}$$

and

$$U_l = 1 \qquad \text{if } l \text{ points in } x \text{ or } z \text{ directions} \tag{30}$$

When l points in y direction eq.(1) is used with

$$A^a = A_0^a \sin(2\pi \frac{y}{aN})| \tag{31}$$

where A_0^a are arbitrary constants.

However, in all the cases the color components are proportional to each other (see Figure 4). Such a system is equivalent to a U(1) lattice gauge theory.

C. Chaotic behavior and instability of magnetic field

If there is more than a single plaquette, they interact through four-body, nearest-neighbor interactions. Due to the energy exchange the fields on each plaquette oscillate with frequency which varies in time and is a function of the

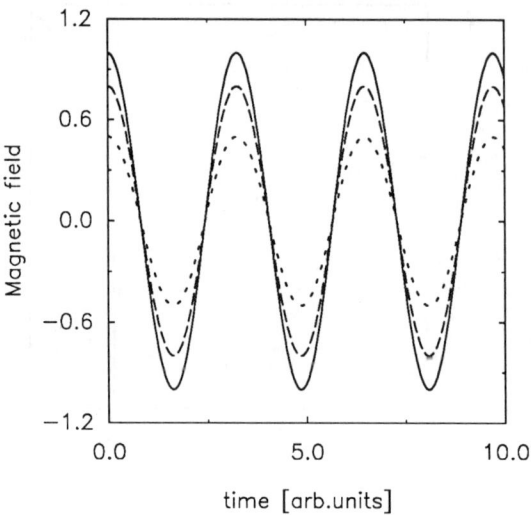

Figure 4: Periodic oscillations of the color components of the magnetic field.

fields on all the neighboring links and plaquettes. We found that in almost all cases the lattice gauge theory has an apparently chaotic behavior. This is not surprising, since the continuum SU(2) gauge theory has been shown to be a non-integrable dynamical system.[11] More detailed study is needed to determine Lyapunov exponents, Poincare surface of section, etc.

It is known from the perturbative theory that a constant magnetic field in the Yang-Mills equations is unstable against small perturbations.[13] The instability is caused by the strong attractive interaction between the chromomagnetic background field and chromomagnetic moment of an elementary field excitation, i.e. a gluon. Within the limits of this theory it is not possible to calculate the value to which a static magnetic field tends after a small fluctuation. We confirm this prediction of the perturbative theory and show that a small fluctuations triggers out-of-phase, non-correlated oscillations of the magnetic field in all plaquettes of the lattice. Different stages of the developing of magnetic field instability are illustrated on Figure 5. Soon after the onset of instability the average magnetic field vanishes (Figure 6).

V. Outlook

We present here a new method to study the real-time evolution in lattice gauge theory which is based on Hamiltonian formulation of the color fields and numerical solution of the equations of motion. It gives information about the

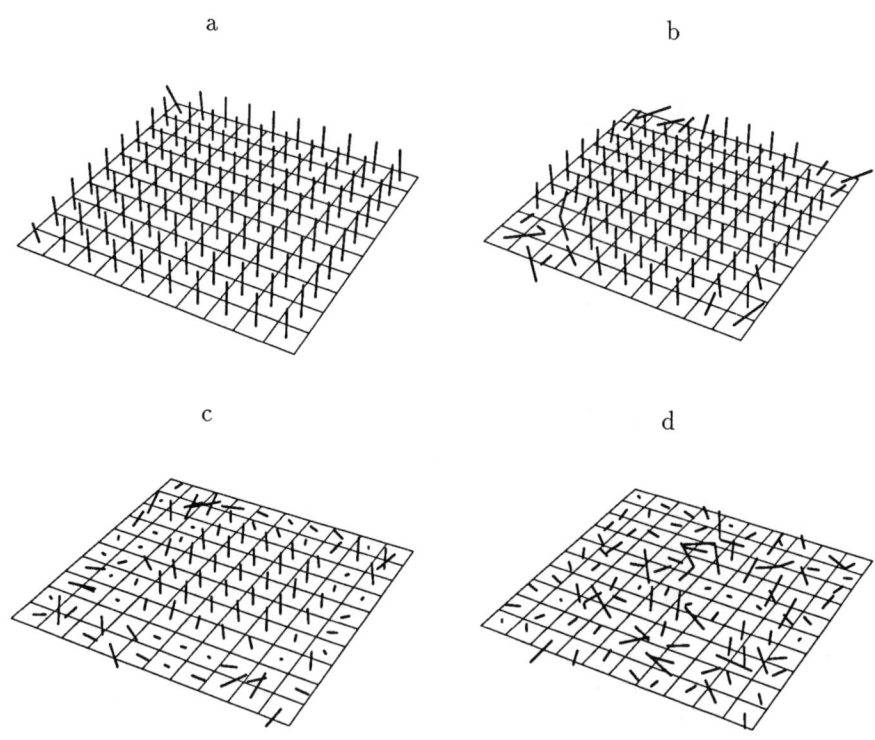

Figure 5: Initial stages of the color magnetic gauge field instability on a $10\times10\times10$ lattice. For better illustration only one lattice layer is shown. The triggering fluctuation of the link variable is taken to occur at the lower left corner. The 3 different SU(2) color field components are plotted in different spatial directions. a) $t = 0.05$; b) $t = 5.0$; c) $t = 10.0$ and d) $t = 20.0$ dimensionless time units.

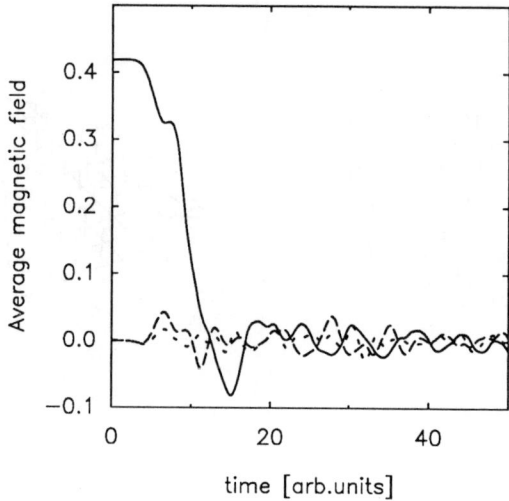

Figure 6: Magnetic gauge field averaged over the whole lattice as a function of time.

deterministic, coherent dynamics in lattice gauge system. The next step to follow will be to couple the system to a thermal bath which will permit to study the full dynamics of an open system. Similar method has been developed recently[5] and has been successfully applied to quantum optics and to ferromagnetic systems.

We believe that in its full extend the method will ultimately be capable of providing information on relaxation times to thermal and chemical equilibrium, and nucleation times in the presence of a phase transition.

Acknowledgements

This work was supported in part by the U.S. Department of Energy (grant DE-FG05-90ER40592) and by the North Carolina Supercomputer Center. We enjoy the fruitful discussions with Dr. T. Biro (Giessen), Dr. R. G. Brown (Duke) and Dr. Michael Ciftan (Duke/ARO).

References

[1] M. Creutz, *Quarks, gluons and lattices*, Cambridge University Press, London, 1983.

[2] H. G. Dosch, Phys. Lett. **B 190**, 177, (1987)

[3] Yu. A. Simonov, Nucl. Physics **B 307**, 512 (1988)

[4] K. Johnson, L. Lellouch and J. Polonyi, preprint

[5] R. G. Brown and M. Ciftan, Phys.Rev. **A 40**, 3080 (1989)

[6] R. G. Brown and M. Ciftan, Condensed Matter Theories 6, (Proceedings of the 1990 Condensed Matter Workshop at Elba, Italy) (in press)

[7] J. Ambjorn, T. Askgaard, H. Porter and M. E. Shaposhnokov, Nucl. Physics **B 353**, 346 (1991)

[8] C. DeTar, *Lattice QCD in 1990's*, invited lecture at Computational Quantum Physics Conference, May 23-25 1991, Nashville, TN

[9] J. Kogut and L. Susskind, Phys. Rev. **D 11**, 395 (1975)

[10] S. A. Chin, O. S. van Roosmalen, E. A. Umland and S. E. Koonin, Phys. Rev. **D 11**, 395 (1975)

[11] S. G. Matinyan, S. G. Savvidi, and N. G. Ter-Arutyunyan-Savvidi, Sov. Phys. JETF **53**, 421 (1981)

[12] A. J. Lichtenberg and M. A. Lieberman, Regular and Stochastic Motion, Springer, New York, 1983

[13] N. K. Nielsen and P. Olesen, Nucl. Physics, **B 144**, 376 (1978)

[14] A. Trayanov and B. Muller, to be published

MONTE CARLO SIMULATION OF A DYNAMICAL FERMION PROBLEM: THE LIGHT $q^2\bar{q}^2$ SYSTEM

G. Grondin
*Department of Physics, University of Toronto,
Toronto, Ontario, Canada M5S 1A7*
and
*Physics Division and Center for Computationally Intensive Physics,
Oak Ridge National Laboratory, Oak Ridge, TN 37831-6373*

We present results from a Guided Random Walk Monte Carlo simulation of the light $q^2\bar{q}^2$ system in a Coulomb-plus-linear quark potential model using an Intel iPSC/860 hypercube. A solvable model problem is first considered, after which we study the full $q^2\bar{q}^2$ system in $(J, I) = (2, 2)$ and $(2, 0)$ sectors. We find evidence for no bound states below the vector-vector threshold in these systems.

I. INTRODUCTION

The possible existence of light four-quark resonances ($q^2\bar{q}^2$, also referred to as baryonia) was first suggested in studies of duality diagrams[1] in the late 1960s. Through the 1970s other models, including the MIT bag model[2] and potential models with truncated color degrees of freedom, predicted that a rich spectrum of four-quark resonances should exist. This was not verified experimentally. The unphysical feature of these models was that they did not allow "fall-apart" decays, so that the four-quark system was not allowed to dissociate into two separate $q\bar{q}$ mesons. The binding implicit in the models held the four quarks together as a single system and generated a tower of excited states.

There is however experimental evidence for two multiquark mesons. The $f_0(975)$ and $a_0(980)$ $J^{PC} = 0^{++}$ resonances were discovered in the early 1960s[3], and could not easily be described as $q\bar{q}$ states. These states were unusual because they were just below $K\bar{K}$ threshold and coupled more strongly to strange final states than to nonstrange states, contrary to expectations for $\frac{1}{\sqrt{2}}(u\bar{u} \pm d\bar{d})$ $^3P_0$ $q\bar{q}$ mesons. Following early suggestions by Jaffe and Johnson[4] that these might be four-quark baryonium states, Weinstein and Isgur[5] used the nonrelativistic quark model[6] in a variational calculation of the four-quark system in spin-0, isospin-0 and -1 sectors, which led them to identify the $f_0(975)$ and $a_0(980)$ with the loosely-bound $K\bar{K}$ states found in this calculation. Other theoretical and experimental work (for example two-photon decays[7]) also supported this conclusion. Weinstein and Isgur also found that the light $q^2\bar{q}^2$ system in the spin-0 sector had no excited metastable states above these "$K\bar{K}$ molecules" other than the two-meson continuum, which explained why no other four-quark states were seen experimentally.

Approximate solutions of QCD and the four-quark problem have of course been attempted using other methods. One very promising approach is lattice QCD, which simulates the full theory in terms of the QCD Lagrangian on a space-time lattice. This lattice formulation can be studied using Monte Carlo methods to estimate the masses of low-lying states such as the π, ρ, and glueballs. However, these algorithms suffer from a nonlocal equivalent action when applied to dynamical fermions; this leads to slow updates and relatively noisy simulations. The accuracy required to measure the small binding energies expected if the four-quark ground state is a "molecule" of two mesons (perhaps a few tens of MeVs) are not yet attainable using unquenched QCD simulations.

Y.G.Liang et al[8] used lattice QCD Monte Carlo to search for light $q^2\bar{q}^2$ resonances below vector-vector thresholds in $(J,I) = (0,0)$, $(0,2)$, $(2,0)$, and $(2,2)$. Although they found possible evidence for a bound state in the $(J,I) = (0,0)$ sector, they noted that its mass equals their $\pi\pi$ threshold energy and hence may be an artifact due to transitions to $\pi\pi$ states. Their lattice was modest in size at only $16^3 \times 24$ sites and their quark masses were heavy compared to the physical situation. This demonstrates some useful properties of simulating a Hamiltonian potential model. In a potential model, we can set various Hamiltonian terms to zero to prevent transition to unwanted states and our quark masses can easily be set to currently accepted values.

Another approach, which we follow in the present work, is to attempt a Monte Carlo simulation of "QCD-inspired" potential models, which are based on phenomenological interquark potentials and nonrelativistic dynamics. Since these models are not equivalent to QCD, care must be taken to insure that the relevant physics is incorporated, following which the model predictions can then be tested by comparison with the experimental spectrum of resonances.

Other Monte Carlo studies of the four-quark system in potential models have appeared in the literature, albeit with important simplifications. Carlson, Heller, and Tjon[9] studied four-quark systems in which the two quarks were heavy (c and b) and the antiquarks were light ($\bar{u}$ and $\bar{d}$). They carried out a Green's Function Monte

Carlo (GFMC) simulation of the Born-Oppenheimer approximation to the bag model and found that the four-quark system was lighter than two separate mesons (*i.e.* a bound state is present). Note however that the four-quark system was again not allowed to decay into two mesons by their Ansatz – this is what originally led to the erroneous prediction of many four-quark resonances. Carlson and Pandeharipande[10] used Variational Monte Carlo and GFMC to study light (u and d) multiquark states in a strong-coupling QCD flux-tube potential model. They found no bound states, not even for six quarks; since the deuteron exists, their results imply that either their QCD flux-tube model is an inadequate description of multiquark states or their solution is inaccurate. A possible source of error is that this model mixes the two distinct color basis states only at fourth order in the flux-tube breaking Hamiltonian. This mixing between color ground-state basis vectors may be too weak, so that one finds no binding in systems with more than three quarks. (This problem does not arise in meson and baryon states, which have unique color states.)

In the next section we briefly describe the QCD-inspired quark potential model and discuss some of the relevant physics of the four-quark system. This is followed by a description of the Guided Random Walk (GRW) Monte Carlo algorithm and the modifications which we implemented for our simulation of dynamical fermions. In the fourth section, a simple model Hamiltonian will be introduced and solved as a test case, following which we quote results for the full four-quark problem.

II. THE HAMILTONIAN

We employ a QCD-inspired quark potential model which has been very successful in describing meson and baryon physics[6,11]. The success of Weinstein and Isgur's description of the $f_0(975)$ and $a_0(980)$ as $K\bar{K}$ bound states also motivates further study of the model as applied to $q^2\bar{q}^2$ states. The Hamiltonian is given by

$$H = \sum_{i=1}^{4}(m_i + \frac{P_i^2}{2m_i}) + \sum_{i<j}(V_{conf}^{ij} + V_{eff}^{ij}), \qquad (1)$$

where

$$V_{conf}^{ij} = -(C + \frac{3}{4}b\, r_{ij})\, \vec{F}_i \cdot \vec{F}_j \qquad (2)$$

contains constant and linear confining terms, and

$$V_{eff}^{ij} = V_{Coul}^{ij} + V_{hyp}^{ij} + V_{so}^{ij} \qquad (3)$$

is a one-gluon-exchange term with

$$V_{Coul}^{ij} = \frac{\alpha_s}{r_{ij}}\, \vec{F}_i \cdot \vec{F}_j \qquad (4)$$

and

$$V_{hyp}^{ij} = -\frac{\alpha_s}{m_i m_j} \left[\frac{8\pi}{3} \vec{S}_i \cdot \vec{S}_j \, \delta^3(r_{ij}) + V_{ten}^{ij} \right] \vec{F}_i \cdot \vec{F}_j. \tag{5}$$

These terms suffice for a surprisingly accurate description of light nonrelativistic hadron spectroscopy. We choose to neglect the usually unimportant tensor and spin-orbit terms, V_{ten}^{ij} and V_{so}^{ij}, because their contribution to a dominantly S-wave ground state is expected to be much smaller than that of the spin-spin term.

Since we are simulating a Schrödinger equation numerically, the delta-function must be smoothed if we are to obtain a physically realistic spectrum. Our replacement

$$\delta^3(r_{ij}) \longrightarrow \frac{\sigma_{hf}^3}{\pi^{3/2}} \exp(-\sigma_{hf}^2 r_{ij}^2) \tag{6}$$

can also be motivated as a "smearing" of the delta-function caused by Zitterbewegung. It is also necessary to regulate the color Coulomb term V_{Coul}^{ij} at contact. We substitute

$$\frac{1}{r} \longrightarrow \begin{cases} \frac{1}{r} & r > h_x \\ \frac{1}{h_x} & 0 \leq r \leq h_x \end{cases} \tag{7}$$

where h_x is the spatial lattice size, which we subsequently extrapolate to zero.

For the present we consider only light u and d quarks in the equal-mass limit. We must construct totally antisymmetric four-quark states, which are products of spatial (ψ), spin (χ), flavor (ϕ), and color (C) wavefunctions. Each carries a subscript S or A, which denotes a symmetric or antisymmetric state under exchange of quarks (labelled 1 and 2) or antiquarks (labelled 3 and 4).

There are two "natural" color bases which we might employ (see Figure 1). The $(|1_{13}1_{24}\rangle, |1_{14}1_{23}\rangle)$ basis corresponds to the "color flux lines" that can be drawn for a system of two separate mesons; this is the more physical basis if we expect the four-quark system to be molecule-like or unbound. These two $|1,1\rangle$ basis states are not orthogonal in color space, and this complication leads to nondiagonal kinetic terms which are not easily simulated by the GRW algorithm. We therefore use the orthogonal basis $(|\bar{3}_{12}3_{34}\rangle, |6_{12}\bar{6}_{34}\rangle)$ in which the fermion symmetry is explicit; interpretation of the results in terms of color-singlet meson states is, unfortunately, rather more difficult in this basis. The relation between the two bases is

$$|1_{13}1_{24}\rangle = \sqrt{\frac{1}{3}} |\bar{3}_{12}3_{34}\rangle + \sqrt{\frac{2}{3}} |6_{12}\bar{6}_{34}\rangle, \tag{8}$$

$$|1_{14}1_{23}\rangle = -\sqrt{\frac{1}{3}} |\bar{3}_{12}3_{34}\rangle + \sqrt{\frac{2}{3}} |6_{12}\bar{6}_{34}\rangle.$$

296 The Light $q^2\bar{q}^2$ System

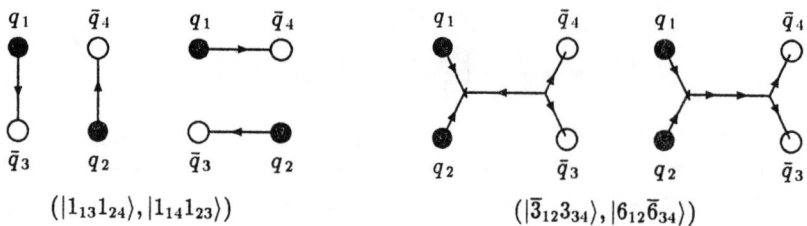

$(|1_{13}1_{24}\rangle, |1_{14}1_{23}\rangle)$ $(|\bar{3}_{12}3_{34}\rangle, |6_{12}\bar{6}_{34}\rangle)$

FIG. 1. Color bases.

From four spin-$\frac{1}{2}$ objects with no orbital excitation we can construct states with total spin $J = 0, 1$ and 2. We neglect orbital angular momentum in listing the possible J values because dominantly S-wave bound states are most likely *a priori*. We shall first consider the $J = 2$ case. These states have a symmetric spin wavefunction χ_S, and are diagonal in the spin-spin Hamiltonian,

$$\vec{S}_i \cdot \vec{S}_j |J = 2, J_m\rangle = \frac{\hbar^2}{4} |J = 2, J_m\rangle. \tag{9}$$

The flavor wavefunctions have total isospin $I = 0$, 1, or 2. Here we consider only $I = 0$ and $I = 2$, which are totally symmetric or antisymmetric under identical fermion exchange. The $I = 1$ states, in contrast, have opposite exchange symmetry for the quarks and antiquarks. This more complicated symmetry requires a more complicated spatial wavefunction to maintain the overall fermion antisymmetry. (The color and spin states are already fixed.)

Finally, the spatial wavefunction $\psi(\vec{x}, \vec{y}, \vec{z})$ can be shown to have definite symmetry in $\vec{x}$ and $\vec{y}$ under identical fermion exchange. Figure 2 shows the definition of these Weinstein-Isgur coordinates; the center-of-mass coordinate can be factored out of the Hamiltonian and need not be considered. The separation between the center-of-mass coordinates of specified $q\bar{q}$ pairs is given by the coordinate $\vec{x}$ (or $\vec{y}$); if the two-meson system is unbound, then the expected $\vec{x}$ or $\vec{y}$ must diverge.

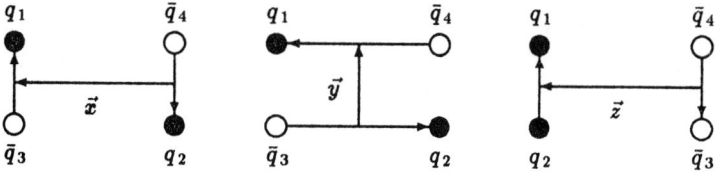

FIG. 2. Spatial coordinates.

To illustrate the spatial symmetry of ψ, consider the $(J,I) = (2,2)$ system; we specialize to the $\rho^+\rho^+$ sector, which has flavor and spin states $|\phi_S\rangle = |uu\bar{d}\bar{d}\rangle$ and $|\chi_S\rangle = |\uparrow\uparrow\uparrow\uparrow\rangle$. One might *a priori* consider writing the state as $\psi(\vec{x},\vec{y},\vec{z})|1_{13}1_{24}\rangle$, but this does not give a correctly symmetrized state. If we apply identical-fermion exchange symmetry to the ground-state wavefunction (which we denote by $|\rho^+\rho^+\rangle$ for simplicity), we can rewrite it as

$$|\rho^+\rho^+\rangle = \big(\psi(\vec{x},\vec{y},\vec{z}) + \psi(-\vec{x},-\vec{y},\vec{z})\big)|1_{13}1_{24}\rangle \pm \big(\psi(\vec{y},\vec{x},\vec{z}) + \psi(-\vec{y},-\vec{x},\vec{z})\big)|1_{14}1_{23}\rangle; \quad (10)$$

the $\pm$ determines the symmetry of the wavefunction under exchange of one pair of quarks or antiquarks. For $|\rho^+\rho^+\rangle$ we must chose the minus sign but in the $(J,I) = (2,0)$ sector we can find an antisymmetric flavor function which allows the positive sign. For the negative case in (10) we can rewrite the state as

$$|\rho^+\rho^+\rangle = \sqrt{\frac{1}{3}}\,\psi_S(\vec{x},\vec{y},\vec{z})|\bar{3}_{12}3_{34}\rangle + \sqrt{\frac{2}{3}}\,\psi_A(\vec{x},\vec{y},\vec{z})|6_{12}\bar{6}_{34}\rangle. \quad (11)$$

This uses the color basis $(|\bar{3}_{12}3_{34}\rangle, |6_{12}\bar{6}_{34}\rangle)$ and introduces the symmetrized combinations

$$\psi_S(\vec{x},\vec{y},\vec{z}) \equiv \big(\psi(\vec{x},\vec{y},\vec{z}) + \psi(-\vec{x},-\vec{y},\vec{z})\big) + \big(\psi(\vec{y},\vec{x},\vec{z}) + \psi(-\vec{y},-\vec{x},\vec{z})\big), \quad (12)$$

$$\psi_A(\vec{x},\vec{y},\vec{z}) \equiv \big(\psi(\vec{x},\vec{y},\vec{z}) + \psi(-\vec{x},-\vec{y},\vec{z})\big) - \big(\psi(\vec{y},\vec{x},\vec{z}) + \psi(-\vec{y},-\vec{x},\vec{z})\big).$$

Equation (12) explicitly shows the symmetry of ψ_S and ψ_A under $(\vec{x},\vec{y})$ interchange, and also shows that $\psi_S(\vec{x},\vec{y},\vec{z}) = \psi_S(-\vec{x},-\vec{y},\vec{z})$ and $\psi_A(\vec{x},\vec{y},\vec{z}) = \psi_A(-\vec{x},-\vec{y},\vec{z})$.

III. THE GUIDED RANDOM WALK ALGORITHM

The guided random walk (GRW) Monte Carlo algorithm was introduced by Barnes, Daniell, and Storey[12] as an alternative to branching algorithms. The earlier versions of GRW considered systems having either continuous[12] or discrete[13] degrees of freedom but did not consider systems with both in the same Hamiltonian (for a summary of both versions see Barnes[14]). The modified GRW algorithm presented here allows the simulation of systems having both types of state.

First consider an unguided RW method based on a Euclidean form of the Schrödinger equation; on taking $it \to \tau$ it becomes a diffusion equation,

$$-\hbar\frac{\partial}{\partial\tau}|\psi(\tau)\rangle = H|\psi(\tau)\rangle. \quad (13)$$

Such a diffusion equation can then be simulated by generating random walks on a space-time lattice (Figure 3), as first proposed by E.Fermi[15], and weighting each walk with an appropriate factor. (This is similar to the e^{-S} weight in path integrals.) In the limit h_x, $h_\tau \to 0$ with h_τ/h_x^2 constant, the weighted RW process can be made equivalent to the diffusion process (13). In a simulation of the diffusion equation (13), the ground-state wavefunction and energy are generated in the large-τ limit, as can be seen from an eigenmode expansion of the initial wavefunction. We assume that the initial wavefunction has a nonzero overlap with the true ground state,

$$|\psi(0)\rangle = \sum_n c_n |\psi_n\rangle, \qquad (14)$$

so that

$$|\psi(\tau)\rangle = \sum_n c_n |\psi_n\rangle e^{-E_n\tau}, \qquad (15)$$

which implies

$$\lim_{\tau \to \infty} |\psi(\tau)\rangle = c_0 e^{-E_0\tau} |\psi_0\rangle + O(e^{-(E_1-E_0)\tau}). \qquad (16)$$

An unweighted random walk simulates a diffusion equation with no potential. To include a potential, each walk is given a weight which is determined by the potential V and the path of the walk. The ground-state energy and wavefunction are found using the expected weight and a weight-factor histogram. In the unguided algorithm these weights have widely scattered values (again see Barnes[14]); the wider the scatter of the weights, the greater the computational effort required to achieve a prescribed statistical accuracy.

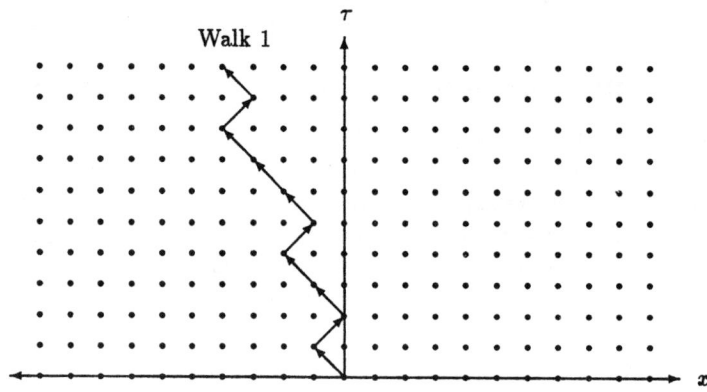

FIG. 3.   A random walk on the lattice.

The GRW algorithm attempts to improve the statistical accuracy by introducing a trial "guiding" wavefunction, which guides the walks into regions where the ground-state ψ_0 is expected to be large. The guiding wavefunction does not introduce a bias in the determination of the ground-state energy or matrix elements, since a generalized weight is introduced which compensates for the effects of the guidance. Formally the algorithm used here is essentially that of Barnes and Daniell[13]; only a few changes are required for our simulation.

The walk is performed on a lattice (with coordinate resolution h_x) in the configuration space $Z = (x_i, \mathcal{S})$; x_i is the $q^2\bar{q}^2$ ($\vec{x}, \vec{y}, \vec{z}$) coordinate space, and $\mathcal{S}$ is the color $\otimes$ isospin $\otimes$ spin state (only two such states are present). At each time step h_τ, the probability to go from Z at time τ to Z' at time $\tau + h_\tau$ is written as

$$p_{Z \to Z'} = r_{Z \to Z'} h_\tau \tag{17}$$

with h_τ determined from conservation of probability,

$$\sum_{Z'} p_{Z \to Z'} = 1. \tag{18}$$

Each walk is assigned a weight factor W_{tot} which is given by

$$W_{tot} = W_{diag} \cdot W_{tran}, \tag{19}$$

$$W_{diag} = \exp\left\{-\int_0^\tau \left(V_0(Z(\tau')) - \sum_{Z' \neq Z} r_{Z \to Z'}\right) d\tau'\right\},$$

$$W_{tran} = \prod_{\substack{transitions \\ Z \to Z' \neq Z}} \left(-\frac{\langle Z'|H_I|Z\rangle}{r_{Z \to Z'}}\right),$$

where $H = H_0 + H_I$, and $H_0(Z) \equiv V_0(Z) = \langle Z|H_0|Z\rangle$. (By definition H_0 is diagonal and H_I is off-diagonal in state space.)

Previously Barnes[14] chose to specialize $r_{Z \to Z'}$ to a particular form in terms of a "guiding wavefunction" ψ_g,

$$r_{Z \to Z'} = -\langle Z'|H_I|Z\rangle \frac{\psi_g(Z')}{\psi_g(Z)}. \tag{20}$$

This form simplifies W_{tran}, and in the "perfect guidance" case in which ψ_g is equal to the true ground-state wavefunction, sets the diagonal weight exactly equal to $\exp(-E_0\tau)$. As the energy is determined from the τ-dependence of W_{tot}, equation (20) makes W_{tran} a function of Z only, and it can therefore be discarded in energy measurements. In practice, the true ground-state wavefunction is not known and is approximated by a parametrized guiding wavefunction ψ_g; we then vary the parameters of ψ_g and search for a minimum of the variance of the diagonal weights. (Zero variance in W_{diag} indicates that ψ_g actually equals an energy eigenstate wavefunction.) In this $q^2\bar{q}^2$ problem, however, the right hand side of (20) is not always positive definite, whereas the stepping probability $r_{Z\to Z'}h_\tau$ must be. To force $r_{Z\to Z'}$ to be positive we instead choose

$$r_{Z\to Z'} = \left| -\langle Z'| H_I |Z\rangle \frac{\psi_g(Z')}{\psi_g(Z)} \right|, \tag{21}$$

which requires

$$W_{tran} = \left| \frac{\psi_g(Z(0))}{\psi_g(Z(\tau))} \right| \prod_{\substack{transitions \\ Z\to Z'\neq Z}} \text{sign}\left(-\langle Z'| H_I |Z\rangle\right). \tag{22}$$

This form allows a dynamical generation of the wavefunction's nodal structure, since both positive and negative weights are generated by the algorithm if some $\langle Z'| H_I |Z\rangle$ matrix elements are positive. It also retains the zero-variance property for $\psi_g = \psi_0$ which $r_{Z\to Z'}$ defined by (20) possessed for negative definite or zero $\langle Z'| H_I |Z\rangle$.

The ground-state energy is estimated from the τ-dependence of the average sign-weighted diagonal weight

$$\left\langle W_{diag}\,\text{sign}(W_{tran})\right\rangle = \frac{1}{N_{rw}} \sum_i^{N_{rw}} \left\{ W_{diag}^i \prod_{\substack{i\,transitions \\ Z\to Z'\neq Z}} \text{sign}(-\langle Z'| H_I |Z\rangle) \right\}; \tag{23}$$

$$\lim_{\tau\to\infty} \left\langle W_{diag}(\tau)\,\text{sign}(W_{tran})\right\rangle \propto \exp(-E_0 \tau). \tag{24}$$

Unbiased matrix elements can in principle be determined from W_{tot} (see Barnes[14]), although we shall only consider energies here.

If we evaluate equation (24) directly, the statistical error in the energy could be large, due to cancellations between negative and positive weights. As the energy is determined from the τ-dependence of the expected weight alone, we can multiply the final histogram of weights by any function of the coordinate Z (which we call a "masking function") to reduce the fraction of negative weights and hence to increase the statistical accuracy in the energy. We normally attempt to approximately reproduce the nodal structure of ψ_0 in the masking function. Note however that no bias will result if the nodal structure of the masking function is not identical to that of ψ_0. (Actually in this problem we do know the exact nodes, which we will discuss subsequently.)

The implementation of the code is parallel, in that we run n copies of the program, each generates a set of walks on its own "node" (CPU), and the resulting weights are summed at the end of the run. It is possible to run one program per node because the algorithm has very small memory requirements; this avoids the complications of internode communication.

IV. RESULTS

Before considering the full Hamiltonian (1), we first consider an analytically-solvable Hamiltonian which is very similar to that introduced by Weinstein and Isgur[5]. Our test problem is defined by

$$\tilde{H}_0 = \left[4m - \frac{\hbar^2}{2m}\left(\nabla_{\vec{x}}^2 + \nabla_{\vec{y}}^2 + \nabla_{\vec{z}}^2\right) + \frac{1}{2}(2k)(x^2+y^2) + \frac{1}{2}(4k)z^2\right]\begin{pmatrix}1&0\\0&1\end{pmatrix}$$
$$+\frac{1}{2}k_n(y^2-x^2)\begin{pmatrix}0&1\\1&0\end{pmatrix}, \qquad (25)$$

whereas the test problem of Weinstein and Isgur[5] instead substitutes

$$V_{conf}^{ij} + V_{eff}^{ij} \longrightarrow -(C + \frac{4}{3}(\frac{1}{2}kr_{ij}^2))\vec{F}_i\cdot\vec{F}_j, \qquad (26)$$

in (1), which produced an intractable Hamiltonian. Our solvable $\tilde{H}_0$ is their (26) without a diagonal term. The energy of a general eigenstate of $\tilde{H}_0$ is

$$E_{ijk} = 4m + \frac{\hbar}{2\sqrt{m}}\left(\sqrt{2k-k_n}(2i_1+2i_2+2i_3+3)\right.$$
$$\left.+\sqrt{2k+k_n}(2j_1+2j_2+2j_3+3) + \sqrt{4k}(2k_1+2k_2+2k_3+3)\right), \qquad (27)$$

where

$$\gamma_1 = \frac{1}{2\hbar}\sqrt{m(2k-k_n)} \qquad (28)$$
$$\gamma_2 = \frac{1}{2\hbar}\sqrt{m(2k+k_n)}$$
$$\gamma_3 = \frac{1}{2\hbar}\sqrt{m(4k)}.$$

If we define

$$H_{i;l}(\vec{x}) \equiv H_{i_1}(\sqrt{2\gamma_l}\,x_1)\,H_{i_2}(\sqrt{2\gamma_l}\,x_2)\,H_{i_3}(\sqrt{2\gamma_l}\,x_3), \qquad (29)$$

where $H_i(x)$ is a Hermite polynomial, then the energy eigenfunction corresponding to (27) is

$$\left|\psi_{ijk}^{\pm}\right\rangle = \begin{bmatrix}H_{i;1}(\vec{x})H_{j;2}(\vec{y})\exp(-\gamma_1 x^2 - \gamma_2 y^2) \mp H_{j;2}(\vec{x})H_{i;1}(\vec{y})\exp(-\gamma_2 x^2 - \gamma_1 y^2)\\ H_{i;1}(\vec{x})H_{j;2}(\vec{y})\exp(-\gamma_1 x^2 - \gamma_2 y^2) \pm H_{j;2}(\vec{x})H_{i;1}(\vec{y})\exp(-\gamma_2 x^2 - \gamma_1 y^2)\end{bmatrix}$$
$$\times H_{k;3}(\vec{z})\exp(-\gamma_3 z^2). \qquad (30)$$

Our test $\tilde{H}_0$ has the same degrees of freedom as the full Hamiltonian (1), and also has terms of either sign off-diagonal, as is expected in a dynamical fermion problem. It also has a "fall-apart" solution; the ground-state wavefunction is independent of the relative-separation coordinate $\vec{x}$ when $2k = k_n$.

Figure 4 shows the Euclidean-time dependence of the extracted ground-state energy with a fixed lattice size; in practice we increase τ until evidence for τ-dependence has disappeared to the required accuracy, which indicates that contributions from excited states have become unimportant. We then extrapolate the energy to zero lattice size at this value of τ, as shown in Figure 5. The algorithm evidently finds the correct ground-state energy to our statistical accuracy of about $0.1\ MeV$. We used $|\psi_{000}^+\rangle$ from (30) as our guiding wavefunction for this simulation; this guiding function equals the true ground state only when $h_x = 0$, and we find accordingly that the variance of the diagonal weights does indeed go to zero as $h_x \to 0$.

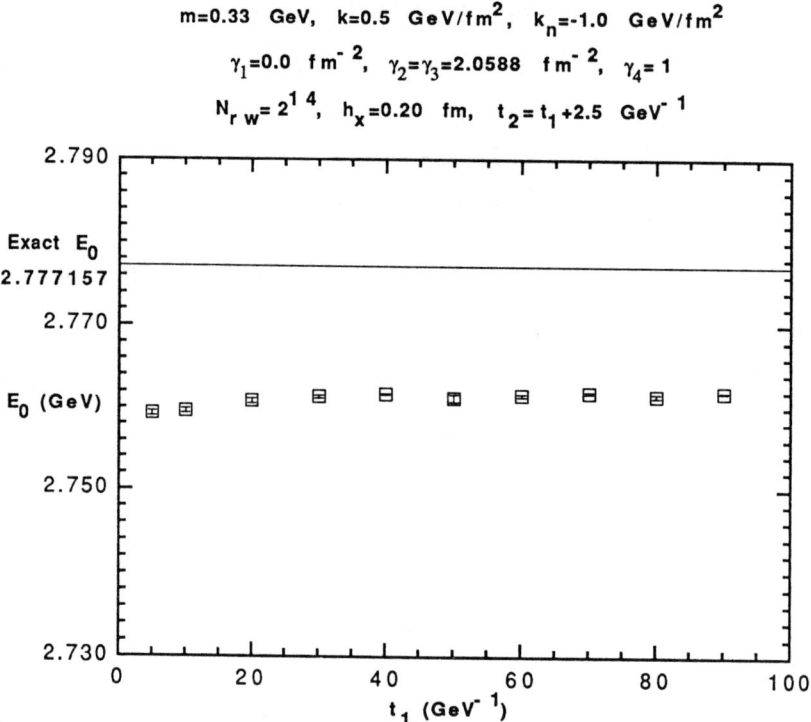

FIG. 4. $E_0(\tau)$ for the analytic test case. See also the h_x^2 extrapolation in Figure 5.

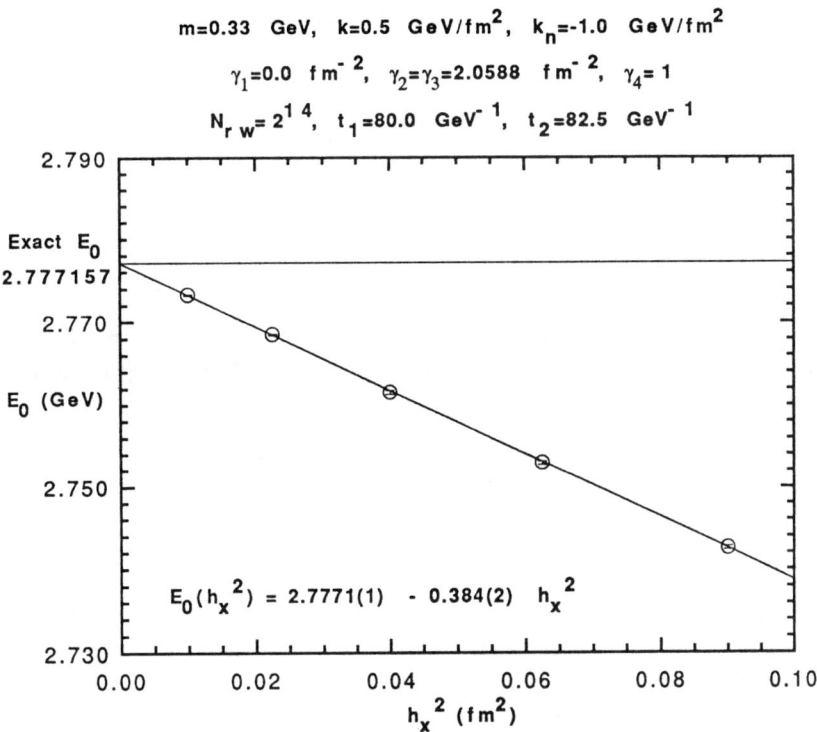

FIG. 5. $E_0(h_x^2)$ for the analytic test case.

One of our principal concerns in solving equation (1) with the GRW algorithm is to insure that ψ_g does not impose an incorrect nodal structure. In a general multifermion problem the nodes are usually unknown. For this problem we can use the eigenfunctions of $\tilde{H}_0$ to exactly determine the nodes of the ground-state eigenfunction of (1).

We will treat (1) as a perturbation $\tilde{H}_I$ about (25); we must restrict the eigenfunction basis set of (30) to eigenfunctions for which $i_1 + i_2 + i_3 + j_1 + j_2 + j_3$ is even, in order to satisfy the constraint $\psi(\vec{x}, \vec{y}, \vec{z}) = \psi(-\vec{x}, -\vec{y}, \vec{z})$. Note that $\tilde{H}_I$ can be written as

$$\tilde{H}_I = \begin{bmatrix} f_1(\vec{x}, \vec{y}, \vec{z}) & f(\vec{x}, \vec{y}, \vec{z}) \\ f(\vec{x}, \vec{y}, \vec{z}) & f_2(\vec{x}, \vec{y}, \vec{z}) \end{bmatrix}, \quad (31)$$

where

$$f_m(\vec{x}, \vec{y}, \vec{z}) = f_m(-\vec{x}, -\vec{y}, \vec{z}) \quad (32)$$

$$= f_m(\vec{y}, \vec{x}, \vec{z})$$

and

$$f(\vec{x}, \vec{y}, \vec{z}) = f(-\vec{x}, -\vec{y}, \vec{z}) \qquad (33)$$
$$= -f(\vec{y}, \vec{x}, \vec{z}).$$

From this form we can easily show that

$$\langle \psi^-_{ijk} | \tilde{H}_I | \psi^+_{lmn} \rangle = 0. \qquad (34)$$

We may now apply perturbation theory in $\tilde{H}_I$ to determine properties of the eigenstates of $\tilde{H}$. Let the eigenstates of $\tilde{H}_0$ be written as $|n^\pm\rangle$ with eigenvalue $E_n(0)$. The eigenstates of $\tilde{H}$ can then be written as

$$E = E(0) + \lambda E(1) + \lambda^2 E(2) + \cdots \qquad (35)$$

and

$$|\Psi\rangle = |\Phi_0\rangle + \lambda|\Phi_1\rangle + \lambda^2|\Phi_2\rangle + \cdots, \qquad (36)$$

where $|\Phi_0\rangle = |\psi^+_{000}\rangle$ and $\langle \Phi_n|\Phi_0\rangle = \delta_{n0}\ \forall n$. (We can also determine $|\psi^-_{000}\rangle$ by changing $+$ to $-$ where appropriate.) We need only determine the ground-state wavefunction, since the GRW algorithm converges to that state. The perturbative results for E and $|\Psi\rangle$ are given by

$$E(n) = \langle \Phi_0 | \tilde{H}_I | \Phi_{n-1} \rangle \qquad (37)$$

and

$$|\Phi_n\rangle = [E(0) - \tilde{H}_0]^{-1}[I - |\Phi_0\rangle\langle\Phi_0|] \left[[\tilde{H}_I - E(1)]|\Phi_{n-1}\rangle \right.$$
$$\left. + E(2)|\Phi_{n-2}\rangle + \cdots + E(n-1)|\Phi_1\rangle \right]. \qquad (38)$$

Using (34), we can easily show that the first- and second-order perturbative contributions to the ground-state wavefunction are

$$|\Phi_1\rangle = \sum_{n \neq 0} \frac{|n^+\rangle \langle n^+ | \tilde{H}_I | \Phi_0 \rangle}{E(0) - E_n(0)} \qquad (39)$$

and

$$|\Phi_2\rangle = \sum_{n \neq 0} \sum_{m \neq 0} \frac{|n^+\rangle \langle n^+ | (\tilde{H}_I - E(1)) | m^+ \rangle \langle m^+ | \tilde{H}_I | \Phi_0 \rangle}{(E(0) - E_n(0))(E(0) - E_m(0))}. \qquad (40)$$

Neither of these has a contribution from the $\{|n^-\rangle\}$ states. One may prove by induction that no $|\Phi_n\rangle$ has a contribution from the $\{|n^-\rangle\}$ states[16]. The ground-state wavefunction may therefore be written as

$$|\Psi\rangle = |0^+\rangle + \sum_{n\neq 0} c_n |n^+\rangle. \qquad (41)$$

Using (30) and letting $S = \frac{1}{2}(\gamma_2 + \gamma_1)$ and $D = \frac{1}{2}(\gamma_2 - \gamma_1)$, we find that the node in the upper component of (41) is determined by

$$0 = 2\sinh(D(x^2 - y^2)) + \sum_{ijk} c_{ijk} \left[H_{i;1}(\vec{x}) H_{j;2}(\vec{y}) \exp(D(x^2 - y^2)) \right.$$
$$\left. - H_{j;2}(\vec{x}) H_{i;1}(\vec{y}) \exp(-D(x^2 - y^2)) \right] H_{k;3}(\vec{z}). \qquad (42)$$

As $D \neq 0$ in general, this equation is only satisfied by $\vec{x} \pm \vec{y} = 0$, or equivalently by $x^2 - y^2 = 0$. If we now consider finding the eigenvectors of $\tilde{H}_I$ alone, we can use the eigenfunctions (30) as the basis eigenvectors, since they span the configuration space and satisfy the symmetry requirement (12) for eigenvectors of $\tilde{H}_I$. The ground-state eigenvector can therefore be written in the form (41), which implies that the nodes of the ground-state eigenfunction of (1) are given by $x^2 - y^2 = 0$. This motivates our choice of the ground-state eigenfunction of $\tilde{H}_0$ as the trial wavefunction ψ_g in the simulation of (1), as this ψ_g has this nodal structure.

The Euclidean-time dependence of the ground-state energy estimate for (1) in the $(J, I) = (2, 2)$ sector is shown in Figure 6. The values of the parameters were determined in a fit to light meson spectroscopy[17], and are similar to the conventional quark model values. The lattice energy extrapolation gives a value for the four-quark ground-state energy which equals that of two free mesons to within $\approx 5\,MeV$. Preliminary results for the ground-state energy in the $(J, I) = (2, 0)$ sector leads to a four-quark ground-state energy equal to that of two free mesons to within $\approx 10\,MeV$. Thus we conclude that there are no bound states in the $I = 2$ ($\rho^+\rho^+$) and $I = 0$ ($\rho\rho + \omega\omega$) sectors of the nonrelativistic quark potential model to within an accuracy of $10\,MeV$. The former result is consistent with the lattice QCD results of Y.G.Liang et al[8] in the $(J, I) = (2, 2)$ sector.

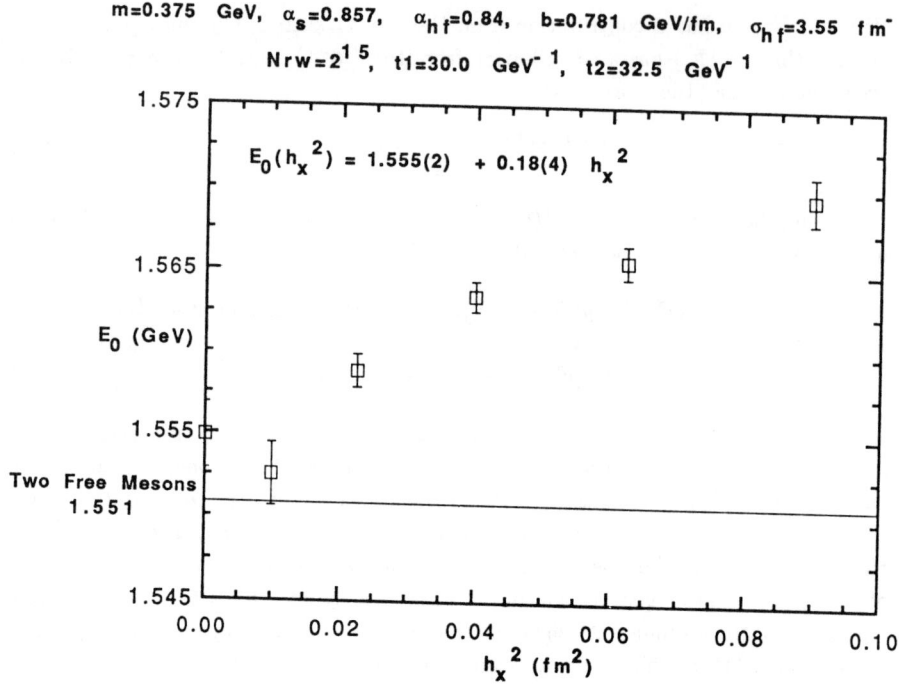

FIG. 6. $E_0(h_x^2)$ in the $(J,I) = (2,2)$ sector ($\rho^+\rho^+$ quantum numbers). The point at the origin is extrapolated from a linear fit in h_x^2.

V. CONCLUSIONS

We have applied an extension of the GRW Monte Carlo algorithm to a dynamical fermion problem, the $q^2\bar{q}^2$ system. The extended algorithm allows the simulation of systems having both continuous and discrete degrees of freedom, and is applicable to Hamiltonians with both positive and negative off-diagonal matrix elements, as characteristically occur in dynamical fermion problems.

We first applied the algorithm to an analytically solvable test problem which has many of the properties of the full four-quark Hamiltonian. From the solution of the test problem we demonstrated that the node of the four-quark ground-state wavefunction is given by $x^2 - y^2 = 0$. Employing this in our choice of guiding wavefunction, we used the Monte Carlo algorithm to estimate the ground-state energies of the $(J, I) = (2, 2)$ and $(2, 0)$ light four-quark systems in a Coulomb-plus-linear quark potential model.

We find evidence for no bound states of two light vector mesons in the channels studied ($J = 2$; $\rho^+\rho^+$, $\rho^-\rho^-$, $\omega\omega$, $\rho \cdot \rho$, ...) in the nonrelativistic quark potential model. The mass of the four-quark state was found to equal that of two free mesons to within our statistical errors of about 5 MeV for $I = 2$ and about 10 MeV for $I = 0$. We hope to improve this measurement in future and to extract an equivalent potential between vector meson pairs, which should be useful in the study of final state interactions in these systems.

VI. ACKNOWLEDGEMENTS

I would like to thank T.Barnes for his advice and instruction throughout this work. I also thank K.Dooley, E.S.Swanson and J.Weinstein for their discussions of four-quark physics and D.Kotchan and M.Kovarik for helpful discussions of the GRW algorithm. I wish to acknowledge the support and facilities provided by the Physics Department of the University of Toronto, Oak Ridge National Laboratory (in particular the Theory Division and the EPM Division), and the Physics Department of the University of Tennessee. This research was supported in part by grants from the Natural Sciences and Engineering Research Council of Canada and by an E.C.Stevens Fellowship and a Walter C.Sumner Memorial Fellowship and also by the Division of Nuclear Physics, U.S. Department of Energy under Contract No. DE-AC05-84OR21400 managed by Martin Marietta Energy Systems Inc., the Physics Department of the University of Tennessee under Contract No. DE-AS05-76ER03956, and the State of Tennessee Science Alliance Center under Contract No. R01-1062-32. Finally, I thank the organizers of the Conference on Computational Quantum Physics (hosted by the Physics Department of Vanderbilt University, Nashville, Tennessee) for extending an invitation to discuss this material with my fellow participants.

---

[1] J.Rosner, Phys. Rev. Lett. 21, 950 (1968); H.Harari, Phys. Rev. Lett. 22, 562 (1969).

[2] R.L.Jaffe, Phys. Rev. Lett. 38, 195 (1977) and 38, 617(E) (1977); K.Saito, Prog. Theor. Phys. 72, 674 (1984).

[3] T.-T.Wang et.al., JETP 13, 323 (1961) for the $f_0(975)$; F.Turkot et.al., Sienna

Conf. 1, 661 (1963) for the $a_0(980)$.

[4] R.L.Jaffe and K.Johnson, Phys. Lett. 60B, 201 (1977); R.L.Jaffe, Phys. Rev. D15, 267, 281 (1977) and D17, 1444 (1978).

[5] J.Weinstein and N.Isgur, Phys. Rev. Lett. 48, 659 (1982); Phys. Rev. D27 588 (1983).

[6] S.Godfrey and N.Isgur, Phys. Rev. D32, 189 (1985) for mesons; S.Capstick and N.Isgur, Phys. Rev. D34, 2809 (1986) for baryons.

[7] T.Barnes, Phys. Rev. B165, 434 (1985).

[8] Y.G.Liang, B.A.Li, K.F.Liu, T.Draper and R.M.Woloshyn, Proceedings of Int. Conf. "Lattice 89", Capri, 1989, Nucl. Phys. B, Proc. Supplement.

[9] J.Carlson, L.Heller, and J.A.Tjon, Phys. Rev. D37, 744 (1988).

[10] J.Carlson and V.R.Pandharipande, Phys. Rev. D43, 1652 (1991).

[11] N.Isgur and G.Karl, Phys. Rev. D18, 4187 (1978).

[12] T.Barnes, G.Daniell, and D.Storey, Nucl. Phys. B265[FS15], 253 (1986).

[13] T.Barnes and G.J.Daniell, Phys. Rev. B37, 3637 (1988).

[14] T.Barnes, "Numerical Solution of High Temperature Superconductor Spin Systems", p.83, *Nuclear and Atomic Physics at One Gigaflop*, ed. C.Bottcher, M.R.Strayer, and J.B.McGrory (Harwood Academic Publishers, Chur, Switzerland, 1989).

[15] E.Fermi, in N.Metropolis and S.Ulam, J. Am. Stat. Assoc. 247, 335 (1949).

[16] G.Grondin, University of Toronto PhD Thesis, in progress.

[17] E.S.Swanson, private communication.

CONCEPTUAL FRAMEWORK FOR HIGH–SPIN HADRONIC PHYSICS

D. V. Ahluwalia and D. J. Ernst
Department of Physics and Center for Theoretical Physics
Texas A&M University
College Station, Texas 77843

Abstract

A formal structure for a calculable and covariant phenomenology of particles with arbitrary spin is presented. As wave equations for particles with high spin have difficulties, the construction proceeds without reference to any specific wave equation. The $2(2j + 1)$ covariant particle/antiparticle spinors for any spin are explicitly constructed. These covariant spinors are then used to construct configuration, as well as momentum space, Feynman–Dyson propagators for arbitrary spin.

Perhaps, one of the first papers on the subject of high–spin matter fields is that of Dirac[1], published eight years after his celebrated paper[2] on spin one–half particles. In this 1936 paper, the first paragraph contains the following observation, "All the same, it is desirable to have the equations ready for a possible future discovery of an elementary particle with a spin greater than a half, or for approximate application to composite particles. Further, the underlying theory is of considerable mathematical interest." Then followed the classic 1939 paper of Wigner[3] showing how the concept of spin is intricately related to the underlying spacetime symmetries. Sixteen years after the 1936 paper of Dirac, in 1952 in a series of three back–to–back letters in *Physical Review*, Anderson et al. reported the existence of an "intermediate state" of spin 3/2, the $\Delta(1232)$. The latest[5] *Review of Particle Properties (1990)* now contains excited composite hadronic states of spin $0 \leq j \leq 15/2$. Modern accelerators will allow the production of these high–spin hadrons in the nuclear medium. An understanding of these processes will require a formalism in which particles of arbitrary spin can be treated in a covariant manner. No totally satisfactory approach yet exists. We outline below an approach[6] to this problem which is internally consistent and produces a calculable phenomenology.

Is such an approach unique? We quote from a 1989 remark by Weinberg[7], "A distinguished nuclear physicist asked me not very long ago what I thought of a proposal to do an experiment on the scattering of a nucleus (I forget which nucleus) with spin–2, which aimed at finding out experimentally which relativistic wave equation that nucleus satisfied. This is all wrong. If you go into the streets of College Park and a passer–by asks you what is the Lorentz transformation of a particle of spin–2, you do not have to ask him if he is referring to a symmetric traceless tensor wave function, or a wave function belonging to some other representation of the homogeneous Lorentz group that contains spin–2. All you have to do is ... tell him that the states transform according to $j = 2$ matrix representation of the rotation group (I promise you that if you do that the pedestrian will not ask you any more questions.). The kinematic classification of particles according to their Lorentz transformation properties is entirely (for finite mass) determined by their familiar representation of the rotation group. It has nothing whatever to do with the choice of one relativistic wave equation rather than another."

In this spirit we first completely do away with all relativistic wave equations. We demonstrate (for details see Ref. 6) that the canonical representation $2(2j+1)$ particle/antiparticle covariant spinors, $u_\sigma(\vec{p})$ and $v_\sigma(\vec{p})$, $\sigma = -j, -j+1, \cdots, j-1, +j$, for any spin can be obtained by the action of $2(2j+1) \times 2(2j+1)$ canonical representation boost matrix

$$M_{CA}(\vec{p}) = \begin{pmatrix} \cosh(\vec{J} \cdot \vec{\varphi}) & \sinh(\vec{J} \cdot \vec{\varphi}) \\ \sinh(\vec{J} \cdot \vec{\varphi}) & \cosh(\vec{J} \cdot \vec{\varphi}) \end{pmatrix}, \tag{1}$$

on the $2(2j+1)$ rest spinors in the form of $2(2j+1)$–element columns

$$u_{+j}(\vec{0}) = \begin{pmatrix} N(j) \\ 0 \\ 0 \\ \vdots \\ 0 \end{pmatrix}, \cdots, v_{-j}(\vec{0}) = \begin{pmatrix} 0 \\ 0 \\ \vdots \\ 0 \\ N(j) \end{pmatrix}. \tag{2}$$

with $N(j) = m^j$. For $j = 1/2$, the result is identical to the standard[8] Dirac spinors (with a slightly different normalization). For $j = 1$, this procedure produces the following covariant spinors:

$$u_{+1}(\vec{p}) = \begin{pmatrix} m + [(2p_z^2 + p_+p_-)/2(E+m)] \\ p_zp_+/\sqrt{2}(E+m) \\ p_+^2/2(E+m) \\ p_z \\ p_+/\sqrt{2} \\ 0 \end{pmatrix}, \qquad (3)$$

$$u_o(\vec{p}) = \begin{pmatrix} p_zp_-/\sqrt{2}(E+m) \\ m + [p_+p_-/(E+m)] \\ -p_zp_+/\sqrt{2}(E+m) \\ p_-/\sqrt{2} \\ 0 \\ p_+/\sqrt{2} \end{pmatrix}, \qquad (4)$$

$$u_{-1}(\vec{p}) = \begin{pmatrix} p_-^2/2(E+m) \\ -p_z p_-/\sqrt{2}(E+m) \\ m + [(2p_z^2 + p_+ p_-)/2(E+m)] \\ 0 \\ p_-/\sqrt{2} \\ -p_z \end{pmatrix}, \qquad (5)$$

$$v_{+1}(\vec{p}) = \begin{pmatrix} p_z \\ p_+/\sqrt{2} \\ 0 \\ m + [(2p_z^2 + p_+ p_-)/2(E+m)] \\ p_z p_+/\sqrt{2}(E+m) \\ p_+^2/2(E+m) \end{pmatrix}, \qquad (6)$$

$$v_0(\vec{p}) = \begin{pmatrix} p_-/\sqrt{2} \\ 0 \\ p_+/\sqrt{2} \\ p_z p_-/\sqrt{2}(E+m) \\ m + [p_+ p_-/(E+m)] \\ -p_z p_+/\sqrt{2}(E+m) \end{pmatrix}, \qquad (7)$$

$$v_{-1}(\vec{p}) = \begin{pmatrix} 0 \\ p_-/\sqrt{2} \\ -p_z \\ p_-^2/2(E+m) \\ -p_z p_-/\sqrt{2}(E+m) \\ m + [(2p_z^2 + p_+ p_-)/2(E+m)] \end{pmatrix} \qquad (8)$$

In the $m \to 0$ limit, only $u_{\pm 1}(\vec{p})$ and $v_{\pm 1}(\vec{p})$ are non–vanishing. For the $(1,0) \oplus (0,1)$ matter field this can be explicitly seen by considering a massless particle travelling along the z–axis (for an arbitrary direction the quantization axis for the angular momentum would have to be chosen accordingly). The explicit results, along with the orthonormality properties and the $m \to 0$ limit, for $j = 3/2$ and $j = 2$ are found in Ref. [6].

The Feynman–Dyson propagators can also be constructed without reference to a relativistic wave equation. For this purpose we define the $(j,0) \oplus (0,j)$ matter field operator

$$\Psi^j(x) = \sum_{\sigma=-j}^{+j} \int \frac{d^3p}{(2\pi)^3} \frac{1}{2\omega_{\vec{p}}} \qquad (9)$$
$$\times \left[u_\sigma(\vec{p}) a(\vec{p},\sigma) \exp(-ip \cdot x) + v_\sigma(\vec{p}) b^\dagger(\vec{p},\sigma) \exp(+ip \cdot x) \right],$$

with

$$\omega_{\vec{p}} = \sqrt{m^2 + \vec{p}^2}, \quad \overline{\Psi}^j(x) \equiv \Psi^{j\dagger}(x) \gamma^{CA}_{oo\ldots o}. \qquad (10)$$

Next we evalutate the vacuum expectation value

$$\langle x | S^j_{FD} | y \rangle \equiv \langle \ | T[\Psi^j(x) \overline{\Psi}^j(y)] | \ \rangle, \qquad (11)$$

called Feynman–Dyson propagator. This is the object which enters naturally in canonical perturbation theory.

Using $\{a_\sigma(\vec{p}), a^\dagger(\vec{p}')\} = (2\pi)^3 2\omega_{\vec{p}} \delta_{\sigma\sigma'} \delta(\vec{p} - \vec{p}')$ etc., for fermions and similar relation for bosons (with anticommutator replaced by commutator), we obtain the configuration space Fynman–Dyson propagator for arbitrary spin. It reads

$$\langle x | S^j_{FD} | y \rangle = \sum_{\sigma=-j}^{+j} \int \frac{d^3p}{(2\pi)^3} \frac{1}{2\omega_{\vec{p}}}$$
$$\times \left[u_\sigma(\vec{p}) \overline{u}_\sigma(\vec{p}) e^{-ip\cdot(x-y)} \theta(x^\circ - y^\circ) + \epsilon v_\sigma(\vec{p}) \overline{v}_\sigma(\vec{p}) e^{+ip\cdot(x-y)} \theta(y^\circ - x^\circ) \right], \qquad (12)$$

with

$$\epsilon = \begin{cases} +1 & \text{for bosons,} \\ -1 & \text{for fermions.} \end{cases} \qquad (13)$$

The momentum space Feynman–Dyson propagator is similarly obtained to be:

$$\langle k' | S^j_{FD} | k \rangle = \int \frac{d^4x}{(2\pi)^3} \frac{d^4y}{(2\pi)^3} e^{ik'\cdot x} e^{-ik\cdot y} \langle x | S^j_{FD} | y \rangle$$
$$= -\frac{i\delta^{(4)}(k'-k)}{(2\pi)^2 2\omega_{\vec{k}}} \sum_{\sigma=-j}^{+j} \left(\frac{u_\sigma(\vec{k}) \overline{u}_\sigma(\vec{k})}{k_o + i\eta - E(\vec{k})} - \epsilon \frac{v_\sigma(-\vec{k}) \overline{v}_\sigma(-\vec{k})}{k_o - i\eta + E(\vec{k})} \right). \qquad (14)$$

We thus have explicit expressions for the spinors and the propagators. If we combine these with a model of the interaction, the theory can be defined to be the perturbation theory given by the usual diagrammatic rules. This allows the construction of a covariant phenomenology without reference to a wave equation and thus circumvents the many problems associated with equations for the higher spin particles.

This work is supported, in part, by the National Science Foundation. The authors would like to thank the Organizing Committee, and especially Sait Umar, for their kind invitation and warm hospitality.

References

1. P. A. M. Dirac, *Proc. Roy. Soc. (London) A* **155** (1936), 447.
2. P. A. M. Dirac, *Proc. Roy. Soc. (London) A* **117** (1928), 610; *ibid.* **118** (1928), 351.
3. E. P. Wigner, *Ann. of Math.* **40** (1939) 149.
4. H. L. Anderson, E. Fermi, E. A. Long, R. Martin and D. E. Nagle, *Phys. Rev.* **85**, 934 (1952); E. Fermi, H. L. Anderson, A. Lundby, D. E. Nagle and G. B. Yodh, *Phys. Rev.* **85**, 935 (1952); H. L. Anderson, E. Fermi, E. A. Long and D. E. Nagle, *Phys. Rev.* **85**, 936 (1952).
5. J. J. Hernandez *et al.*, Phys. Lett B **239** (1990) 1.
6. D. V. Ahluwalia, "Relativistic Quantum Field Theory of High–Spin Matter Fields: A Pragmatic Approach For Hadronic Physics," *Ph. D thesis*, Texas A&M University, unpublished (available upon request from the author), (1991).
7. S. Weinberg, *Nucl. Phys. B (Proc. Suppl.)*, **6**, (1989) 67.
8. J. D. Bjorken and S. D. Drell, "Relativistic Quantum Mechanics," McGraw–Hill Book Co., New York, 1964.

LIST OF PARTICIPANTS

Dharam V. Ahluwalia
Department of Physics
Texas A&M University
College Station, TX 77843-4242

Thomas L. Ainsworth
Department of Physics
Texas A&M University
College Station, TX 77843-4242

Charles J. Albert
Department of Physics
Texas A&M University
College Station, TX 77843-4242

Nigel R. Badnell
Department of Physics
Auburn University
Auburn, AL 36849-5311

Ted Barnes
Physics Division
Oak Ridge National Laboratory
Oak Ridge, TN 37831-6373

Sayoko J. Blodgett-Ford
National Institute of Standards
 & Technology
Gaithersburg, MD 20899

Christopher Bottcher
Physics Division
Oak Ridge National Laboratory
Oak Ridge, TN 37831-6373

James Boyle
Department of Physics
University of Virginia
Charlottesville, VA 22901

Eric Braaten
Dept. of Physics & Astronomy
Northwestern University
Evanston, IL 60208

Tomas Brage
Dept. of Computer Science
Vanderbilt University
Nashville, TN 37235

Aurel Bulgac
Cyclotron Laboratory
Michigan State University
East Lansing, MI 48824-1321

Ziyong Cai
Dept. of Computer Science
Vanderbilt University
Nashville, TN 37235

Alan Calder
Dept. of Physics & Astronomy
Clemson University
Clemson, SC 29631

Bing-Qing Chen
106-38
California Institute of Technology
Pasadena, CA 91125

Jasson Chen
Department of Physics
Texas A&M University
College Station, TX 77843

S. A. Chin
Department of Physics
Texas A&M University
College Station, TX 77843

Elbio R. Dagotto
Institute of Theoretical Physics
University of California
Santa Barbara, CA 93106

David J. Dean
Physics Division
Oak Ridge National Laboratory
Oak Ridge, TN 37831-6373

Carleton E. DeTar
Department of Physics
University of Utah
Salt Lake City, UT 84112

Jerry P. Draayer
Dept. of Physics & Astronomy
Louisiana State University
Baton Rouge, LA 70803

List of Participants

David J. Ernst
Department of Physics
Texas A&M University
College Station, TX 77843

Charlotte F. Fischer
Dept. of Computer Science
Vanderbilt University
Nashville, TN 37235

Gregory R. J. Grondin
Physics Division
Oak Ridge National Laboratory
Oak Ridge, TN 37831-6373

William B. Herrmannsfeldt
SLAC
P. O. Box 4349
Stanford, CA 94309

Malvin H. Kalos
Cornell Theory Center
Cornell University
Ithaca, NY 14853-3801

Hugh P. Kelly
Department of Physics
University of Virginia
Charlottesville, VA 22901

Kate Kirby
Atomic & Molecular Physics Division
Harvard-Smithsonian Center
Cambridge, MA 02138

Rubin H. Landau
Department of Physics
Oregon State University
Corvallis, OR 97331

Richard A. Matzner
Department of Physics
University of Texas
Austin, TX 78712

Gregory Miecznik
Dept. of Computer Science
Vanderbilt University
Nashville, TN 37235

John C. Morrison
Department of Physics
University of Louisville
Louisville, KY 40292

Berndt Müller
Department of Physics
Duke University
Durham, NC 27706

Volker E. Oberacker
Dept. of Physics & Astronomy
Vanderbilt University
Nashville, TN 37235

C. Edward Oliver
Office of Laboratory Computing
Oak Ridge National Laboratory
Oak Ridge, TN 37831-6259

Michael S. Pindzola
Department of Physics
Auburn University
Auburn, AL 36849

Mark J. Rhoades-Brown
RHIC Project, Bldg. 1005
Brookhaven National Laboratory
Upton, NY 11973

Peter Rochford
Dept. of Physics & Astronomy
Louisiana State University
Baton Rouge, LA 70803

Michael R. Strayer
Physics Division
Oak Ridge National Laboratory
Oak Ridge, TN 37831-6373

Atanas Trayanov
Department of Physics
Duke University
Durham, NC 27706

Herbert Uberall
Department of Physics
Catholic University
Washington, DC 20064

A. Sait Umar
Dept. of Physics & Astronomy
Vanderbilt University
Nashville, TN 37235

Michel Vallières
Dept. of Physics & Atmos. Science
Drexel University
Philadelphia, PA 19104

Jack C. Wells
Physics Division
Oak Ridge National Laboratory
Oak Ridge, TN 37831-6373

Michael F. Werby
NOARL
Code 221
Stennis Space Center
MS 39529-5004

Peter Winkler
Department of Physics
University of Nevada
Reno, NV 89557

J.-S. Wu
Physics Division
Oak Ridge National Laboratory
Oak Ridge, TN 37831-6373

Bin Zhou
Department of Physics
University of Pittsburgh
Pittsburgh, PA 15260

R. Zingarelli
NOARL
Code 221
Stennis Space Center
MS 39529-5004

AUTHOR INDEX

A

Ahluwalia, D. V., 309
Albert, C. J., 255

B

Badnell, N. R., 207
Bottcher, C., 47, 215, 255
Braaten, E., 244
Brage, T., 94
Bulgac, A., 23

C

Chen, C. M., 109
Chin, S. A., 35

D

Dagotto, E., 172
Dean, D. J., 159
DeTar, C., 1
Draayer, J. P., 59

E

Ernst, D. J., 109, 255, 309

G

Geiger, K., 228
Griffin, D. C., 207
Grondin, G., 292

H

Herrmannsfeldt, W. B., 142

K

Kalos, M. H., 122
Kelly, H. P., 131
Kirby, K. P., 149
Krotschek, E., 35
Kusnezov, D., 23

L

Landau, R. H., 83
Lu, Zi-M., 267

M

Matzner, R. A., 12
Müller, B., 228, 280

O

Oberacker, V. E., 215

P

Park, S. C., 59
Pindzola, M. S., 207

R

Rochford, P., 59

S

Shakin, C. M., 47
Sloan, J. H., 23
Strayer, M. R., 47, 159, 215, 255

T

Trayanov, A., 280

U

Umar, A. S., 159, 215

V

Vallières, M., 267

W

Wells, J. C., 215
Werby, M. F., 180
Winkler, P., 72
Wu, J.-S., 47, 159, 215

Y

Yan, Y. T., 142
Yuan, J.-M., 267

Z

Zhang, S., 122
Zhen, S.-Q., 59

AIP Conference Proceedings

		L.C. Number	ISBN
No. 198	Astrophysics in Antarctica (Newark, DE, 1989)	89-46421	0-88318-398-6
No. 199	Surface Conditioning of Vacuum Systems (Los Angeles, CA, 1989)	89-82542	0-88318-756-6
No. 200	High T_c Superconducting Thin Films: Processing, Characterization, and Applications (Boston, MA, 1989)	90-80006	0-88318-759-0
No. 201	QED Stucture Functions (Ann Arbor, MI, 1989)	90-80229	0-88318-671-3
No. 202	NASA Workshop on Physics From a Lunar Base (Stanford, CA, 1989)	90-55073	0-88318-646-2
No. 203	Particle Astrophysics: The NASA Cosmic Ray Program for the 1990s and Beyond (Greenbelt, MD, 1989)	90-55077	0-88318-763-9
No. 204	Aspects of Electron–Molecule Scattering and Photoionization (New Haven, CT, 1989)	90-55175	0-88318-764-7
No. 205	The Physics of Electronic and Atomic Collisions (XVI International Conference) (New York, NY, 1989)	90-53183	0-88318-390-0
No. 206	Atomic Processes in Plasmas (Gaithersburg, MD, 1989)	90-55265	0-88318-769-8
No. 207	Astrophysics from the Moon (Annapolis, MD, 1990)	90-55582	0-88318-770-1
No. 208	Current Topics in Shock Waves (Bethlehem, PA, 1989)	90-55617	0-88318-776-0
No. 209	Computing for High Luminosity and High Intensity Facilities (Santa Fe, NM, 1990)	90-55634	0-88318-786-8
No. 210	Production and Neutralization of Negative Ions and Beams (Brookhaven, NY, 1990)	90-55316	0-88318-786-8
No. 211	High-Energy Astrophysics in the 21st Century (Taos, NM, 1989)	90-55644	0-88318-803-1
No. 212	Accelerator Instrumentation (Brookhaven, NY, 1989)	90-55838	0-88318-645-4
No. 213	Frontiers in Condensed Matter Theory (New York, NY, 1989)	90-6421	0-88318-771-X 0-88318-772-8 (pbk.)
No. 214	Beam Dynamics Issues of High-Luminosity Asymmetric Collider Rings (Berkeley, CA, 1990)	90-55857	0-88318-767-1
No. 215	X-Ray and Inner-Shell Processes (Knoxville, TN, 1990)	90-84700	0-88318-790-6
No. 216	Spectral Line Shapes, Vol. 6 (Austin, TX, 1990)	90-06278	0-88318-791-4
No. 217	Space Nuclear Power Systems (Albuquerque, NM, 1991)	90-56220	0-88318-838-4

No. 218	Positron Beams for Solids and Surfaces (London, Canada, 1990)	90-56407	0-88318-842-2
No. 219	Superconductivity and Its Applications (Buffalo, NY, 1990)	91-55020	0-88318-835-X
No. 220	High Energy Gamma-Ray Astronomy (Ann Arbor, MI, 1990)	91-70876	0-88318-812-0
No. 221	Particle Production Near Threshold (Nashville, IN, 1990)	91-55134	0-88318-829-5
No. 222	After the First Three Minutes (College Park, MD, 1990)	91-55214	0-88318-828-7
No. 223	Polarized Collider Workshop (University Park, PA, 1990)	91-71303	0-88318-826-0
No. 224	LAMPF Workshop on (π, K) Physics (Los Alamos, NM, 1990)	91-71304	0-88318-825-2
No. 225	Half Collision Resonance Phenomena in Molecules (Caracus, Venezuela, 1990)	91-55210	0-88318-840-6
No. 226	The Living Cell in Four Dimensions (Gif sur Yvette, France, 1990)	91-55209	0-88318-794-9
No. 227	Advanced Processing and Characterization Technologies (Clearwater, FL, 1991)	91-55194	0-88318-910-0
No. 228	Anomalous Nuclear Effects in Deuterium/Solid Systems (Provo, UT, 1990)	91-55245	0-88318-833-3
No. 229	Accelerator Instrumentation (Batavia, IL, 1990)	91-55347	0-88318-832-1
No. 230	Nonlinear Dynamics and Particle Acceleration (Tsukuba, Japan, 1990)	91-55348	0-88318-824-4
No. 231	Boron-Rich Solids (Albuquerque, NM, 1990)	91-53024	0-88318-793-4
No. 232	Gamma-Ray Line Astrophysics (Paris–Saclay, France, 1990)	91-55492	0-88318-875-9
No. 233	Atomic Physics 12 (Ann Arbor, MI, 1990)	91-55595	0-88318-811-2
No. 234	Amorphous Silicon Materials and Solar Cells (Denver, CO, 1991)	91-55575	0-88318-831-7
No. 235	Physics and Chemistry of MCT and Novel IR Detector Materials (San Francisco, CA, 1990)	91-55493	0-88318-931-3
No. 236	Vacuum Design of Synchrotron Light Sources (Argonne, IL, 1990)	91-55527	0-88318-873-2
No. 237	Kent M. Terwilliger Memorial Symposium (Ann Arbor, MI, 1989)	91-55576	0-88318-788-4
No. 238	Capture Gamma-Ray Spectroscopy (Pacific Grove, CA, 1990)	91-57923	0-88318-830-9
No. 239	Advances in Biomolecular Simulations (Obernai, France, 1991)	91-58106	0-88318-940-2
No. 240	Joint Soviet-American Workshop on the Physics of Semiconductor Lasers (Leningrad, USSR, 1991)	91-58537	0-88318-936-4

No. 241	Scanned Probe Microscopy (Santa Barbara, CA, 1991)	91-76758	0-88318-816-3
No. 242	Strong, Weak, and Electromagnetic Interactions in Nuclei, Atoms, and Astrophysics: A Workshop in Honor of Stewart D. Bloom's Retirement (Livermore, CA, 1991)	91-76876	0-88318-943-7
No. 243	Intersections Between Particle and Nuclear Physics (Tucson, AZ, 1991)	91-77580	0-88318-950-X
No. 244	Radio Frequency Power in Plasmas (Charleston, SC, 1991)	91-77853	0-88318-937-2
No. 245	Basic Space Science (Bangalore, India, 1991)	91-78379	0-88318-951-8
No. 246	Space Nuclear Power Systems (Albuquerque, NM, 1992)	91-58793	1-56396-027-3 1-56396-026-5 (pbk.)
No. 247	Global Warming: Physics and Facts (Washington, DC, 1991)	91-78423	0-88318-932-1
No. 248	Computer-Aided Statistical Physics (Taipei, Taiwan, 1991)	91-78378	0-88318-942-9
No. 249	The Physics of Particle Accelerators (Upton, NY, 1989, 1990)	92-52843	0-88318-789-2
No. 250	Towards a Unified Picture of Nuclear Dynamics (Nikko, Japan, 1991)	92-70143	0-88318-951-8
No. 251	Superconductivity and its Applications (Buffalo, NY, 1991)	92-52726	1-56396-016-8
No. 252	Accelerator Instrumentation (Newport News, VA, 1991)	92-70356	0-88318-934-8
No. 253	High-Brightness Beams for Advanced Accelerator Applications (College Park, MD, 1991)	92-52705	0-88318-947-X
No. 254	Testing the AGN Paradigm (College Park, MD, 1991)	92-52780	1-56396-009-5
No. 255	Advanced Beam Dynamics Workshop on Effects of Errors in Accelerators, Their Diagnosis and Corrections (Corpus Christi, TX, 1991)	92-52842	1-56396-006-0
No. 256	Slow Dynamics in Condensed Matter (Fukuoka, Japan 1991)	92-53120	0-88318-938-0
No. 257	Atomic Processes in Plasmas (Portland, ME, 1991)	91-08105	0-88318-939-9
No. 258	Synchrotron Radiation and Dynamic Phenomena (Grenoble, France 1991)	92-53790	1-56396-008-7
No. 259	Future Directions in Nuclear Physics with 4π Gamma Detection Systems of the New Generation (Strasbourg, France 1991)	92-53222	0-88318-952-6
No. 260	Computational Quantum Physics (Nashville, TN 1991)	92-71777	0-88318-933-X
No. 261	Rare and Exclusive B&K Decays and Novel Flavor Factories (Santa Monica, CA 1991)	92-71873	1-56396-055-9